Silicon in Agriculture

Yongchao Liang • Miroslav Nikolic
Richard Bélanger • Haijun Gong
Alin Song

Silicon in Agriculture

From Theory to Practice

Springer

Yongchao Liang
Zhejiang University
Hangzhou, China

Chinese Academy of Agricultural Sciences
Beijing, China

Richard Bélanger
Laval University
Québec, Canada

Alin Song
Chinese Academy of Agricultural Sciences
Beijing, China

Miroslav Nikolic
University of Belgrade
Belgrade, Serbia

Haijun Gong
Northwest A&F University
Yangling, China

ISBN 978-94-017-9977-5 ISBN 978-94-017-9978-2 (eBook)
DOI 10.1007/978-94-017-9978-2

Library of Congress Control Number: 2015938303

Springer Dordrecht Heidelberg New York London

Printed on acid-free paper

Springer Science+Business Media B.V. Dordrecht is part of Springer Science+Business Media (www.springer.com)

This book is dedicated to the late Professors:
Volker Römheld (University of Hohenheim, Germany)
and
Tongsheng Ma and Ruixing Ding (Nanjing Agricultural University, China)

Preface

For over 150 years, plant scientists have recorded the beneficial effects of silicon (Si) in enhancing plant resistance and/or tolerance to various biotic and abiotic stresses. Interestingly, Si is the only mineral element known to effectively mitigate multiple abiotic stresses including salinity, drought, flooding, freezing, high temperature, ultraviolet radiation and mineral nutrient deficiency/toxicity stress. In recent years, Si has become more globally accepted as an agriculturally important addition. Application of Si fertilizers have been shown to significantly enhance plant resistance to diseases and pests, thus contributing to increased food safety, higher production with lower input costs and reduced negative impacts on environmental health. However, it was not until the late 1980s that Si began to attract the attention of a broader group of plant scientists. Over the past decade, rapid progress has been made in cloning and functional characterization of the genes responsible for Si transport and distribution in higher plants. Progress has also been made, in unravelling the molecular basis of Si-mediated resistance against both biotic and abiotic stresses. Nevertheless, the mechanisms involved in the manifold roles of Si in plant biology still remain poorly understood and in need of further research. Moreover, with the exception of rice and sugarcane, less attention has been paid to the application of Si-based fertilizers for general agricultural practice.

In the present book, we provide an overview of the current knowledge on Si in agriculture, as it relates to plants, soils and fertilizers. Finally, we discuss the future interdisciplinary research that is needed to further the knowledge and potential uses of Si in agriculture and in broader environmental sciences. The book is organized into 13 chapters. Dr. Yongchao Liang, from Zhejiang University, China, authored Chaps. 1, 2, 3, 6 and 10. He also collaborated with Dr. Alin Song from Chinese Academy of Agricultural Sciences on Chaps. 5 and 12 and with Dr. Haijun Gong from Northwest A&F University, China, on Chap. 11. Dr. Miroslav Nikolic from University of Belgrade, Serbia, wrote Chaps. 4, 8 and 13. Dr. Richard Bélanger from Laval University, Canada, wrote Chap. 9, and Dr. Haijun Gong wrote Chap. 7. Drs. M. Nikolic, R. Bélanger and Y. Liang reviewed and finalized all the chapters.

This book is primarily intended for graduate students and researchers in various fields of agricultural, biological and environmental sciences, as well as for experts and consultants in the agronomic and fertilizer industry. This book aims to serve as an important reference for those engaged in crop nutritional physiology and molecular biology, crop production, crop protection, environmental protection, soil biogeochemistry, soil and plant nutrient management, fertilizer production and agricultural extension. We hope that it will also contribute to stimulating discussions on the need for future research to broaden our understanding of the roles of Si in plant biology and, especially, plant stress physiology.

We would like to express our sincere gratitude to Dr. Nina Nikolic from the University of Belgrade, Serbia, for her critical reading of all the chapters and her useful comments. Last but not least, our thanks are given to Ms. Abbey Huang from Springer for her instructions and assistance during the publication process.

Hangzhou, China Yongchao Liang
Beijing, China
Belgrade, Serbia Miroslav Nikolic
Québec, Canada Richard Bélanger
Yangling, China Haijun Gong
Beijing, China Alin Song
March, 2015

Acknowledgment of Copyright Approvals

Permission for use of selected published material was obtained from the following companies or organizations:

Elsevier Science Publishers
John Wiley and Sons
National Academy of Sciences
Nature Publishing Group
Oxford University Press
Science Press, China
Springer Science + Business Media
Taylor & Francis
The American Phytopathological Society

The authors gratefully acknowledge the support and assistance of the above publishers with the use of pertinent data and information. All other references cited in this book are also acknowledged with gratitude.

Contents

About the Authors

Yongchao Liang is currently a full professor at the College of Environmental and Resource Sciences of the Zhejiang University, China; a deputy director of the Scientific Committee for Soil and Plant Nutrition of the Society of Soil Sciences of China; a member of the Academic Committee of the Chinese Academy of Agricultural Sciences; an editor of Plant and Soil and Applied and Environmental Soil Science; and an editorial board member of Journal of Plant Nutrition. He obtained his B.Sc. degree in Soil Science and Agrochemistry from the Zhejiang University, in 1983, and his M.Sc. degree in Soil Science from the Nanjing Agricultural University, Jiangsu, China, in 1987. From 1987 he worked at the Nanjing Agricultural University and obtained his Ph.D. degree and professorship in 1999. He did a postdoc at the University of Hohenheim, Stuttgart, Germany, as an Alexander von Humboldt Research Fellow (2002–2003). From 2004 he worked at the Institute of Agricultural Resources and Regional Planning of the Chinese Academy of Agricultural Sciences as a First-Class Subject Elitist for Plant Nutrition and Fertilizer, and moved back to the Zhejiang University in 2014. In 2006 he was officially appointed by the Ministry of Education, People's Republic of China, as the special-term professor under the Changjiang Scholars Award Program. His broad research interest includes plant silicon biology and related stress-resistance mechanisms, adaptations of higher plants to stressful environments, soil contamination and remediation, soil biology and biochemistry, and fertilization management. His most accomplished work on plant silicon biology ranges from Si uptake and transport to abiotic and biotic stresses. He and his co-workers have demonstrated for the first time that silicon-enhanced antioxidant defense capacity is one of the major mechanisms of resistance to salinity stress and a universal mechanism of Si-enhanced resistance to abiotic stress. They concluded that both active and passive Si-uptake components coexist in different crop species with their relative contribution depending on both the genotype and the external Si concentrations. He organized the 5th International Conference on Silicon in Agriculture in Beijing, China (2011).

Miroslav Nikolic graduated with honours in 1990 (Viticulture major) and obtained his M.Sc. degree in Plant Physiology in 1995 from the University of Belgrade, Serbia. He did his Ph.D. in the form of a sandwich program at the University of Hohenheim, Stuttgart, Germany, supported by the German Academic Exchange Service (DAAD), and received his Ph.D. degree in Agricultural Sciences from the University of Belgrade in 1999. He completed two postdocs in plant nutritional physiology, the first with Volker Römheld at the University of Hohenheim, working on iron dynamics in the leaf apoplast, and then with Roberto Pinton and Zeno Varanini at the University of Udine, Italy, working on root-to-shoot nitrate translocation and the anion uptake by leaf cells. Since 2003 he has been at the Institute for Multidisciplinary Research, University of Belgrade, becoming a full research professor in 2007. He currently serves as a consulting editor for the international journal Plant and Soil. Miroslav developed the method for silicon uptake tracing by radioactive germanium-68 and with his co-workers indicated an active uptake of silicon in barley a few years before the related root transporters were identified. Current research of his group of Plant Nutrition centers on plant response mechanisms to mineral stresses, from nutrient deficiency to the excess of different minerals, with a special focus towards understanding the role(s) of silicon in stress alleviation. Miroslav's team was the first one to demonstrate that silicon promotes root acquisition and shoot mobilization of iron through up-regulation of the expression of key genes involved in these processes. Recently, his group has been working on elucidation *in planta* mechanism of how silicon modulates phosphorus bioavailability in the rhizosphere of crops growing under acidic conditions.

Richard Bélanger obtained his bachelor degree in forest biology in 1984 from Laval University in Quebec, Canada. He went on to complete his Ph.D. in Plant Pathology at the College of Environmental Science and Forestry, State University of New York, Syracuse, in 1988. He was first hired as a research associate in 1988 to work on disease problems in greenhouse crops for the Horticulture Research Center at Laval University. He became an assistant professor in 1991 at the same university. He is now a full professor in plant pathology and holder of a Canada Research Chair in plant protection. His research endeavors have concentrated on the development of biological and non-chemical approaches to control plant diseases. Along those lines, sustained efforts have been devoted to the study of the prophylactic properties of silicon (Si) against plant pathogens with a special emphasis towards understanding the elusive properties of Si as it interacts with plants. His group was the first one to propose that the mode of action of Si extended beyond a mechanical barrier through the stimulation of plant defense reactions. This concept is now widely recognized and has been validated in a large number of plant–pathogen and plant–insect interactions. Additional efforts have led to the first full transcriptomic analyses of plants' responses to Si amendments under conditions of pathogen stress. These results led to the conclusions that Si benefits were significantly more manifest when a plant was exposed to a stress. In recent activities, Bélanger's team has been investigating Si transporters and absorption in plants in order to optimize the application of Si through a better understanding of the genetic predisposition of plants to absorb the element.

Haijun Gong completed his undergraduate study in Biology at Lanzhou University, China, in 1999. During his Ph.D. period at Lanzhou University, he investigated the plant responses to gradual abiotic stresses, mainly about plasma membrane energizing systems. He also studied the effect of silicon on drought tolerance in wheat plants and observed the alleviative effect of silicon on oxidative damages. After he graduated from Lanzhou University in 2004, he did one year of visiting research in the lab of Professor Timothy J. Flowers at the University of Sussex, UK, and studied the effect of silicon on transpirational bypass flow and sodium transport in rice. After that, he did two years of postdoctoral research in Zhejiang University, China, and three years of postdoctoral research with CSIRO, Australia, working on oilseed rape herbicide and chloride exclusion mechanisms in grapevine, respectively. In 2010, Haijun Gong accepted a position in the College of Horticulture at Northwest A&F University, China, where he continued his research in stress physiology. At present, he is mainly working on the mechanisms of silicon-mediated stress tolerance in tomato and cucumber plants.

Alin Song received her Ph.D. degree in Plant Nutrition from Nanjing Agricultural University in 2009. She is currently an associate professor of plant nutrition at Institute of Agricultural Resources and Regional Planning, Chinese Academy of Agricultural Sciences. She has been working on silicon nutrition, resistance mechanism to heavy metal and biogeochemical cycling of silicon and carbon for over 10 years. Her main findings are: 1) silicon-enhanced resistance to cadmium in *Brassica chinensis* is attributed to Si-suppressed cadmium uptake and transport and Si-enhanced antioxidant defense capacity, and 2) the alleviation of zinc toxicity by silicon is related to zinc transport, antioxidative reactions and key genes expression in *Oryza sativa*. These findings are both theoretically and practically important in terms of *Brassica chinensis* or *Oryza sativa* safety and yield because heavy metal (Cd/Zn) contamination in soil is becoming more and more serious in some parts of China and in the world due to repeated agricultural use of metal-containing agrochemicals and compost. In biogeochemical cycling of silicon and carbon aspect, her finding was: The potential for carbon bio-sequestration in China's paddy rice is impacted by slag-based silicate fertilizer; thus regulating Si supply during rice growth may serve as an effective tool in improving PhytOC production rate and mitigating global warming. These results show that silicon is beneficial for plant growth especially when plants are exposed to heavy metal stress, while providing a basis for better management of cropland systems to maximize phytolith C sequestration and aid in reducing atmospheric carbon loads which have been tied to global climate change.

Chapter 1
History and Introduction of Silicon Research

Abstract Although silicon (Si) is not yet listed among the essential elements for the growth of higher plants, it has been well documented to play an important role in providing beneficial effects on growth and yield, especially in plants under stressful environments. From a practical perspective, the use of slag-based silicate fertilizers in agriculture can be dated back to the Middle Ages in Europe. Over the last decade, the discovery of specific Si transporters in rice roots has allowed great progress in the understanding of Si uptake by plants at the molecular level. In the same manner, important advancements have been made in dissecting the molecular mechanisms by which Si enhances plant resistance to fungal and bacterial diseases and insect pest damage. In contrast, more efforts are needed to explain at the molecular level the numerous reports showing Si benefits against abiotic stresses. In this chapter, a brief review is presented focusing on the most important historical points and general introduction of worldwide Si research.

Keywords Abiotic stress • Biotic stress • Silicon uptake and transport • Slag

1.1 General

Silicon (Si) is a tetravalent metalloid and, together with its chemical analogue germanium (Ge), is a crucial semiconductor widely used in circuits for modern electronics. Si is a ubiquitous element and the second most abundant after oxygen in soil, comprising approximately 28 % of the Earth's crust. Despite this, most sources of Si in soil are present as crystalline aluminosilicates, which are insoluble and not directly available for plants (Richmond and Sussman 2003). The plant-available form of Si is monosilicic acid (H_4SiO_4), which is present in the soil solution bathing the plant roots at concentrations normally ranging from 0.1 to 0.6 mM (Gunnarsson and Arnórsson 2000), roughly two orders of magnitude higher than the concentrations of phosphorus (P) in soil solutions (Epstein 1994, 1999). However, Si is contained in varying amounts in all terrestrial plants, ranging from 0.1 to 10 % of shoot dry weight.

Y. Liang et al., *Silicon in Agriculture*, DOI 10.1007/978-94-017-9978-2_1

Despite its ubiquity and abundance in both soils and plants, Si has not yet been considered an 'essential' mineral element of any terrestrial higher plant taxon, with the exception of the members of Equisetaceae, nor has it received the attention of plant physiologists (Epstein 1994). This is partly because convincing evidence that Si can meet the definition of essentiality as established by Arnon and Stout in the 1930s is still scarce, and it is currently impossible to purify nutrient solution cultures to be completely free of Si. Consequently, Si is not deliberately included in the formulation of any of the commonly used hydroponics nutrient solutions, where it might occur from a purified water, chemicals and contamination by air dusts. Thus, Epstein (1994, 1999, 2001) reiterates that plants grown in conventional nutrient solutions are, to an important extent, experimental artefacts, because in reality plants rooting in soils are exposed to ample supply of Si from growth media and its omission from solution cultures may lead to distorted results in experiments on inorganic plant nutrition, growth and development and responses to environmental stress. Accordingly, Epstein's lab prescribed adding 1 mM Na_2SiO_3 for preparation of a modified Hoagland solution to eliminate experimental artefacts mentioned above (Epstein 1994). Epstein (1999) considered Si a 'quasi-essential' element for many of plant species for which its absolute essentiality has not been established. According to the definition of 'quasi-essential' elements, an element is defined as quasi-essential if it is ubiquitous in plants and if a deficiency of it can be severe enough to result in demonstrable adverse effects or abnormalities with respect to growth, development, reproduction or viability (Epstein 1999). More recently, however, Epstein and Bloom (2005) have further modified the near universally accepted definition of essentiality of elements established by Arnon and Stout (1939). Based on this newly modified definition, an element is essential if it fulfils either one or both of the two following criteria: (1) the element is part of a molecule that is an intrinsic component of the structure or metabolism of the plant, and (2) the plant can be so severely deficient in the element that it exhibits abnormalities in growth, development or reproduction, i.e. 'performance', compared to plants with lower deficiency. Following this new definition, Si is undoubtedly an essential element for higher plants.

Based on the current knowledge documented in the literature, it is clear that Si does show its significance for the life of plants and the performance of crops in the following aspects, but definitely not confined to these: (1) essentiality (diatoms (Bacillariophyta), horsetails or scouring rushes (Equisetaceae)); (2) enhancement of growth, yield and quality; (3) promotion of mechanical strength or plant erectness and hence resistance to lodging; (4) better light interception and hence promotion of photosynthesis; (5) resistance to insufficient sunshine or shading; (6) effects of plant surface properties; (7) resistance to plant diseases caused by fungi, bacteria, viruses and nematodes; (8) resistance to herbivores ranging from phytophagous insects to mammals; (9) resistance to excess metal toxicity; (10) resistance to salinity stress; (11) inhibition of transpiration and hence resistance to drought stress and efficient water use; (12) resistance to high temperature and chilling or freezing stress; (13) resistance to UV radiation or monochromic exposure; (14) enhancement of root oxidizing power and root activities and hence alleviation of reduced toxicity under low Eh; (15) effects on enzyme activities; (16) alleviation of stress from other minerals [e.g. potassium (K), phosphorus (P) and iron (Fe) deficiency; nitrogen (N), P and boron (B) excess; etc.]; (17) promotion of nodule formation in legume plants and hence promotion of N_2 fixation (Nelwamondo and Dakora 1999; Nelwamondo et al. 2001; Mali and Aery 2008); and (18) promotion of formation of phytolith-occluded carbon and hence having implications in carbon bio-sequestration of atmospheric CO_2 and global climate change (Song et al. 2012, 2013).

Over the last century, especially the last decade, numerous studies have focused on a better understanding of the mechanism(s) involved in (1) the effects of Si on agricultural and horticultural crop productivity and quality; (2) Si-mediated alleviation of various forms of abiotic and biotic stress; (3) Si uptake, transport and accumulation; (4) soil Si biogeochemistry and its bioavailability; (5) analytical methods of Si in soil, plant and fertilizer; and (6) Si fertilizer manufacturing, application and management. The international Si community has grown through a history of more than one decade since the first International Conference on Silicon in Agriculture was successfully held in Florida, USA (1999), followed by the second in Tsuruoka, Japan (2002); the third in Uberlandia, Brazil (2005); the fourth in KwaZulu-Natal, South Africa (2008); the fifth in Beijing, China (2011); and the sixth in Stockholm, Sweden (2014). All these conferences provided a forum for the distinguished scientists, colleagues, students, fertilizer producers and consumers to present their most recent findings and achievements and to exchange their valuable experiences with their international partners, thus putting forward the science of Si in agriculture at an international level. Rapid and ample progress has been achieved since the last two to three decades in all the aspects mentioned above (e.g. Horst and Marschner 1978a, b; Horiguchi and Morita 1987; Chérif et al. 1992a, b; 1994; Hodson and Evans 1995; Liang 1999; Epstein 1999; Rogalla and Römheld 2002; Rodrigues et al. 2003, 2004; Fauteux et al. 2005; Gong et al. 2005; Guo et al. 2005; Liang et al. 2005; Nikolic et al. 2007; Chain et al. 2009; Ghareeb et al. 2011a; Detmann et al. 2012; Liu et al. 2013a, b; Pavlovic et al. 2013; Ye et al. 2013; He et al. 2015), while the fastest and greatest progress has been made in mechanisms of Si uptake and transport in higher plants, pioneering work performed by Ma's group on isolation, identification and functional analysis of a series of genes responsible for Si transport in rice (*Oryza sativa*), maize (*Zea mays*), barley (*Hordeum vulgare*) and pumpkin (*Cucurbita moschata*) (Ma et al. 2006, 2007a; Yamaji and Ma 2009; Mitani et al. 2009, 2011; for more details, see Chapter 4). More recently, He et al. (2013, 2015) have provided more convincing and direct evidence showing that Si is naturally present as a constituent of the cell walls, where it is firmly bound to the cell wall matrix (hemicellulose) rather than occurring within intra- or extracellular silica deposition, as determined by using inductively coupled plasma mass spectrometry (ICP-MS) and X-ray photoelectron spectroscopy (XPS). This species of Si, linked with the cell wall matrix, improves the structural stability of cell walls during their expansion and subsequent cell division. Maintaining cell shape is thereby enhanced, which may be crucial for the function and survival of cells. These authors have provided further evidence indicating that organosilicon is present in plant cell walls, which broadens our understanding of the chemical nature of 'anomalous Si' in plant biology (He et al. 2013, 2015).

In this book, the current and updated knowledge of Si in soil, plant and fertilizer will be summarized and overviewed, and future research needs for Si in agriculture and in eco-environmental systems will also be discussed. In this chapter a review is presented on the history and general introduction of worldwide Si research.

1.2 Silicon Research and Application in Europe

Field application of slag-based fertilizer in Europe can be traced back to the Middle Ages (fifth to fifteenth centuries) when the medieval Cistercians in monasteries of European countries such as Dürstel in Switzerland, Noraskog in Sweden, Champagne in France and Märkische Sauerland in Germany, who were known to have been skilled metallurgists, extracted iron (Fe) using iron ore deposits and used the by-product of iron-making industry, phosphate-rich (also silicate- and calcium-rich) slag, from their furnaces as an agricultural fertilizer. The blast furnace, discovered in the twelfth century, is ranked fourth among the top 10 inventions of the Middle Ages. Like manure and cinder, iron-making slag has been used extensively for centuries as fertilizer or liming materials in maize, potato and sugar beet to improve crop yield and to rectify soil pH in Germany, Belgium, France and Britain.

Early studies of Si in plants in Europe showed that the concentration of Si was found to be between 10 and 20 times higher in species of Poaceae than in non-monocotyledonous species (de Saussure 1804). This may be the earliest recording of extremely wide variation of Si content in higher plants. In the book *Elements of Agricultural Chemistry* edited by Davy (1819), the greatest chemist of his age, the roles of Si in providing protection against insect damage were commented as saying that 'the siliceous epidermis of plants serves as support, protects the bark from the action of insects, and seems to perform a part in the economy of these feeble vegetable tribes (Grasses and Equisetables) similar to that performed in the animal kingdom by the shell of crustaceous insects'. Struve (1835) in Germany was also one of the pioneers to investigate the form of silica deposited on the cell walls of higher plants (phytoliths). He demonstrated for 'tabashir' in bamboo stems that silicified parts remain intact after ashing but dissolved in caustic potash solution. Later, more research was done on silica bodies in Cauto bark (*Moquilea* or *Hirtella* species, Chrysobalanaceae) and *Calamus* parenchyma (Cruger 1857) and in *Zea* and *Saccharum* (Wiesner 1867).

Liebig (1840), a German chemist and agricultural scientist (Fig. 1.1), first recommended sodium silicate as a Si fertilizer to improve crop productivity after he conducted the pioneering greenhouse experiments to explore silicate effect on sugar beet growth. He concluded that the true purpose of a fertilizer is to supply ammonia and such salts as potassium silicate, calcium phosphate and magnesium phosphate to plants. By using the solution culture technique around the 1860s in Germany, Sachs (1865) determined which elements are indispensable to plant nutrition and commented in his book (a chapter 'Nutrients') that Si does not belong among the

Fig. 1.1 Justus von Liebig (1803–1873)

indispensable elements in the same sense as potassium (K) and P but that Si is widely distributed in plants. This is the first paper excluding Si as an essential element, but stressing its wide distribution in plants.

Since 1856, a long-term field experiment at the Rothamsted Experimental Station in England has been conducted to test sodium silicate effect on grass and barley growth and yield. Sodium silicate, both as a fresh application and as a residue, continued to give substantial yield increases in the period 2002–2005 on plots lacking P or K but had no effect on plots receiving these nutrients (Rothamsted Research 2006).

Over the last century, numerous researches on the physiological role of Si and its agronomic effects have been carried out across Europe. Some of the most impressive work was done by Marschner and his co-workers, who criticized the claims of the essentiality of Si by Japanese scientists (Marschner et al. 1990). For instance, Miyake and Takahashi (1978, 1983) reported that Si-deprived cucumber (*Cucumis sativus*) and tomato (*Lycopersicon esculentum*) grown hydroponically exhibited typical deficiency symptoms, leading to an implication of Si as an essential element for higher plants. However, with regard to the experiments with cucumber, Marschner et al. (1990) concluded that the Si deficiency symptoms observed by Miyake and Takahashi (1983) were actually zinc (Zn) deficiency symptoms, because the concentrations of Zn in the nutrient solution were low (0.1 μM or less) and those of P were excessive. Marschner et al. (1990) could rectify the so-called Si deficiency symptoms by lowering the concentration of P in the nutrient solution or raising that of Zn. Thus, the observation by Miyake and Takahashi (1983) is an instance of Si-mediated amelioration of an unfavourable environment and that, according to the classical definition for essentiality established by Arnon and Stout (1939), does not qualify the element for 'essential' status.

Numerous European studies have been performed on Si-mediated resistance to metal toxicity stress including aluminium (Al) (e.g. Hodson and Evans 1995; Wang et al. 2004), manganese (Mn) (e.g. Horst and Marschner 1978a, b; Iwasaki et al. 2002a, b; Rogalla and Römheld 2002; Führs et al. 2009; Dragisic Maksimovic et al. 2012) and cadmium (Cd) (e.g. Vaculík et al. 2012).

Interestingly, Si has been found to alleviate Fe deficiency symptoms in cucumber as demonstrated for the first time by Nikolic and his colleagues, by enhancing remobilization of Fe from the root apoplast and subsequent xylem transport (Pavlovic et al. 2013; for more details, see Chap. 8). Pioneering research work has been done by Wydra and his colleagues showing that Si is also very effective in suppressing bacterial wilt in tomato, a Si non-accumulating (Si-excluder) plant species (Dannon and Wydra 2004; Diogo and Wydra 2007; Ghareeb et al. 2011a, b). The interactive mechanisms of Si and bacterial wilt in tomato were explained at both physiological and molecular levels (also see Chap. 9).

Another important finding has recently been reported by Fleck et al. (2011). In their work, they showed that Si reduced radical oxygen and oxidation power in rice, thus increasing suberization of exodermis and lignification of sclerenchyma. These changes increased with Si supply and were related to a higher transcription of genes associated with suberin and lignin synthesis.

1.3 Silicon Research in Asia

1.3.1 Japan

Japanese scientists started to investigate the role of Si in higher plants, especially in rice plants at the beginning of the twentieth century (reviewed by Ma and Takahashi 2002). The earlier research mainly focused on field observations of the relationship between disease severity and leaf Si content. By collecting rice plants from 13 different regions in Western Japan and comparing the chemical composition of the rice plants infected with blast with that of healthy ones grown in the same paddy field, Onodera (1917) first published a paper entitled 'Chemical studies on rice blast disease' in a scientific journal, concluding that blast diseased plants contained less Si than healthy ones obtained from the same field. Therefore, a close relationship existed between Si content in leaves and rice blast incidence, and the natural Si content found in rice tissue depended on the paddy field in which the plants had been grown. Although Onodera did not intend to study the physiological role Si played in rice blast resistance, his pioneering work stimulated further research on Si in Japan. Subsequently, more papers were published dealing with the effects of genotype and silicate fertilizer (Kawashima 1927; Ma and Takahashi 2002 and references therein) on rice blast severity both under controlled environments and in the field. Kawashima (1927) first demonstrated that both rice blast resistance and Si content in rice plants were significantly increased by application of Si under controlled conditions. He also found that Si content in rice straw and husks was proportional to the amount of Si amended to the soil and the severity of blast on panicles was inversely proportional to the amount of Si in rice tissues. Similar results were reported by Ito and Hayashi (1931) and Miyake and Ikeda (1932). Inokari and Kubota (1930) demonstrated that the incidence of blast was significantly reduced by application of Si fertilizer to Si-deficient paddy fields derived from peat soils.

To dramatically reduce the intensity of blast, application of 1.5–2.0 t ha^{-1} of various Si sources to Si-deficient paddy soils was recommended in Japan (Suzuki 1935). It was also shown that application of Si fertilizer was more effective in alleviating brown spot and blast in rice under excess nitrogen stress.

Rice is a staple food in Japan and other Asian countries. Monocropping with rice results in leaching and deficiency of nutrients and available Si. Soil degradation and low productivity thus occur in such rice fields (Ma and Takahashi 2002). To solve this problem, the Ministry of Agriculture, Forestry and Fisheries of Japan started nationwide trials using slag in rice fields at different experimental stations in 1952. The results indicate that, in addition to supplementing base ions (e.g. P, K, Fe, Mn and Zn), slag contains an important source of Si important for the improvement of the productivity in the degraded paddy soils. Consequently, Si was first among the fertilizers listed as a nutrient in the 1950s. The first national standard for slag-based Si fertilizer was established in 1955 in Japan. This is one of the most important contributions to the field of Si research and application by Japanese scientists. The extensive use of slag-based Si fertilizer to the degraded Si-deficient paddy soils also contributed significantly to sustained rice productivity and food safety after World War II. The annual consumption of slag-based Si fertilizer averages 1 million tons with its peak at 3 million tons. In addition, after Imaizumi and Yoshida (1958) made an intensive investigation on the Si-supplying capacity of paddy soils by using various solvents including hot hydrochloric acid, ammonium oxalate (at pH 3.0), 2 % sodium carbonate, 0.002 N sulphuric acid (at pH 3.0), saturated carbonic acid water (at pH 3.8) and sodium acetate (at pH 4.0), they found the best relationship between the amount of Si extracted by the acetate buffer and the amount of Si in rice straw. This finding served as the basis to define a method to assess a soil Si-supplying power that has been widely used both in Japan and in the rest of the world.

As for investigations on the roles Si plays in plant biology, Takeoka et al. (1979) postulated the famous 'window' hypothesis that the presence of epidermal silica bodies facilitates the transmission of light through the epidermis to the photosynthetic mesophyll or to stem cortical tissue, consequently increasing photosynthesis and plant growth. This hypothesis, however, was not supported by the later findings of Kaufman et al. (1979, 1981, 1985) and finally was disproved by Agarie et al. (1996).

By using IR and UV absorption spectroscopy, Inanaga et al. (1995) provided some evidence that Si exists in association with some organic compounds in the cell walls of rice seedlings. They hypothesized that in rice, Si may bind with the lignin–carbohydrate complexes or phenol–carbohydrate complexes in a similar way to calcium (Inanaga and Okasaka 1995; Inanaga et al. 1995). If Si is present in association with organic compounds in the cell walls of rice plants, it may compete with calcium for binding sites on the cell wall. This hypothesis has recently been confirmed by He et al. (2013, 2015).

From efforts initiated in the early 2000s, Ma and his co-workers made a breakthrough in Si transport in plants that led to the identification, characterization, cloning and functional analysis of a series of Si transporter genes (*Lsi1*, *Lsi2* and *Lsi6*) responsible for Si uptake, translocation and distribution first in rice, then in barley (*Hordeum vulgare*) and maize (*Zea mays*) and more recently in pumpkin

(Ma et al. 2006, 2007a, b, 2011; for more details, see Chap. 4). These Si genes first identified in higher plants bear no similarity to the Si transporters previously identified in diatoms (Hilderbrand et al. 1997).

1.3.2 China

China has a long history of contributions to agriculture with more than 5000 years of agricultural civilization. According to China's first agricultural encyclopaedia, transcribed in Latin letters as 'Qi Min Yao Shu' (English translation 'Important Arts for the People's Welfare'; see http://www.chinaculture.org/created/2005-04/30/content_68368.htm) by Jia Sixie in the late years of the Northern Wei Dynasty (386–534), applying organic manure such as straw, manure and composting to cropland is the best way to improve soil fertility and sustain crop productivity through recycling of nutrients including Si. Also, wood or crop straw burning in China has been a rather common practice for cooking in rural areas; the returning of the burning ash and/or cinder containing large quantities of mineral elements including Si and K to crop fields as a soil amendment is an extremely effective supplement of Si nutrient removed by crop harvest. However, the effects of steel or iron slag-based Si fertilizer on crop growth, yield and quality were not studied until the 1950s with the paper published by Zhu and Chen (1963) in the scientific journal *Acta Pedologia Sinica*. Starting in 1958, Zhu and Chen (1963) conducted extensive field trials with different blast furnace slags in different types of soils differing in pH (pH 6.5–7.6) and soil texture and in varying crops including rice, wheat (*Triticum aestivum*), maize, millet (*Setaria italica*), sorghum (*Sorghum bicolor*), soybean (*Glycine max*), cotton (*Gossypium*), potato (*Solanum tuberosum*), sugar beets (*Beta vulgaris*), sunflower (*Helianthus annuus*) and pumpkin (*Cucurbita moschata*) in Liaoning, Jilin and Heilongjiang provinces of Northeast China. The results showed that, on average, application of slag increased rice yield by 10 % and upland crop yield by 5 % with distinct residual effects. Application of slag increased not only crop yield but also crop quality. For example, oil content of soybean and sunflower, starch content of potato and rice, sugar content of beets, protein content of wheat and cotton yield before frost were significantly raised by application of blast furnace slag. Since then, numerous studies have focused on Si-supplying capacity in low-pH oxisols and ultisols located in tropical and subtropical areas of south China including Chinese Taiwan and the responses of different crops such as rice, sugarcane, wheat, maize and other vegetable and horticultural crops (Lian 1976; Qin 1979; He et al. 1980; Ma et al. 1985).

Most of the research work conducted in China before 1990 focused mainly on Si-supplying power and crop responses to slag-based Si fertilizers in acid soils in tropical and subtropical regions. The positive effect of Si on crops was restricted to Si-deficient (critical plant-available SiO_2 content of 105–120 mg kg^{-1}) paddy soils derived from acid soils, while those developed from calcareous soils with high pH were thought to be Si sufficient with plant-available SiO_2 content of over 200 mg kg^{-1} since application of Si fertilizers to these soils did not affect crops.

Nevertheless, Liang et al. (1994) first demonstrated that Si deficiency still occurred in rice grown on calcareous paddy soils in which plant-available Si was not sufficient although the acetate-extractable Si was much higher than the critical value. The acetate buffer (pH 4.0) extraction method seems to overestimate Si-supplying power in calcareous soils with considerable amount of the extracted Si unavailable to plants. Positive rice growth and yield responses were obtained with Si applied to those soils.

Investigations into the physiological and biochemical roles of Si in stimulating plant growth, increasing crop yield and enhancing resistance to lodging, plant disease and pest damage and various forms of environmental stress have been ongoing since the 1970s. Many papers have been published in domestic and international journals over the last decades. One of the highlights in Si research in China is the hypothesis proposed by Liang and his colleagues (e.g. Liang 1999; Liang et al. 2003, 2007, 2008; Song et al. 2009, 2011) that Si-enhanced antioxidant defence capacity is a universal and major strategy for Si-mediated tolerance to abiotic stress such as salinity, freezing, drought and heavy metals in plants. Progress has also been made in mechanisms of Si uptake and transport by dicots such as cucumber. It has been demonstrated for the first time that Si uptake and transport is an active process in cucumber (Liang et al. 2005). Furthermore, Liang et al. (2006) concluded that both active and passive Si-uptake components coexist in rice, maize, sunflower and wax gourd, with their relative contribution being dependent upon the plant species and external Si concentrations. The contribution of passive component cannot be overlooked in rice and maize, especially at a higher external Si concentration, while the active component is also important in sunflower and wax gourd, especially at a lower external Si concentration. Furthermore, xylem loading of Si is an active process in rice, maize, sunflower and wax gourd, regardless of external Si concentrations.

By analysing single cells of the suspended rice cell culture in the absence and presence of Si using a combination of physical techniques including atomic force microscopy (AFM), He et al. (2013, 2015) have recently provided more convincing and direct evidence demonstrating that Si is naturally present as a constituent of the cell walls rather than occurring within intra- or extracellular silica deposition. More importantly, it seems that a majority of Si occurred firmly bound to the polysaccharide matrix of cell walls, such as hemicellulose, not as free silicate/silicic acid or silica, which is less readily liberated by weak alkali or acid or even enzymes. This wall-bound form of Si also plays a significant role in maintaining cellular integrity and structure of the walls, thereby highlighting the biochemical and structural role of Si at the trace level in preserving cell shape and mechanical properties to ensure subsequent wall expansion and cell division, which may be crucial for the function and survival of cells.

Rapid and ample progress has recently been achieved by Ye et al. (2013) who have first demonstrated that apart from the roles Si may play in passive or constitutive defences, Si primes jasmonate (JA)-mediated inducible antiherbivore defence responses and increases rice resistance to the leaffolder caterpillar, and Si accumulation in rice leaves is mediated by the JA pathway, suggesting a strong interaction between Si and JA in rice defence against insect herbivores. This interaction is considered to be a possible widespread mechanism by which Si enhances plant resistance against biotic stresses in Si-accumulating plant species (Ye et al. 2013).

1.3.3 *Other Southeast Asian Countries*

In addition to the extensive research on soil, plant and fertilizer Si in Japan and China, numerous researches targeting raising crop yield by using slag-based Si fertilizers were also conducted in South Korea, North Korea, Thailand, Malaysia, Indonesia, the Philippines, Sri Lanka, Ceylon, India, Pakistan, etc. Application of slag-based Si fertilizers to paddy soils is a rather common practice in these countries.

South Korea has a long history of research on the effects of Si on rice growth and yield, which has started since the 1960s (Kim and Choi 2002; Lee et al. 2005). Korean paddy soils are generally characterized by low pH and low available Si content. The soil pH and available Si content averaged 5.6 and 72 mg SiO_2 kg^{-1}, respectively, which were lower than the optimum ranges of pH (6.0–6.5) and available Si (130–180 mg SiO_2 kg^{-1}).

The first silicate fertilizer in Korea from 1963 to 1990 was the grounded wollastonite. Another important source of Si fertilizer is blast furnace slag, a by-product of iron and steel industry. This silicate fertilizer has been provided to farms every 4 years and subsidized by the government since 1996. With successive use of silicate fertilizer, the average Si content in soil was increased from 75 mg SiO_2 kg^{-1} in the 1970s to 118 mg SiO_2 kg^{-1} in 2003.

The beneficial effect of Si on rice yield has been extensively studied in Korea. For instance, a 26-year field experiment was done from 1975 to 2000 on the Si fertilization of 1.5 Mg ha^{-1} to the clay loam paddy fields. The average rice yield for every 5-year interval was observed to continually increase with silicate fertilizer applied at 1.5 Mg ha^{-1} as compared to a NPK treatment (Kim and Choi 2002).

In the Philippines, scientists in IRRI (International Rice Research Institute) conducted a series of field experiments to investigate the effects of different forms of Si sources, rates and methods of application on agronomic and entomological aspects of flooded rice and obtained encouraging results (IRRI 1965, 1966, 1967). Positive rice yield responses to Si fertilizers have also been reported in Sri Lanka (Rodrigo 1964; Takijima et al. 1970), Thailand (Takahashi et al. 1980), Indonesia (Burbey et al. 1988), and India (Datta et al. 1962; Sadanandan and Varghes 1968).

1.4 Silicon Research in America and Australia

Although the first patents for using slag as a fertilizer were obtained by Zippicotte in the USA in 1881 (Zippicotte 1881), Si was not approved as a plant-beneficial substance by the Association of American Plant Food Control Officials (AAPFCO) until 2012. Indeed, the first analysis for Si content in soils available for plants was conducted on the Hawaiian Islands in 1898 (Maxwell 1898). Before the 1950s, numerous researches were conducted in the USA to demonstrate whether Si was an essential element for higher plants, but no conclusion was drawn because no convincing evidence was obtained. The earlier investigations showed that Si was a beneficial or essential element by some plants (Sommer 1926; Lipman 1938; Raleigh 1939). For example, Sommer (1926) reported that the growth of rice and of millet was improved by the addition of Si to the nutrient solution. Lipman (1938) found that sunflower and barley definitely benefited from addition of Si to the culture medium, especially in terms of seed production. Raleigh (1939) reported that Si was an indispensible element for the growth of beet plant. However, such experimental results of earlier investigations into the essentiality of Si were difficult to interpret because these experiments were performed with paraffin-coated glass containers or asphalt-painted iron containers. It was then impossible and is now still impossible to create a Si-free experimental environment to verify whether Si is essential or not for higher plants because of contamination from chemical and water impurities and from dust. In a solution culture experiment, Woolley (1957) grew tomato plants with double distilled water (0.00085 g L^{-1} Si) and purified salts to exclude Si from the environment to determine whether it is an essential element for tomato. The Si-deprived plants with only 4.2 mg kg^{-1} of Si in the shoots grew as well as those deliberately supplied with the element. Physiologically speaking, Si can perhaps be considered as a micronutrient, the inadvertent supply of which through environmental contamination meeting the requirements of solution-cultured plants.

An early and important discovery into the significance of Si in plant biology was made by Williams and Vlamis (1957a, b), who demonstrated that addition of Si to a nutrient solution supplied with excess Mn prevented Mn toxicity symptoms (dark brown necrotic spots in the leaves) in barley, however with no effect on the total leaf Mn content. They concluded that Si caused Mn to be more evenly distributed instead of being concentrated in discrete necrotic spots.

In the 1960s, two most impressive review articles on Si were published, which stimulated and deepened the research work related to Si in agriculture and plant biology. In 1967, Jones and Handreck (1967) from Australia published a thorough review article, which covered several aspects of Si ranging from Si soil chemistry and availability through plant Si uptake and the factors affecting Si uptake to Si-alleviated manganese toxicity to animal Si nutrition via ingesting pasture plants. In 1969, Lewin and Reimann (1969) published a review article 'Silicon and Plant Growth', in which they discussed in detail chemical forms, uptake and distribution of Si in plants and its effect on plant growth, increased Mn and Fe toxicity and water loss by transpiration under Si deficiency, as well as the interactions of Si with P and B. In recent years, the most important review articles regarding plant Si nutrition are arguably the two reviews by Epstein (1994, 1999), which have greatly stimulated worldwide research on Si at chemical, physiological and molecular aspects. Considering the importance of Si in plant growth and the reality of presence of Si at an appreciable concentration in the medium to which plant roots are normally exposed, Epstein (1994, 1999) first discussed the reasons why Si should not have been omitted from conventional culture solutions and why the experimental plants grown in Si-deprived nutrient solutions should be considered to be experimental artefacts. Also inclusion of Si should be recommended in commercial hydroponics.

Since the 1990s, great progress has been made in Canada and the USA regarding the mode of action of Si against plant diseases. Earlier reports were mostly associated with the mechanical barrier hypothesis whereby fungal penetration was believed to be impeded by silica opals or phytoliths deposited on the cell walls (Wagner 1940; Heath and Stumpf 1986; Carver et al. 1987, 1994), and for a long time, this mode of action prevailed but stood uncontested. This hypothesis, however, was challenged by Samuels et al. (1991a, b) and particularly by Bélanger's group (Chérif et al. 1992a, b, 1994) who first linked this beneficial effect of Si to the elicitation of defence mechanisms. Indeed, they found that the activities of some pathogenesis-related proteins (enzymes) such as peroxidase (POD), polyphenol oxidase (PPO) and chitinase in cucumber infected with *Pythium* spp. were significantly stimulated by Si. Subsequently, Bélanger and his co-workers (Fawe et al. 1998, 2001) provided some direct evidence showing that Si played more than a mechanical role *in planta* as Si was shown to be involved in enhancing cucumber resistance to *P. xanthii* by stimulating phytoalexin production within the plant. Thus, Si was hypothesized to be actively involved in triggering defence mechanisms in response to fungal attack in cucumber. This hypothesis was further supported by Datnoff and his co-workers who reported that the fungitoxic compounds identified as momilactones, known as rice phytoalexins, were detected in blast-infected rice supplied with Si (Rodrigues et al. 2003, 2004). Furthermore, evidence was provided that Si-alleviated powdery mildew infection in wheat was attributable to a cellular reaction to fungal presence (Bélanger et al. 2003). Later on, Fauteux et al. (2005) suggested that Si could act as a potentiator of plant defence responses or as an activator of strategic signalling proteins to interact with several key components of plant stress signalling systems, ultimately leading to induced resistance against pathogenic fungi. However, the molecular mechanisms of how such priming is associated with Si still remain unclear. The association between Si and priming in plant–pathogen interactions has recently been corroborated by Vivancos et al. (2015), but at the same time their work suggests that mechanisms other than SA-dependent plant defence priming are involved. These latest results may be helpful in defining a unifying theory explaining the elusive and debated mode of action of Si in plant–pathogen interactions (see Chap. 9).

1.5 Silicon Research in Africa

The first discovery that applications of silica may benefit sugarcane growth can be dated back to 1937 in Mauritius by D'Hotman De Villiers (Meyer and Keeping 2001). Application of Si-containing crushed basalt at rates varying from 200 to 400 t ha^{-1} could lead to cumulative yield increases between 30 and 60 t ha^{-1}. Since then, many field trials on application of silicate materials to highly weathered entisols, ultisols and oxisols low in soluble Si have resulted in significant responses in both cane and sugar yields in several African countries including Mauritius and South Africa (Bishop 1967; Cheong and Halais 1970; Du Preez 1970; Meyer and Keeping 2001). The more recent research on Si in Africa has been mainly focused on the influence of Si on the resistance of sugarcane to the stalk borer *Eldana saccharina* (Keeping and Meyer 2006; Smith et al. 2007; Kvedaras et al. 2007, 2009; see Chap. 10).

References

Agarie S, Agata W, Uchida H, Kubota F, Kaufman P. Function of silica bodies in the epidermal system of rice (*Oryza sativa* L.): testing the window hypothesis. J Exp Bot. 1996;47:655–60.

Arnon DI, Stout PR. The essentiality of certain elements in minute quantity for plants with special reference to copper. Plant Physiol. 1939;14:371–5.

Bélanger RR, Benhamou N, Menzies JG. Cytological evidence of an active role of silicon in wheat resistance to powdery mildew (*Blumeria graminis* f. sp. *tritici*). Phytopathology. 2003;934:402–12.

Bishop RT. Aluminium and silica relationships in growth failure areas. Proceedings of the South African Sugar Technologists Association; 1967. pp. 190–196.

Burbey A, Rizaldi B, Yulizar A. Response of upland rice to potassium and silicate application on Ultisol. Pemberitaan Penelitian Sukarami. 1988;15:26–31 (In Indonesian, with English summary).

Carver TLW, Zeyen RJ, Ahlstrand GG. The relationship between insoluble silicon and success or failure of attempted primary penetration by powdery mildew (*Erysiphe graminis*) germlings on barley. Physiol Mol Plant P. 1987;31:133–48.

Carver TLW, Ingerson-Morris SM, Thomas BJ, Gay AP. Light-mediated delay of primary haustorium formation by *Erysiphe graminis* f sp *avenae*. Physiol Mol Plant P. 1994;45:59–79.

Chain F, Côté-Beaulieu C, Belzile F, Menzies JG, Bélanger RR. A comprehensive transcriptomic analysis of the effect of silicon on wheat plants under control and pathogen stress conditions. Mol Plant Microbe Inter. 2009;22:1323–30.

Cheong YWY, Halais P. Needs of sugar cane for silicon when growing in highly weathered latosols. Exp Agric. 1970;6:99–106.

Chérif M, Benhamou N, Menzies JG, Bélanger RR. Silicon induced resistance in cucumber plants against *Pythium ultimum*. Physiol Mol Plant P. 1992a;41:411–25.

Chérif M, Benhamou N, Menzies JG, Bélanger RR. Studies of silicon distribution in wounded and *Pythium ultimum*-infected cucumber plants. Physiol Mol Plant P. 1992b;41:371–85.

Chérif M, Asselin A, Bélanger RR. Defense responses induced by soluble silicon in cucumber roots infected by *Pythium* spp. Phytopathology. 1994;84:236–42.

Cruger H. Westindische Fragmente, 9. El Cauto. Bot Zeitung. 1857;15:281–308.

Dannon EA, Wydra K. Interaction between silicon amendments, bacterial wilt development and phenotype of *Ralstonia solanacearum* in tomato genotypes. Physiol Mol Plant Pathol. 2004;64:233–43.

Datta NP, Shinde JE, Kamath MD, Datta SK. Effect of sodium silicate on the uptake of soil and fertilizer phosphorus by wheat, rice and berseem. Indian J Agric Sci. 1962;32:219–27.

Davy H. The elements of agricultural chemistry. Hartford: Hudson and Co.; 1819.

de Saussure T. Recherches chimiques sur la végétation. Paris: Nyon; 1804.

Detmann KC, Araújo L, Martins SCV, et al. Silicon nutrition increases grain yield, which, in turn, exerts a feed-forward stimulation of photosynthetic rates via enhanced mesophyll conductance and alters primary metabolism in rice. New Phytol. 2012;196:752–62.

Diogo RVC, Wydra K. Silicon-induced basal resistance in tomato against *Ralstonia solanacearum* is related to modification of pectic cell wall polysaccharide structure. Physiol Mol Plant Pathol. 2007;70:120–9.

Dragisic Maksimovic J, Mojovic M, Maksimovic V, Römheld V, Nikolic M. Silicon ameliorates manganese toxicity in cucumber by decreasing hydroxyl radical accumulation in the leaf apoplast. J Exp Bot. 2012;63:2411–20.

Du Preez P. The effect of silica on cane growth. Proc S Afr Sug Technol Ass. 1970;44:83–8.

Epstein E. The anomaly of silicon in plant biology. Proc Natl Acad Sci U S A. 1994;91:11–7.

Epstein E. Silicon. Annu Rev Plant Physiol Plant Mol Biol. 1999;50:641–64.

Epstein E. Silicon in plants: facts vs. concepts. In: Datnoff LE, Snyder GH, Korndöfer GH, editors. Silicon in agriculture. Amsterdam: Elsevier Science; 2001. p. 1–15.

Epstein E, Bloom AJ. Mineral nutrition of plants: principles and perspectives. 2nd ed. Sunderland: Sinauer Associates Inc; 2005.

Fauteux F, Rémus-Borel W, Menzies JG, Bélanger RR. Silicon and plant disease resistance against pathogenic fungi. FEMS Microbiol Lett. 2005;249:1–6.

Fawe A, Abou-Zaid M, Menzies JG, Bélanger RR. Silicon-mediated accumulation of flavonoid phytoalexins in cucumber. Phytopathology. 1998;88:396–401.

Fawe A, Menzies JG, Chérif M, Bélanger RR. Silicon and disease resistance in dicotyledons. In: Datnoff LE, Snyder GH, Korndöfer GH, editors. Silicon in agriculture. Amsterdam: Elsevier; 2001. p. 159–70.

Fleck AT, Nye T, Repenning C, Stahl F, Zahn M, Schenk MK. Silicon enhances suberization and lignification in roots of rice (*Oryza sativa*). J Exp Bot. 2011;62:2001–11.

Führs H, Götze S, Specht A, Erban A, Gallien S, Heintz D, Van Dorsselaer A, Kopka J, Braun H-P, Horst WJ. Characterization of leaf apoplastic peroxidases and metabolites in *Vigna unguiculata* in response to toxic manganese supply and silicon. J Exp Bot. 2009;60:1663–78.

Ghareeb H, Bozsó Z, Ott PG, Repenning C, Stahl F, Wydra K. Transcriptome of silicon-induced resistance against Ralstonia solanacearum in the silicon non accumulator tomato implicates priming effect. Physiol Mol Plant Pathol. 2011a;75:83–9.

Ghareeb H, Bozsó Z, Ott PG, Wydra K. Silicon and *Ralstonia solanacearum* modulate expression stability of housekeeping genes in tomato. Physiol Mol Plant Pathol. 2011b;75:176–9.

Gong HJ, Zhu XY, Chen KM, Wang S, Zhang CL. Silicon alleviates oxidative damage of wheat plants in pots under drought. Plant Sci. 2005;169:313–21.

Gunnarsson I, Arnórsson S. Amorphous silica solubility and the thermodynamic properties of H_4SiO_4 degrees in the range of 0 degrees to 350 degrees C at P-sat. Geochim Cosmochim Acta. 2000;64:2295–307.

Guo W, Hou YL, Wang SG, Zhu YG. Effect of silicate on the growth and arsenate uptake by rice (*Oryza sativa* L.) seedlings in solution culture. Plant Soil. 2005;272:173–81.

He DY, Zang HL, Zhang XP. The effect of silicate slags on rice in paddy soils derived from red earth. Acta Pedol Sin. 1980;17:355–63 (In Chinese with English abstract).

He CW, Wang LJ, Liu J, Liu X, Li XL, Ma J, Lin YJ, Xu FS. Evidence for 'silicon' within the cell walls of suspension-cultured rice cells. New Phytol. 2013;200:700–9.

He CW, Ma J, Wang LJ. A hemicellulose-bound form of silicon with potential to improve the mechanical properties and regeneration of the cell wall of rice. New Phytol. 2015;206:1051–62.

Heath MC, Stumpf MA. Ultrastructural observations of penetration sites of the cowpea rust fungus in untreated and silicon-depleted French bean cells. Physiol Mol Plant P. 1986;29:27–39.

Hilderbrand M, Volcani BE, Gossmann W, Schroeder J. A gene family of silicon transporters. Nature. 1997;385:688–9.

Hodson MJ, Evans DE. Aluminium/ silicon interactions in higher plants. J Exp Bot. 1995;46:161–71.

Horiguchi T, Morita S. Mechanism of manganese toxicity and tolerance of plants. 6. Effect of silicon on alleviation of manganese toxicity of barley. J Plant Nutr. 1987;10:2299–310.

Horst WJ, Marschner H. Effect of silicon on manganese tolerance of bean plants (*Phaseolus vulgaris* L.). Plant Soil. 1978a;50:287–303.

Horst WJ, Marschner H. Einfluβ von Silizium auf den Bindungszustand von Mangan im Blattgewebe von Bohnen (*Phaseolus vulgaris*). Z Pflanzenernaehr Bodenkd. 1978b; 141:487–97.

Imaizumi K, Yoshidai S. Edaphological studies on the silicon supplying power of paddy soils. Bull Natl Inst Agric Sci Tokyo. 1958;B8:261–304.

Inanaga S, Okasaka A. Calcium and silicon binding compounds in cell walls of rice shoots. Soil Sci Plant Nutr. 1995;41:103–10.

Inanaga S, Okasaka A, Tanaka S. Does silicon exist in association with organic compounds in rice plant? Soil Sci Plant Nutr. 1995;41:111–7.

Inokari G, Kubota T. On the effect of soil dressing and silica application to peatland paddy fields. J Sapporo Soc Agric Sci. 1930;98:484–5.

IRRI, International Rice Research Institute. Annual report 1964, Los Banos, Laguna, Philippines; 1965.

IRRI, International Rice Research Institute. Annual report 1965, Los Banos, Laguna, Philippines; 1966.

IRRI, International Rice Research Institute. Annual report 1966, Los Banos, Laguna, Philippines; 1967.

Ito S, Hayashi H. On the relation of silica supply to rice blast. J Sapporo Soc Agric Sci. 1931;103:460–1.

Iwasaki K, Maier P, Fecht M, Horst WJ. Effects of silicon supply on apoplastic manganese concentrations in leaves and their relation to manganese tolerance in cowpea (*Vigna unguiculata* (L.) Walp.). Plant Soil. 2002a;238:281–8.

Iwasaki K, Maier P, Fecht M, Horst WJ. Leaf apoplastic silicon enhances manganese tolerance of cowpea (*Vigna unguiculata*). J Plant Physiol. 2002b;159:167–73.

Jones LHP, Handreck KA. Silica in soils, plants and animals. Adv Agron. 1967;19:107–49.

Kaufman PBL, Takeoka Y, Carlson TJ, Bigelow WC, Jones JD, Moore EH, Ghosheh NS. Studies on silica deposition in sugarcane (*Saccharum* spp.) using scanning electron microscopy, energy-dispersive X-ray analysis, neutron activation analysis, and light microscopy. Phytomorphology. 1979;29:185–93.

Kaufman PBL, Dayanandan P, Takeoka Y, Bigelow WC, Jones JD, Iler R. Silica in shoots of higher plants. In: Simpson TL, Volcani BE, editors. Silicon and siliceous structures in biological systems. New York: Springer; 1981. p. 409–49.

Kaufman PB, Dayanandan P, Franklin CI, Takeoka Y. Structure and function of silica bodies in the epidermal system of grass shoots. Ann Bot. 1985;55:487–507.

Kawashima R. Influence of silica on rice blast disease. Jpn J Soil Sci Plant Nutr. 1927;1:86–91.

Keeping MG, Meyer JH. Silicon-mediated resistance of sugarcane to *Eldana saccharina* Walker (Lepidoptera: Pyralidae): effects of silicon source and cultivar. J Appl Entomol. 2006;130:410–20.

Kim C-B, Choi J. Changes in rice yield, nutrients use efficiency and soil chemical properties as affected by annual application of slag silicate fertilizer. Korean J Soil Sci Fertil. 2002;35:280–9.

Kvedaras OL, Keeping MG, Goebel FR, Byrne MJ. Larval performance of the pyralid borer *Eldana saccharina* Walker and stalk damage in sugarcane: influence of plant silicon, cultivar and feeding site. Int J Pest Manag. 2007;53:183–94.

Kvedaras OL, Byrne MJ, Coombes NE, Keeping MG. Influence of plant silicon and sugarcane cultivar on mandibular wear in the stalk borer *Eldana saccharina*. Agr Forest Entomol. 2009;11:301–6.

Lee CH, Yang MS, Chang KW, Lee YB, Chung KY, Kim PJ. Reducing nitrogen fertilization level of rice (*Oryza sativa* L.) by silicate application in Korean paddy soil. Korean J Soil Sci Fert. 2005;38:194–201.

Liebig J. Organic chemistry in its application to agriculture and physiology. From the manuscript of the author by Lyon Playfair. London: Taylor & Walton; 1840.

Lewin J, Reimann BEF. Silicon and plant growth. Annu Rev Plant Physiol. 1969;20:289–304.

Lian S. Silica fertilization of rice. In: The fertility of paddy soils and fertilizer applications for rice. Taipei: Food Fertilizer Technology Center; 1976. p. 197–220.

Liang YC. Effects of silicon on enzyme activity, and sodium, potassium and calcium concentration in barley under salt stress. Plant Soil. 1999;209:217–24.

Liang YC, Ma TS, Li FJ, Feng YJ. Silicon availability and response of rice and wheat to silicon in calcareous soils. Commun Soil Sci Plant Anal. 1994;25:2285–97.

Liang YC, Chen Q, Liu Q, Zhang WH, Ding RX. Exogenous silicon (Si) increases antioxidant enzyme activity and reduces lipid peroxidation in roots of salt-stressed barley (*Hordeum vulgare* L.). J Plant Physiol. 2003;160:1157–64.

Liang YC, Si J, Römheld V. Silicon uptake and transport is an active process in *Cucumis sativus*. New Phytol. 2005;167:797–804.

Liang YC, Hua HX, Zhu YG, Zhang J, Cheng CM, Römheld V. Importance of plant species and external silicon concentration to active silicon uptake and transport. New Phytol. 2006;172:63–72.

Liang YC, Sun WC, Zhu YG, Christie P. Mechanisms of silicon-mediated alleviation of abiotic stresses in higher plants: a review. Environ Pollut. 2007;147:422–8.

Liang YC, Zhu J, Li ZJ, Chu GX, Ding YF, Zhang J, Sun WC. Role of silicon in enhancing resistance to freezing stress in two contrasting winter wheat cultivars. Environ Exp Bot. 2008;64:286–94.

Lipman CB. Importance of silicon, aluminum, and chlorine for higher plants. Soil Sci. 1938;45:189–98.

Liu JG, Zhang HM, Zhang YX, Chai TY. Silicon attenuates cadmium toxicity in *Solanum nigrum* L. by reducing cadmium uptake and oxidative stress. Plant Physiol Biochem. 2013a;68:1–7.

Liu J, Ma J, He CW, Li XL, Zhang WJ, Xu FX, Lin YJ, Wang LJ. Inhibition of cadmium ion uptake in rice (*Oryza sativa*) cells by a wall-bound form of silicon. New Phytol. 2013b;200:691–9.

Ma JF, Takahashi E. Soil, fertilizer, and plant silicon research in Japan. Amsterdam: Elsevier; 2002.

Ma TS, Qian ZR, Zhang ZF, Shu JM, Wang HY. Study on the paddy soils V. Preliminary study on the availability of silica in paddy soil of Jiangsu hilly area and the effect of slag silica fertilizer. J Nanjing Agric Univ. 1985;4:67–73 (In Chinese with English abstract).

Ma JF, Tamai K, Yamaji N, Mitani N, Konishi S, Katsuhara M, Ishiguro M, Murata Y, Yano M. A silicon transporter in rice. Nature. 2006;440:688–91.

Ma JF, Yamaji N, Mitani M, Tamai K, Konishi S, Fujiwara T, Katsuhara M, Yano M. An efflux transporter of silicon in rice. Nature. 2007a;448:209–12.

Ma JF, Yamaji N, Tamai K, Mitani N. Genotypic difference in Si uptake and expression of Si transporter genes in rice. Plant Physiol. 2007b;145:919–24.

Ma JF, Yamaji N, Mitani-Ueno N. Transport of silicon from roots to panicles in plants. Jpn Acad Ser B. 2011;87:377–85.

Mali M, Aery NC. Silicon effects on nodule growth, dry-matter production, and mineral nutrition of cowpea (*Vigna unguiculata*). J Plant Nutr Soil Sci. 2008;171:835–40.

Marschner H, Oberle H, Cakmak L, Romheld V. Growth enhancement by silicon in cucumber (*Cucumis sativus*) plants depends on imbalance in phosphorus and zinc supply. Plant Soil. 1990;124:211–9.

Maxwell W. Lavas and soils of the Hawaiian islands. Honolulu: Hawaiian Sugar Planters' Association; 1898. p. 189.

Meyer JH, Keeping MG. Past, present and future research of the role for sugarcane in southern Africa. In: Datnoff LE, Snyder GH, Korndörfer GH, editors. Silicon in agriculture. Amsterdam: Elsevier; 2001. p. 257–75.

Mitani N, Chiba Y, Yamaji N, Ma JF. Identification and characterization of maize and barley Lsi-2-like silicon efflux transporters reveals a distinct silicon uptake system from that in rice. Plant Cell. 2009;21:2133–42.

Mitani N, Yamaji N, Ago Y, Iwasaki K, Ma JF. Isolation and functional characterization of an influx silicon transporter in two pumpkin cultivars contrasting in Si accumulation. Plant J. 2011;66:231–40.

Miyake K, Ikeda M. Influence of silica application on rice blast. Jpn J Soil Sci Plant Nutr. 1932;6:53–76 (In Japanese).

Miyake Y, Takahashi E. Silicon deficiency of tomato plants. Soil Sci Plant Nutr. 1978;24:175–90.

Miyake Y, Takahashi E. Effect of silicon on the growth of solution-cultured cucumber plant. Soil Sci Plant Nutr. 1983;29:71–83.

Nelwamondo A, Dakora FD. Silicon promotes nodule formation and nodule function in symbiotic cowpea (*Vigna unguiculata*). New Phytol. 1999;142:463–7.

Nelwamondo A, Jaffer MA, Dakora FD. Subcellular organization of N2-fixing nodules of cowpea (Vigna *unguiculata*) supplied with silicon. Protoplasma. 2001;216:94–100.

Nikolic M, Nikolic N, Liang Y, Kirkby EA, Römheld V. Germanium-68 as an adequate tracer for silicon transport in plants. Characterization of silicon uptake in different crop species. Plant Physiol. 2007;143:495–503.

Onodera I. Chemical studies on rice blast. J Sci Agric Soc. 1917;180:606–17.

Pavlovic J, Samardzic J, Masimović V, Timotijevic G, Stevic N, Laursen KH, Hansen TH, Husted S, Schjoerring JK, Liang Y, Nikolic M. Silicon alleviates iron deficiency in cucumber by promoting mobilization of iron in the root apoplast. New Phytol. 2013;198:1096–107.

Qin RC. Effect of silicon fertilizer on disease resistance and yield increase in rice. Zhejiang Agric Sci. 1979;1:12–5 (In Chinese).

Raleigh GJ. Evidence for the essentiality of silicon for growth of the beet plant. Plant Physiol. 1939;14:823–8.

Richmond KE, Sussman M. Got silicon? The non-essential beneficial plant nutrient. Curr Opin Plant Biol. 2003;6:268–72.

Rodrigo DM. Response of rice to silica. Trop Agric. 1964;120:219–26.

Rodrigues FÁ, Benhamou N, Datnoff LE, Jones JB, Bélanger RR. Ultrastructural and cytochemical aspects of silicon-mediated rice blast resistance. Phytopathology. 2003;93:535–46.

Rodrigues FÁ, McNally DJ, Datnoff LE, Jones JB, Labbé C, Benhamou N, Bélanger RR. Silicon enhances the accumulation of diterpenoid phytoalexins in rice: a potential mechanism for blast resistance. Phytopathology. 2004;94:177–83.

Rogalla H, Römheld V. Role of leaf apoplast in silicon-mediated manganese tolerance of *Cucumis sativus* L. Plant Cell Environ. 2002;25:549–55.

Rothamsted Research. Classical and other long-term experiments, datasets and sample archive. Lawes Agricultural Trust Co Ltd; 2006. pp. 31–4.

Sachs J. Handbuch der Experimental Physiologie der Pflanzen. Leipzig: Verlag von Wilhelm Engelmann; 1865. p. 514.

Sadanandan AK, Varghes EJ. Studies on the silicate nutrition of rice in the laterite soil of Kerala. I. Effect on growth and yield. Madras Agric J. 1968;11:261–4.

Samuels AL, Glass ADM, Ehret DL, Menzies JG. Distribution of silicon in cucumber leaves during infection by powdery mildew fungus (*Sphaerotheca fuliginea*). Can J Bot. 1991a;69:140–6.

Samuels AL, Glass ADM, Ehret DL, Menzies JG. Mobility and deposition of silicon in cucumber plants. Plant Cell Environ. 1991b;14:485–92.

Smith MT, Kvedaras OL, Keeping MG. A novel method to determine larval mandibular wear of the African stalk borer, *Eldana saccharina* Walker (Lepidoptera: Pyralidae). Afr Entomol. 2007;15:204–8.

Sommer AL. Studies concerning essential nature of aluminum and silicon for plant growth. Univ Calif Publ Agric Sci. 1926;5:2.

Song AL, Li ZJ, Zhang J, Xue GF, Fan FL, Liang YC. Silicon-enhanced resistance to cadmium toxicity in *Brassica chinensis* L. is attributed to Si-suppressed cadmium uptake and transport and Si-enhanced antioxidant defense capacity. J Hazard Mater. 2009;172:74–83.

Song AL, Li P, Li ZJ, Fan FL, Nikolic M, Liang YC. The alleviation of zinc toxicity by silicon is related to zinc transport and antioxidative reactions in rice. Plant Soil. 2011;344:319–33.

Song Z, Liu H, Si Y, Yin Y. The production of phytoliths in China's grasslands: implications to the biogeochemical sequestration of atmospheric CO_2. Glob Chang Biol. 2012;18:3647–53.

Song Z, Liu H, Li B, Yang X. The production of phytolith-occluded carbon in China's forests: implications to biogeochemical carbon sequestration. Glob Chang Biol. 2013;19:2907–15.

Struve GA. De silicia in plantis nonnullis. Berolini: Diss; 1935.

Suzuki H. The influence of some environmental factors on the susceptibility of the rice plant to blast and Helminthosporium diseases and on the anatomical characters of the plant III. Influence of differences in soil moisture and in amounts of fertilizer and silica given. J Coll Agric. 1935;13:277–331.

Takahashi J, Kanareugsa C, Somboondumrongkul J, Prasittikhet J (1980) The effect of silicon, magnesium and zinc on the yield of rice. Proceedings of the Symposium on Paddy Soils; Nanjing, China; 1980. p. 82–3.

Takeoka Y, Kaufman PB, Matsumura O. Comparative microscopy of idioblasts in lemma epidermis of some C3 and C4 grasses (Poaceae) using SUMP method. Phytomorphology. 1979;29:330–7.

Takijima Y, Wijayaratna HMS, Seneviratne CJ. Nutrient deficiency and physiological disease of lowland rice in Ceylon. III. Effect of silicate fertilizers and dolomite for increasing rice yields. Soil Sci Plant Nutr. 1970;16:11–6.

Vaculík M, Landberg T, Greger M, Luxová M, Stoláriková LA. Silicon modifies root anatomy, and uptake and subcellular distribution of cadmium in young maize plants. Ann Bot. 2012;110:433–43.

Vivancos J, Labbé C, Menzies JG, Bélanger RR. Silicon-mediated resistance of Arabidopsis against powdery mildew involves mechanisms other than the SA-dependent defense pathway. Mol Plant Pathol. 2015. doi:10.1111/mpp.12213.

Wagner F. The importance of silicic acid for the growth of some cultivated plants, their metabolism, and their susceptibility to true mildews. Phytopathol Z. 1940;12:427–79.

Wang YX, Stass A, Horst WJ. Apoplastic binding of aluminum is involved in silicon-induced amelioration of aluminum toxicity in maize. Plant Physiol. 2004;136:3762–70.

Wiesner J. Einleitung in die technische Mikroskopie. Vienna; 1867.

Williams DE, Vlamis J. Manganese toxicity in standard culture solutions. Plant Soil. 1957a;8:183–93.

Williams DE, Vlamis J. The effect of silicon on yield and manganese-54 uptake and distribution in the leaves of barley plants grown in culture solutions. Plant Physiol. 1957b;32:404–9.

Woolley JT. Sodium and silicon as nutrients for the tomato plant. Plant Physiol. 1957;32:317–21.

Yamaji N, Ma JF. A transporter at the node responsible for intervascular transfer of silicon in rice. Plant Cell. 2009;21:2878–83.

Ye M, Song Y, Long J, Wang R, Baerson SR, Pan Z, Zhu-Salzman K, Xie J, Cai K, Luo S, Zeng R. Priming of jasmonate-mediated antiherbivore defense responses in rice by silicon. Proc Natl Acad Sci U S A. 2013;110:3631–9.

Zhu Q, Chen EF. The properties of iron and steel slags and their effects on crops grown on different types of soils. Acta Pedol Sin. 1963;11:70–83.

Zippicotte J. Fertilizer. U.S. Patent No. 238240. Official Gazette of the United States Patent Office; 1881. p. 19, 9, and 496.

Chapter 2
Analysis of Silicon in Soil, Plant and Fertilizer

Abstract The first soil testing of plant-available silicon (Si) was not conducted until 1898 on Hawaiian soils. However, numerous procedures have since been developed for determination of Si content in a wide variety of materials including soils, plants and fertilizers. This chapter reviews current analytical procedures that are widely used for analysis of both total Si in soils, plants and fertilizers and plant-available Si in soils and fertilizers.

Keywords Plant available • Silicon fertilizer • Silicon analysis • Soil • Total silicon

2.1 General

Although silica bodies were isolated from various plant species in the early nineteenth century (Davy 1819; Struve 1835), the first soil test for plant-available Si was not conducted until 1898 on Hawaiian soils (Maxwell 1898). Since then, numerous procedures for determination of Si content in a wide variety of materials including soils, plants and fertilizers have been developed (Snyder 2001; Sauer et al. 2006). However, apart from the review articles by Snyder (2001) and Sauer et al. (2006), no reviews have been published with focus on comparing the published extraction techniques or methodologies for chemically and physically analysing plant materials for Si and for determining total as well as plant-available Si in soils and fertilizers. The classical methods for determining total Si content of various materials involve conversion of insoluble silicates into sodium silicate through high-temperature fusion with sodium carbonate or hydroxide or other sodic bases (Snyder 2001) followed by determination of Si by a variety of methods including gravimetry, colorimetry and absorption/emission spectroscopy. In general, all the analytical methods involve two major steps; the first one is to dissolve Si contained in the insoluble silicates and extract or isolate Si from the materials, and the second one is to gauge Si based on gravimetric methods, spectrometric methods or microscopic observation. X-ray fluorescence spectrometry (XRF) as a nondestructive technique for multielemental analysis of soil and plant materials showed even higher measurement accuracy for Si over the destructive methods based on alkaline fusion or acid

Y. Liang et al., *Silicon in Agriculture*, DOI 10.1007/978-94-017-9978-2_2

digestion (e.g. Ramsey et al. 1995; Reidinger et al. 2012); this technique, however, has not been routinely used so far. In this chapter, current methods that can be widely used are reviewed.

2.2 Analysis of Total Silicon in Soil

2.2.1 *Gravimetric Methods*

Gravimetric methods are classic accurate methods that are suitable for determination of total Si in a wide variety of solid materials including soils, waste materials, sewage sludge, plants, organic manures and fertilizers, etc. They require basic common and simple laboratory equipment except for the expensive platinum wares. On the other hand, they are time consuming and tedious and thus not suitable for routine analysis of a large number of samples especially if only an analysis of Si is required (Snyder 2001). Therefore, rapid and robust methods suitable for routine analyses are needed in modern laboratories. The principle and detailed procedure are listed in Table 2.1.

2.2.2 *Spectroscopic Methods*

The rapid development of modern spectroscopic techniques has made it possible to use more rapid and robust techniques suitable for routine analysis of large samples. However, conversion of the silicates contained in the solid samples into a soluble form of Si by using sodium carbonate (Na_2CO_3), lithium metaborate ($LiBO_2$) or sodium hydroxide (NaOH) is needed before using such spectrometric techniques (Snyder 2001). The principle and detailed procedure are listed in Table 2.2.

2.3 Analysis of Total Si in Plant Material

2.3.1 *Gravimetric Methods*

To analyse Si in an organic matrix, such as plant material, the organic matter should be first removed by oxidation at a high temperature (550 °C), and then the non-Si elements are solubilized by 6 M HCl followed by filtration through ashless filter paper that retains the Si precipitates. The paper is then ignited and weighed. Afterwards, HF is used to evolve Si so that the weight loss is assumed to be Si (Snyder 2001). Yoshida et al. (1976) used a gravimetric method to determine Si and other mineral elements in rice straw following the removal of organic matter by acid digestion.

Table 2.1 The principle and procedure for gravimetric methods

Principle	Procedure
Total Si includes all Si forms existing in soil, which can be dissolved by strong alkali fusion or acid digestion methods (Snyder 2001). The earliest procedures for gravimetric analysis of soil Si are based on chemical reactions that produce either losses or gains in weight. The classical Si analyses like those carried out early in the twentieth century by the Bureau of Chemistry and Soils of the US Department of Agriculture were done gravimetrically by fusing the soil samples with sodium carbonate (Na_2CO_3). The SiO_2 was then dissolved with HCl and precipitated, dehydrated and weighed (Robinson 1930). The typical weight loss methods as used by Robinson (1945) utilize HF to evolve Si contained in soils as SiF_4 gas to gauge Si in soils. Silicon is determined by dissolving the sample in diluted acid and subsequently dehydrating it to precipitate SiO_2. Because a variety of other insoluble oxides are also formed in this process, the precipitate weight is not a direct measure of the amount of Si in the sample. The treatment of the solid residue with HF results in the formation of volatile SiF_4. The decrease in weight following the loss of SiF_4 provides an indirect measure of the amount of Si in the original sample	1. Transfer exactly 1.0 g of air-dried, finely powdered soil sample to a platinum crucible, add approximately 5.0 g of anhydrous Na_2CO_3, and mix thoroughly to a uniform colour
	2. Cover additional 1.0 g of Na_2CO_3 evenly on the surface of the mixture and transfer the crucible to a muffle furnace
	3. Heat the muffle furnace to gradually raise the temperature to 950 °C for about 30 min until a tranquil melt forms
	4. Take out the crucible, swirl it gently and return it to the furnace for another 5 min
	5. Take out the crucible again and keep it on a porcelain plate on the bench
	6. After cooling, dissolve the residue in dilute HCl (1:1), add enough distilled water to completely immerse the crucible and then transfer the crucible to a large casserole
	7. Cover a watch glass on the crucible and leave it overnight on a steam bath
	8. Evaporate the solution to dryness on the steam bath and heat the residue, which contains SiO_2 and other solids, in an oven at 110 °C for 1 h
	9. Moisten the residue with HCl and repeat the dehydration. Remove any acid-soluble materials from the residue by adding 50 mL of distilled water and 5 mL of concentrated HCl
	10. Bring the solution to a boil and filter through ashless filter paper (Whatman No. 40). Wash the residue with hot 2 % (v/v) HCl followed by hot water
	11. Evaporate the filtrate to dryness twice and, following the same procedure, treat to remove any acid-soluble materials
	12. Combine the two precipitates, and dry and ignite to a constant weight at 1,200 °C
	13. After cooling in a desiccator, add 2 drops of 50 % (v/v) H_2SO_4 and 30 mL of 6 M HF. Remove the volatile SiF_4 by evaporating to dryness on a hot plate
	14. Finally, bring the residue to constant weight by igniting it at 1,200 °C
	15. Calculate the Si content based on the weight loss

According to Young (1971)

Table 2.2 The principle and procedure for spectroscopic methods

Principle	Procedures and reagents
Silicon in soil and also in a wide variety of other solid materials can be dissolved by fusion with strongly alkaline chemicals such as Na_2CO_3, NaOH, lithium metaborate ($LiBO_2$) or lithium tetraborate (LiB_4O_7). Because the fusion of samples with Na_2CO_3 involves the use of very expensive platinum crucibles at a high temperature (1,000 °C), relatively rapid fusion of soil samples with NaOH, another alkaline flux, was later recommended in inexpensive Ni crucibles at a relatively low temperature (Kilmer 1965). After cooling, the flux can be dissolved by acid (HCl or H_2SO_4), and Si in the solution can be quantified by a variety of spectroscopic techniques such as AAS or ICP (Snyder 2001). Both $LiBO_2$ and $Li_2B_4O_7$ are also used to dissolve Si compounds in soils because both of these chemicals are acidic compounds that can easily fuse silicates in soil. In either case, the finely ground soil is intimately mixed with the Li and Na fluxes with the mixture fused in a platinum or graphite crucible over a burner or in a muffle furnace. Clear glasses are formed with the Li fluxes and the opaque light-coloured fusion cakes are formed with the sodium fluxes. The fused material is dissolved to form a clear solution for subsequent analysis (Jones and Dreher 1996; Meyer and Bloom 1993)	(a) Sample preparation by NaOH fusion method
	1. Transfer 5 mL portions of 30 % NaOH solution to a series of 75-mL nickel crucibles cleaned with diluted HCl prior to use
	2. Evaporate the solutions to dryness over gas burners or electric heaters. Slight spattering can be ignored
	3. Accurately weigh (to the nearest 0.1 mg) 50 mg of each sample of silicate powder and two 50 mg portions of standard sample (feldspar). As each portion is weighed, transfer them to a crucible containing the fused NaOH
	4. Cover and heat the crucibles to dull redness for about 5 min. Remove each crucible from the heat and swirl the melt around the sides. Allow the melts to cool
	5. Transfer the crucibles plus contents to a series of 1-L plastic beakers
	6. Add about 980 mL of distillate water to each beaker (the exact amount is not important except that the same amount must go into each beaker). Allow to stand for at least 1 h
	7. Stir the contents of each beaker using a plastic stirring rod
	8. Add 20 mL of diluted HCl (1:1) to each beaker, stirring each at the time of addition
	9. Use nichrome tongs to lift each nickel crucible out and examine it; there should be no adhering melt. If any melt remains, allow some more time for dissolution with frequent stirring. When all the melts are dissolved, remove the crucibles and set them aside for washing with HCl before storing them. When the crucibles are removed, the solutions are ready to be used for spectroscopic determination of Si
	(b) Determination of Si by molybdenum blue method
	1. Transfer 8 mL of the reagent blank solution, 8 mL of each standard solution and 8 mL of each sample solution to 150-mL plastic beakers
	2. Add 25 mL of acidified water to each beaker
	3. Add 25 mL of the molybdate reagent to each beaker and leave it for 10 min
	4. Add 23 mL of the tartaric acid solution to each beaker
	5. Add 25 mL of the reducing solution to each beaker and leave it for at least 45 min
	6. Determine the percent transmission for each solution at 640 nm using the reagent blank as the reference, or read calculated Si concentration using an automated spectrophotometer

	Reagents[a]:
	1. Ammonium molybdate solution: dissolve 6.0 g of $(NH_4)_6Mo_7O_{24} \cdot 4H_2O$ in 1,000 mL of distilled water
	2. Tartaric acid solution: dissolve 16 g of $H_2C_2H_4O_6$ in 1,000 mL of distilled water
	3. Acidified water: add 5 mL of 1:1 H_2SO_4 to 950 mL of distilled water
	4. Reducing solution: place 0.28 g of sodium sulphite, 3.6 g of sodium bisulphite and 0.06 g of 1-amino-2 napthol-4-sulphonic acid into a 1,000-mL bottle, add approximately 950 mL of distilled water, stir to dissolve, and fill the bottle up to the mark with distilled water

[a]Note: Highly consistent results are readily obtained when these solutions are prepared fresh, not more than 48 h before use

2.3.2 *Spectrometric Methods*

Spectrometric methods used widely for Si determination normally include colorimetric, atomic absorption spectrometric (AAS) and inductively coupled plasma spectrometric (ICP) assays of Si following alkaline- or acid-based digestion of plant materials. More recently, XRF, particularly portable X-ray fluorescence spectrometry (P-XRF), has been reported to be a much faster, safer, nonconsumptive and potentially more accurate method to determine Si concentrations in plant material (Reidinger et al. 2012). XRF works on the principle of excitation of inner orbital electrons by an X-ray radiation source. As the excited electrons relax to the ground state, they fluoresce, thereby ejecting photons of energy and wavelength characteristic of the atoms present. Despite several advantages of XRF over digestion-based techniques, such as its nonconsumptive nature and its higher measurement accuracy, XRF is still not routinely used for measurement of Si in plants partly due to the higher purchasing price of XRF instruments than that for equipment typically used in digestion-based techniques such as AAS or ICP (Reidinger et al. 2012).

2.3.2.1 Procedures for Sample Preparation (Table 2.3)

Table 2.3 Selected procedures for sample preparation

Method	Procedures for sample preparation
(a) Lithium metaborate fusion method (Meyer and Bloom 1993)	1. Tare 10-ml graphite crucibles and weigh 300 mg of $LiBO_2$
	2. Add 100 mg of plant tissues carefully to ensure that the sample is not in contact with the wall of the crucible
	3. Place the crucibles into a muffle furnace and heat slowly up to 485 °C over a period of 5 h to ash the plant tissues and continue to heat for 24 h at 485 °C
	4. Remove the crucibles from the oven and raise oven temperature to 950 °C
	5. Carefully mix plant ash with $LiBO_2$ with a small clean spatula
	6. Transfer the crucibles into the furnace for approximately 15 min or until the temperature returned to 950 °C using large long-handled tongs and leather gloves
	7. Remove the crucibles and place them on a heat-resistant slab to cool. At this time, the fused $LiBO_2$ and plant ash will form a small solid spherical bead
	9. Transfer the solid spherical bead into a high-density polyethylene plastic bottle containing 50 mL or 100 mL of 2 M HCl. Larger volumes of acid can be used if the sample is expected to contain larger amounts of silica
	10. Cap the bottles and place them on a shaker for 2 h
	11. Make appropriate dilutions and determine the Si content by the molybdenum blue method

(continued)

Table 2.3 (continued)

Method	Procedures for sample preparation
(b) Autoclave-induced digestion (AID) method (Elliott and Snyder 1991)	1. Place 100 mg of straw samples milled to pass a 20-mesh screen and dried to a moisture content of <10 % into a 250-mL polyethylene tube
	2. Add 2 mL 50 % H_2O_2 and 4.5 mL 50 % (w/w) NaOH
	3. Digest the resulting suspension in an autoclave at 138 kPa for 1 h
	4. Dilute the digested sample to 50 mL with deionized/distilled water
	5. Determine Si content colorimetrically
(c) Acid digestion and alkaline dissolution method (Nayar et al. 1975)	1. Prepare 50-mL Corning glass conical flasks which are thoroughly cleaned with hot alkali followed by acids and distilled water
	2. Weigh 100 mg finely ground, oven-dried (70 °C) plant samples into the flask containing 5 mL concentrated HNO_3 or a mixture of 5 mL concentrated HNO_3, 1 mL 70 % $HClO_4$ and 0.5 mL concentrated H_2SO_4
	3. Place the flask on a hot plate for acid digestion
	4. Wait for about 30 min till the brown fumes ceases and the volume of the acid is reduced to about 2 mL (overheating and drying should be avoided)
	5. Transfer the resultant solutions carefully with repeated washings into tall stainless-steel beakers containing 1–1.5 g of anhydrous Na_2CO_3 in suspension so that there is sufficient alkali in excess after neutralization of the acid
	6. Boil the alkali suspension in the stainless-steel beaker for 3–5 min to ensure complete dissolution of silica
	7. After cooling make up the resultant solution to 250 mL for colorimetric analysis of the dissolved Si
(d) Oven-induced digestion (OID) method (Kraska and Breitenbeck 2010)	1. Weigh 100 mg of dry and ground tissue samples into a 50-mL polyethylene screw-cap centrifuge tube previously washed with 0.1 M NaOH, rinsed with distilled water and dried
	2. Add 5 drops of octyl alcohol to reduce foaming prior to adding H_2O_2 and NaOH
	3. Add 2 mL of 30 % H_2O_2 to the samples
	4. Place the tightly capped tube in a convection oven at 95 °C
	5. Remove the tube and add 4 mL of 50 % NaOH to the hot samples after 30 min
	6. Gently vortex the sample tube and tightly cap, and then return to the oven (95 °C)
	7. Remove the sample tube after 4 h and add 1 mL of 5 mM NH_4F to facilitate the formation of monosilicic acid prior to quantitative transfer to 50-mL volumetric flask
	8. Dilute to final volume with distilled water for determination of the dissolved Si

(continued)

Table 2.3 (continued)

Method	Procedures for sample preparation
(e) Tiron extraction method (Guntzer et al. 2010)	1. Weigh 0.5 g of plant sample milled to pass a 20-mesh screen into a polypropylene tube
	2. Add 30 mL of a 0.1 M Tiron (4,5-dihydroxy-1,3-benzene-disulfonic acid disodium salt, $C_6H_4Na_2O_8S_2$) solution buffered at pH 10.5 to the plant sample in the tube
	3. Place the tube in a water bath at 85 °C and horizontally shake for 1 h with the cap loosely closed to prevent evaporation
	4. Cool the tube
	5. Take 10 mL of the extraction solution and add 10 mL of 30 % H_2O_2 to destroy Tiron to reduce the matrix effect during ICP-OES analysis
	6. Shake the new solution at 85 °C in the water bath for one more hour until it was colourless
	7. Cool and filter the solution at a 0.2-μm cellulose nitrate membrane before ICP-OES analysis
(f) Hydrofluoric acid extraction method (Saito et al. 2005)	1. Weigh 500 mg of dried and milled plant samples into a 50-mL polypropylene or polyethylene bottle
	2. Add 10 mL of the extraction solution (1.5 M HF + 0.6 M HCl)
	3. Stopper the bottle, stir the content of the bottle to immerse the plant tissues in the HF solution, and let it stand for 30 min at 30 °C (1 h at 18 °C) with occasional stirring (roughly every 10 min)
	4. Add 40 mL of distilled water, stir to homogenize the content, and let plant materials settle for 20 min
	5. Take the aliquot (0.1 mL) of the clear supernatant for determination of the dissolved Si

2.3.2.2 Procedures for Si Determination

Colorimetric Methods

Silicon (Si) can be determined colorimetrically, using light absorption spectrometer either by the blue Si molybdenum method at lower Si concentrations or by the yellow Si molybdenum method at higher Si concentrations. Si in solution reacts with ammonium molybdate forming a yellow silicomolybdate complex. Silicomolybdate complex can be finally converted into a reduced silicomolybdate complex which is blue, using a reducing solution such as ammonia ferrous sulphate, ascorbic acid or a mixture of sodium sulphite, sodium bisulphite and 1-amino-2-naphthol-4-sulfonic acid. Oxalic acid or tartaric acid is added to remove phosphate interferences. These two methods are similar in principle; however, the blue molybdenum method is more popular because of its higher sensitivity (Table 2.4).

Table 2.4 Procedures and reagent preparation for colorimetric determination of Si using molybdenum blue method

Reagents	Procedure
1. Si standard solution of 50 mg L^{-1} Si by diluting 1,000 mg L^{-1} Si standard with addition of 2.5 mM HCl to adjust the pH of the solution in the range between 2 and 4	(1) Transfer 1 mL sample solution to a 50-mL volumetric flask
2. 0.5 M B solution and 0.1 M B solution form H_3BO_3 (store in plastic bottles)	(2) Add 30 mL 20 % acetic acid
3. Stock 0.5 M Mo solution from $Na_2MoO_4 \cdot 2H_2O$ (store in a polypropylene bottle)	(3) Add 10 mL ammonium molybdate solution (54 g L^{-1}, pH 7.0)
4. Stock H_2SO_4 solution (0.8 M H_2SO_4+0.5 M B) from concentrated H_2SO_4 and H_3BO_3	(4) Shake up to mix thoroughly and keep for 5 min
5. Working Mo solution (0.25 M Mo+0.4 M H_2SO_4+0.25 M B) freshly prepared before use by combining 1 volume of the stock H_2SO_4 solution, and 1 volume of the stock Mo solution (stable for about 1 month at 5 °C)	(5) Immediately add 5 mL 20 % tartaric acid and 1 mL reducing solution containing 8 g L^{-1} Na_2SO_3, 1.6 g L^{-1} 1-amino-2-naphthol-4-sulfonic acid and 100 g L^{-1} $NaHSO_3$
6. 0.5 M citric acid (stock citric) solution with addition of 250 mg L^{-1} of benzoic acid as antiseptic	(6) Adjust to 50 mL with 20 % acetic acid and wait for 30 min
7. 0.1 M citric acid (working citric) solution from 0.5 M citric acid	(7) Measure the absorbance at 650 nm
8. 1 M tartaric acid	

According to Kilmer (1965) and Hallmark et al. (1982)

AAS and ICP Methods

In addition to colorimetric determination of Si using a light absorption spectrometer, dissolved Si can also be determined by AAS using a nitrous oxide–acetylene flame (Eaton et al. 1995) or by ICP (Jones and Dreher 1996).

2.4 Analysis of Soluble Silicon in Soil

As the total Si content is not related to the concentration of soluble Si in soils and can provide little information on soil Si availability to plants, various chemical extractants (Table 2.5; Berthelsen and Korndörfer 2005) are developed to assess Si-supplying power or plant-available Si. Currently, the chemical extracts reported to be useful in assaying plant-available Si are water (at varying soil: water ratios), sodium acetate buffer (pH 4.0), ammonium acetate (pH 4.8), dilute HCl or H_2SO_4, ammonium oxalate (pH 3.0), dilute citric acid or acetic acid, dilute $CaCl_2$, dilute phosphate plus acetate (pH 3.5), sodium phosphate buffer (pH 6.2), dilute Na_2CO_3, etc. (Gillman and Bell 1978; Fox et al. 1967; Fox and Silva 1978; Haysom and Chapman 1975; Imaizumi and Yoshidai 1958; Haysom and Kingston 2001; Kato

Table 2.5 Methods used to determine soluble and extractable soil Si

Extractant	Soil–solution ratio (or recommended weights and volumes)	Method	Suggested critical level	Reference
H_2O	Pre-wet air-dry soil at a matric suction of 0.1 bar	Incubate at 25 °C for 1 day and centrifuge at 900 g (RCF) for 1 h		Gillman and Bell (1978); Menzies and Bell (1988)
H_2O	Saturated paste		2 mg kg^{-1}	Fox and Silva (1978)
H_2O	10 g:100 mL	Continuous shaking for 4 h and centrifuge at 24,000 g (RCF)	<0.9 mg kg^{-1} (deficient)	Fox et al. (1967); Elawad et al. (1982)
			<2.0 mg kg^{-1} (marginal)	
			8.0 mg kg^{-1}	
H_2O	10 g:60 mL	'Incubation method' – shake, degas, seal bottle, incubate at 40 °C for 1 week without shaking		Takahashi and Nonaka (1986); Nonaka and Takahashi (1988, 1990)
H_2O	1:4	Supernatant method – shake, degas, fill to replace all air space, seal bottle, incubate at 30 °C for 4 weeks		Sumida (1992)
Phosphate acetate (pH 3.5) [500 mg L^{-1} P as Ca $(H_2PO_4)_2$ and 0.1 M NH_4OAC]	10 g:100 mL	Continuous shaking for 4 h and centrifuge	<50 mg kg^{-1} (deficient)	Fox et al. (1967)
			50–150 mg kg^{-1} (marginal – adequate)	
0.04 M sodium phosphate buffer (pH 6.2)	1 g:10 mL	Continuous shaking for 24 h at 40 °C		Kato and Sumida (2000)
Modified Truog [0.01 M H_2SO_4 containing 3 g L^{-1} $(NH_4)_2SO_4$]	1 g:100 mL	Continuous shaking for 30 min	<40 mg kg^{-1} (deficient)	Fox et al. (1967)
			40–100 mg kg^{-1} (marginal – adequate)	
0.1 M HCl		Intermittent shaking over 5 h at 40 °C		Nayer et al. (1977)

0.5 M NH_4OAC (pH 4.5–4.8)	5 g:100 mL	Continuous shaking for 1 h	<20 mg kg^{-1} (deficient) 20–40 mg kg^{-1} (marginal – adequate)	Fox et al. (1967), Wong et al. (1970), Ayres (1966)
			<50 mg kg^{-1} (deficient)	
0.5 M NH_4OAC (pH 4.5–4.8)	2.5 g:50 mL	2.5 g soil leached with 10 * 5 mL aliquots of extractant		Bishop (1967)
1.0 M acetate buffer, pH 4.0	10 g:100 mL	Intermittent shaking over 5 h at 40 °C		Imaizumi and Yoshidai (1958)
0.01 M $CaCl_2$	1:10	Continuous shaking for 16 h and centrifuged	<20 mg kg^{-1} (deficient – marginal)	Haysom and Chapman (1975)
0.01 M $CaCl_2$	1:25	Shaking for 7 days, with a few drops of chloroform on a reciprocal shaker at 25 °C		Wickramasinghe (1994)
0.005 M H_2SO_4	1:200	Continuous shaking for 16 h and centrifuged	<100 mg kg^{-1} (deficient – marginal)	Hurney (1973)
0.5 M acetic acid	10:25 (v/v)	2 h shaking following standing overnight	<24 mg mL^{-1} (deficient)	Barbosa et al. (2001), Snyder (2001)
0.025 M Citric acid	1:10	Intermittent shaking over 5 h at 30 °C		Ueda and Yamaoka (1959)
0.1 M Citric acid	1:50	2 h shake, rest O/N, 1 h shake, centrifuge		Acquaye and Tinsley (1965)
KH_2PO_4–KOH pH 6.5	1:10	Incubation at 40 °C without shaking 24 h		Liu (2002)

and Sumida 2000; Korndörfer et al. 2001; Nayer et al. 1977; Nonaka and Takahashi 1988, 1990). In general, the most successful extractants are acid rather than neutral or alkaline solutions (Imaizumi and Yoshidai 1958), and dissolution is further increased by chelating agents due to decreased Si sorption resulting from the lower concentration of Al and Fe in solution (Berthelsen and Korndörfer 2005). Other factors such as the method of equilibration, soil–solution ratio, temperature and pH of extractant solution are also important.

2.4.1 Acetate Buffer Method

The acetate buffer method proposed by Imaizumi and Yoshida (1958) is one of the most widely used analytical methods to assay the Si-supplying capacity in soils (Table 2.6).

By comparing various chemicals including hot hydrochloric acid, ammonium oxalate buffer at pH 3.0, 2 % sodium carbonate, 0.002 N sulphuric acid at pH 3.0, saturated carbonic acid water at pH 3.8 and sodium acetate buffer at pH 4.0 as extractants, Imaizumi and Yoshida (1958) found that more soil Si was extracted by acid than by alkaline chemicals. They further concluded that CO_2-saturated water (pH 3.8) was best suited for estimating the Si uptake by rice plants followed by the acetate buffer, and the amount of SiO_2 extracted by CO_2-saturated water was correlated nicely with that extracted by acetate buffer (pH 4.0). Considering that preparing the acetate buffer is easier, and a more stable pH can be reached during extraction, and that the acetate-extractable SiO_2 is well correlated with the amount of Si taken up by rice, Imaizumi and Yoshida (1958) suggested the use of acetate buffer as an extractant to estimate Si-supplying power in soil. Up to now, this acetate buffer extractant has been used to assess the Si-supplying capacity in soils in Japan, Chinese Taiwan (Lian 1976), China (Qin 1979; He et al. 1980; Zang et al. 1982; Ma et al. 1985; Liang et al. 1994), South Korea (Park 2001), Korea (Lian 1976), Malaysia and Thailand (Kawaguchi 1966), Ceylon (Takijima et al. 1970), India

Table 2.6 The procedure of acetate buffer method

Extractant	Procedure
1.0 M acetate buffer (pH 4.0)	(1) Put 10 g of air-dried soils (<2 mm) into a 200-mL flask
	(2) Add 100 mL 1.0 M sodium acetate buffer which is prepared by diluting 49.2 mL acetic acid and 14.8 g anhydrous sodium acetate to 1 L and adjusting to pH 4.0 with acetic acid or sodium acetate
	(3) Put the flask into a water bath at 40 °C for incubation for 5 h with intermittent shaking (once every hour)
	(4) Filtrate
	(5) Measure Si colorimetrically at 650 nm

According to Imaizumi and Yoshidai (1958)

(Nayer et al. 1977), Australia (Haysom and Kingston 2001) and Brazil (Korndörfer et al. 2001).

As the acetate buffer method involves extracting Si in soil for 5 h at 40 °C and is thus time consuming and laborious, the modified procedure in Korea involves extracting Si for 90 min at 60 °C, which is reported to be more feasible for routine analysis of soil-available Si in laboratories (Park 2001).

However, it is also reported that the acetate buffer method overestimated the Si-supplying power in soils previously fertilized with slag-based calcium silicate fertilizers (Takahashi 1981; Takahashi and Nonaka 1986; Nonaka and Takahashi 1990) and in calcareous soils (Liang et al. 1994), because this acetate buffer is strong enough to dissolve some non-available Si from the residual calcium silicate fertilizers, and it extracts additional portions of Si which may be carbonate bound but not plant available in the calcareous soils (Liang et al. 1994).

2.4.2 *Water Extraction Method*

To overcome overestimation problems, Nonaka and Takahashi (1988, 1990) proposed a method for measuring water-soluble Si in rice paddy soils (Table 2.7).

For soils previously fertilized with slag-based calcium silicate fertilizers, Si extracted by this new method was generally correlated better with Si in rice straw than the Si extracted by the acetate buffer method. However, this method has not been widely used for routine lab testing or for commercial use since the 2-week period between soil sampling, and reporting of results is a serious disadvantage (Savant et al. 1997; Snyder 2001). Sumida (1992) developed two additional soil incubation methods. One involves incubation of soil under flooded conditions for 4 weeks at 30 °C using a soil–water ratio of 1:4. The other method requires 5 days of incubation of soil with the addition of external silicic acid in a range from 0 to 100 mg SiO_2 kg^{-1} at 30 °C using soil–water ratio of 1:10. They serve the purpose to investigate the impacts of silicate fertilizer management on the characteristics of Si dissolution and adsorption of various paddy soils. Although these methods provide

Table 2.7 The procedure of water extraction method

Extractant	Procedure
Distilled water	(1) Put 10 g of air-dried soil sample (<2 mm) into a 100-mL polyethylene cylindrical bottle
	(2) Add 60 mL distilled water to the bottle
	(3) Shake and degas
	(4) Put the bottle in an incubator at 40 °C for 1 week without further shaking
	(5) Filtrate
	(6) Measure Si in the supernatant colorimetrically at 650 nm

According to Nonaka and Takahashi (1988, 1990)

the most suitable the Si-supplying capacity indices in paddy soils previously amended with calcium silicate or organic manure and in the soils with varying clay minerals and textures, they are not suitable for use in a routine soil testing laboratory because of their longer time requirement and/or complexity (Savant et al. 1997; Snyder 2001).

2.4.3 *$CaCl_2$ Extraction Method*

Water extracts have often been used to assay water-soluble or readily soluble Si (Elawad et al. 1982; Fox et al. 1967; Fox and Silva 1978; Gillman and Bell 1978; Menzies and Bell 1988; Nonaka and Takahashi 1988, 1990; Takahashi and Nonaka 1986). It is reported that the water-extractable Si was significantly correlated with Si uptake by plants (Fox et al. 1967; Medina-Gonzales et al. 1988; Takahashi and Nonaka 1986). However, some researchers find that the incubation method is generally not a suitable method since the low ionic strength of the solution may cause dispersion (Elgawhary and Lindsay 1972; Lindsay 1979). Generally, as drying of soil samples results in changes in equilibrium between soluble and solid Si compounds, and as soluble Si compounds (mainly monosilicic acids) adsorbed on soil particles are dehydrated during soil drying, extracting Si with water as an extractant requires a longer period of time (Table 2.5).

Considering the resulting dispersion effect caused by the low ionic strength in the water incubation solution, Elgawhary and Lindsay (1972) recommended the use of 0.02 M $CaCl_2$ as the reactive media to equalize ionic strengths and facilitate ready flocculation of colloidal Si. According to the soil chemical equilibrium principle, Si extracted with diluted $CaCl_2$ may correspond more closely to the levels of $Si(OH)_4$ expected from solubility predictions (Lindsay 1979). Accordingly, a diluted $CaCl_2$ solution was recommended by Haysom and Chapman (1975) as an extractant to assess Si-supplying power in soils under sugarcane (Table 2.8).

The critical value is 100 mg Si kg^{-1} soil, below which Si deficiency for sugarcane is expected. They concluded that although Si extracted with either 0.5 M NH_4OAC or 0.005 M H_2SO_4 was well correlated with cane yield, 0.01 M $CaCl_2$ was the best

Table 2.8 The procedure of water extraction method

Extractant	Procedure
0.01 M $CaCl_2$	(1) Put 2 g of air-dried soil sample (<2 mm) into a 50-mL polyethylene tube
	(2) Add 20 mL 0.01 M $CaCl_2$ to the tube
	(3) Shake for 16 h
	(4) Centrifuge at 2,000 rpm for 10 min
	(5) Measure Si in the supernatant colorimetrically at 650 nm

According to Nonaka and Takahashi (1988, 1990)

extractant. This method was also adopted by Wickramasinghe (1994) with some modifications. After having compared a number of extraction methods over a wide range of soil types from north Queensland in Australia, Berthelsen (2000) concluded that 0.01 M $CaCl_2$-extractable Si provided a measure of the readily available Si present in the soil solution, while NH_4OAC- and acetic acid-extractable Si fractions were likely to contain rather simple polymers affected by changes in pH, CEC and the ratio of soluble Si–Al in the soil solution. Berthelsen et al. (2003) also reported that the Si contents in sugarcane were significantly related to readily soluble soil Si extracted by 0.01 M $CaCl_2$ but were not correlated with the level of soil Si extracted by a stronger acid extractant (e.g. 0.005 M H_2SO_4).

2.4.4 *Citric Acid Extraction Method*

Ueda and Yamaoka (1959) measured soil Si-supplying capacity in degraded paddy soils where nutrients such as bases, Fe, Mn and Si were leached out by using 0.025 M citric acid, 2 % Na_2CO_3, oxalic acid and oxalate buffer and 1 M acetate buffer to extract plant-available Si. Their results showed that the straw Si was significantly correlated with citric acid-extractable Si and with acetate buffer-extractable Si, but was not well correlated to the Si extracted by other extractants. They thus proposed using citric acid as an extractant to assess the soil Si-supplying power. Later, numerous researches all over the world confirmed that 0.025 M citric acid was the most suitable extractant to assay the plant-available Si in soils. For example, by comparing 1.0 M acetate buffer extractant with three other extractants, i.e. distilled water, 0.2 M HCl and 0.025 M citric acid, Nayer et al. (1977) reported that the extracting power was in the order of 0.2 M HCl > 0.025 M citric acid > 1.0 M acetate buffer at pH 4 > distilled water, with only some exceptions in some soils. They further found that citric acid-extractable Si had the strongest correlation with the Si taken up by rice plants. Zhang and Zang (1982) also compared the differences and similarities in the analytical methods for determining plant-available Si in 28 Si-fertilized paddy soils derived from coastal marine deposits, granite, basalt, red sandstone, quaternary red earth and alluvial deposits in Guangdong, Jiangxi, Zhejiang and Jiangsu provinces of South China. They used acetate buffer, 0.01 M H_2SO_4, 0.2 M HCl and 0.025 M citric acid with intermittent shaking at 30 °C for 5 h or 1 h and at 15 °C for 15 min to extract the plant-available Si in all the soils used. The results showed that the Si amounts extracted by these extractants were all significantly correlated with the straw Si content and with the relative increase of rice yield in Si-fertilized plots. However, it was roughly calculated that the amount of Si extracted by the acetate buffer and 0.01 M H_2SO_4 was less and by 0.2 M HCl was more than the amount of Si removed by rice during the growing season and that the amount of Si extracted by citric acid was approximately equal to that removed by rice.

2.4.5 *Sulphuric Acid-Extractable Method*

More recently, Hong et al. (2011) also studied the effect of five extracting methods (0.025 M citric acid with continuous shaking at 30 °C for 2 h, 1 % citric acid with intermittent shaking at 30 °C for 5 h, 1 M acetate buffer with intermittent shaking at 40 °C for 5 h, 0.01 M H_2SO_4 with continuous shaking at 20 °C for 30 min and 0.2 M HCl without shaking at 40 °C for 16 h) on extraction of Si in neutral and alkaline paddy soils sampled from Heilongjiang province, Northeast China. They found that, on the average, the concentration of Si measured by 0.2 M HCl extraction was 39-fold higher than by the other methods, and the concentration of Si determined by 0.01 M H_2SO_4 extraction was almost equal to that by 0.025 M citric acid extraction. They concluded that both 0.01 M H_2SO_4 and 0.025 M citric acid were the most suitable extractants to assess the Si-supplying power in neutral and especially in alkaline soils and argued that the 'too strong' acidity and 'too long' incubation period might be the major reasons for the unexpectedly higher content of Si solubilized by the 0.2 M HCl extractant.

Kanamugire et al. (2006) compared the correlation coefficient between the amount of Si accumulated by sorghum and sugarcane in 5 acid soils and the amount of Si extracted by six different extractants (0.025 M H_2SO_4, 0.01 M H_2SO_4, 0.5 M acetic acid, 0.01 M $CaCl_2$, water and 0.5 M NH_4OAC) and found that the correlations between the total Si taken up by sorghum and sugarcane and Si extracted from soils by the extraction methods used were all statistically significant, with the 0.025 M H_2SO_4 extractant showing the best correlation with total Si uptake by plants. They also proposed that a positive response to the application of Si to be likely when the soil test level using 0.025 M H_2SO_4 is below 45 mg Si kg^{-1} for sandy soils, below 65 mg Si kg^{-1} for loamy sands to sandy clay loams and 100 mg Si kg^{-1} for clayey soils.

2.4.6 *Acetic Acid Extraction Method*

Barbosa et al. (2001) developed a soil testing method using acetic acid as an extractant for determining plant-available Si in paddy and sugarcane soils in South Florida, and this method has been well accepted (Table 2.9). Critical values for soil Si test ranges of low (<7 mg Si L^{-1}), medium (7–24 mg Si L^{-1}) and high (>24 mg Si L^{-1}) were established (Snyder 2001).

2.4.7 *Phosphate Extraction Method*

Khalid et al. (1978) proposed extracting water-soluble Si by shaking 3 g of soil with 30 mL distilled water for 4 h to measure the 'intensity factor' and using a phosphate extractant to gauge plant-available Si as the 'capacity factor'. Briefly, 3 g of soil are

Table 2.9 The procedure of acid extraction method

Extractant	Procedure
0.5 M acetic acid	(1) Scoop 10 cm^3 of air-dried and sieved (<2 mm) soil into a 75-mL plastic extraction bottle
	(2) Add 25 mL of 0.5 M acetic acid to the bottle and let the mixture stand overnight (approximately 20 h)
	(3) Shake on a reciprocating shaker (120 rpm) for 50 min
	(4) Filter and collect the extract into plastic containers
	(5) Measure Si in the supernatant by ICP-OES
	(6) Calculate Si on a soil volume basis (mg Si L^{-1}) by multiplying mg Si L^{-1} in soil extract with 2.5

According to Barbosa et al. (2001)

shaken for 4 h with 30 mL 0.1 M acetic acid containing 50 mg P L^{-1} as $Ca(H_2PO_4)_2$ at pH 3.5 (adjusted with NH_4OH). The results obtained from 5-year field trials showed that the amount of Si taken up by rice seedlings was higher than extracted by the acetic acid plus phosphate (Khalid et al. 1978). Kato and Sumida (2000) proposed a phosphate buffer method for measuring plant-available Si, in which 5 g soil is shaken for 24 h with 50 mL of 0.04 M phosphate buffer solution at pH 6.2 made by titrating 0.04 M Na_2HPO_4 with 0.04 M NaH_2PO_4 to pH 6.2. After centrifugation and filtration, the Si concentration in the supernatant is measured colorimetrically. Kato and Sumida (2000) reported a better correlation between the amount of Si taken up by rice with soil Si extracted by the phosphate buffer than by the acetate buffer method. They concluded that the phosphate buffer method was better for assessing the Si-supplying power in paddy soils than the traditional acetate buffer method because the proposed method did not overestimate Si availability in soils previously treated with calcium silicate fertilizers.

2.4.8 *Alkaline Extraction Method*

Alkaline wet chemical dissolution extracts such as NaOH, a strong base, and Na_2CO_3, a weak base, are frequently used to extract Si in soils. The NaOH- or Na_2CO_3-extractable Si is considered to represent the fraction of amorphous Si in soils (Foster 1953; Follett et al. 1965) and amorphous biogenically derived Si in sediments (DeMaster 1981; Mortlock and Froelich 1989; Koning et al. 2002). The principle of these alkaline-extractable methods is based on the fact that the solubility of amorphous Si increases with increasing pH values (Iler 1979). While NaOH is more often used in soils than Na_2CO_3, the latter is the most commonly used method to extract amorphous biogenically derived Si in aquatic sediments (Conley 1998). Foster (1953) used 0.5 M NaOH to extract amorphous soil Si at soil–solution ratio of 1 g : 50 mL in a boiling water bath for 4 h. Since then, this NaOH-based extraction has become a standard method to assay amorphous Si in soils. As NaOH also partly dissolves the silicate minerals, Hashimoto and Jackson (1960) proposed using 0.5 M NaOH to extract amorphous Si within 2.5 min to reduce the dissolution of silicate minerals.

Some extractants can dissolve or extract different components from certain siliceous soils. For example, dithionite–citrate–bicarbonate (DCB) (Weaver et al. 1968) and ammonium oxalate extracts (Wang and Schuppli 1986) which are originally prepared for determining crystalline and amorphous Fe also are often used for analysis of crystalline and amorphous Si.

Although numerous extractants have been proposed to assay the soil Si-supplying power all over the world, the most widely used extract is 1.0 M acetate buffer at pH 4.0 proposed by Imaizumi and Yoshidai (1958) as there are threshold or critical values below which plant response to Si fertilizers can be expected. The critical value of acetate buffer-extracted Si, below which positive response of rice to slag-based Si fertilizers can be expected, is 105 mg SiO_2 kg^{-1} in Japan, 100 mg SiO_2 kg^{-1} in Korea, 40 mg SiO_2 kg^{-1} in Chinese Taiwan (Lian 1976) and 95 mg SiO_2 kg^{-1} in China (Zang et al. 1982).

In fact, no extraction has been found to work equally well on all soils (Snyder 2001). The Si content extracted by all of the discussed extractants can only be considered as a reference index to estimate the Si-supplying power in soils, but it does not measure the exact amount of plant-available Si and should be used to make diagnosis for Si deficiency in plants together with plant Si content, the truely important variable. Research is also warranted on testing whether the Si extracted by any of the established extractants is well correlated with Si uptake by plants.

2.5 Analysis of Plant-Available Silicon in Fertilizers

Procedures for determination of total Si in soils such as alkaline fusion method can apply to analysis of Si in fertilizers. However, total Si content in Si fertilizers, which is not related to Si availability, can provide little information on plant-available Si, which is most important to assessing the quality and Si-supplying power of the fertilizers. Therefore, a powerful method for assaying plant-available Si content is needed to assess the Si-supplying capacity of the fertilizer of interest. Unfortunately, it is difficult to select a universally suitable method or extractant for assessing the Si-supplying capacity of all types of materials, all soils and soil conditions (Gascho 2001; Berthelsen and Korndörfer 2005). While a number of chemical extractants have been developed to estimate soluble or extractable Si in silicate materials, often the correlations between the Si contents in a fertilizer and a response of the Si concentrations in plants on fertilized fields are rather low (Berthelsen and Korndörfer 2005). Actually, a lot of physical and chemical properties of the materials such as pH, molar ratio of $CaO–SiO_2$ and mineral particle size can greatly influence Si solubility and plant availability (Ma and Takahashi 2002).

Up to recently, many researchers have used different extractants to assay the plant-available Si in solid fertilizers. These extractants include HCl+HF, Na_2CO_3

Table 2.10 Quantities of total and extractable Si in the waste materials tested[a]

Material	Total Si (%)	NH_4 acetate-extractable Si (g kg^{-1})	HCl-extractable Si (g kg^{-1})	Water/cation exchange resin Si (g kg^{-1})	Na_2CO_3–NH_4NO_3-extractable Si (g kg^{-1})	Na_2CO_3–NH_4NO_3 + 5 d equilibration-extractable Si (g kg^{-1})
Ca silicate	24.2	1.8 (0.78)	1.9 (0.78)	28.3 (11.7)	58.8 (24)	64.5 (27)
BF slag 1	17.3	29.9 (17.2)	49.9 (28.8)	7.2 (4.2)	3.1 (1.8)	17.2 (9.9)
BF slag 2	14.8	27.1 (18.3)	66.2 (44.7)	1.9 (1.3)	1.7 (1.1)	7.2 (4.9)
BF slag 3	16.1	31.5 (19.6)	58.1 (36.1)	1.8 (1.1)	0.17 (0.11)	0.67 (0.41)
Steel slag	5.4	25.7 (47.6)	38.3 (70.1)	11.3 (20.9)	3.63 (6.7)	4.6 (8.5)
Processing mud	6.8	33.3 (49.0)	47.9 (70.4)	3.2 (4.7)	0.14 (0.21)	0.37 (0.54)
Fly ash	29.1	0.01	0.02	0.02	0.07 (0.02)	0.30 (0.10)

From Haynes et al. (2013)
[a]Percentage of total Si present in extractable form is shown in parentheses

(10 g L^{-1}) + NH_4NO_3 (16 10 g L^{-1}), citric acid (50 g L^{-1}), HCl (0.5 mM) and ammonium citrate (Buck et al. 2011; Korndörfer and Pereira 2011; Sebastian et al. 2013). Based on the correlation coefficients, the best extractant for available Si in solid fertilizers was Na_2CO_3 + NH_4NO_3, while for liquid fertilizers, HCl + HF was found to be superior for total Si assessment (Buck et al. 2011; Korndörfer and Pereira 2011). It is shown that the 5-day Na_2CO_3–NH_4NO_3 soluble Si extraction method can be applied to quantify the plant-available Si in solid fertilizer products at levels ranging from 0.2 to 8.4 % Si, and the single-laboratory validation of the 5-day Na_2CO_3–NH_4NO_3 soluble Si extraction method has been approved by the Association of American Plant Food Control Officials for testing nonliquid Si fertilizer products (Sebastian et al. 2013) and the Association of American Plant Food Control Officials (AAPFCO) (Korndörfer and Pereira 2011).

Given in Table 2.10 are quantities of total and extractable Si in the waste materials tested (Haynes et al. 2013). The data listed in Table 2.10 clearly indicate that the total Si was over 20 % in fly ash and Ca silicate and between 14 % and 18 % in the BF slags, while it ranged from 5.4 % in the steel slag to 6.8 % in the processing mud. The water/exchange resin method extracted 11.7 % of total Si from Ca silicate, but ammonia acetate and HCl extracted < 1 %. The two Na_2CO_3–NH_4NO_3 methods extracted the greatest proportions of total Si from the Ca silicate (i.e. 24 % and 27 %). In the BF slags, approximately 17 %–20 %, 28 %–44 %, 1 %–5 %, 0.17 %–3.1 % and 0.67 %–17 % of total Si were extracted by ammonium acetate, HCl, water/exchange resin, Na_2CO_3–NH_4NO_3 and Na_2CO_3–NH_4NO_3 plus a 5-day equilibration, respectively. Although the total Si was lower in steel slag and processing mud than in the BF-slag samples, the proportions of ammonia acetate- and HCl-extractable Si were much greater in steel slag and processing mud than in the BF-slag samples. This was also true for the water/exchange resin extraction for steel slag. For the Na_2CO_3–NH_4NO_3 method, adding 5 days of equilibration prior to Si analysis could extract greater amount of Si from all the materials tested. Haynes et al. (2013) draw a conclusion that the BF slags tested are the most effective as sources of Si fertilizer and that, in slag-amended soils, $CaCl_2$ and NH_4OAC are the most reliable soil test extractants.

2.5.1 Acid Extraction Methods

The earliest widespread commercial Si fertilizer used for rice crop in Japan is slag-based calcium silicate (Ma and Takahashi 2002). Si extraction with 0.5 M HCl at 30 °C for 1 h is proposed as an 'official' method of gauging Si availability in slags in Japan (NIAES 1987; Ma and Takahashi 2002). In China, this procedure is also a commonly used method to gauge the plant-available Si in Si materials (Wang et al. 1995; Li et al. 2004). Wang et al. (1995) reported that both 0.5 M HCl and 2 % citric acid could be satisfactorily used to assess plant-available Si content in fertilizers. The amount of Si extracted by 0.5 M HCl was higher than that by 2 % citric acid. The amounts of plant-available Si extracted from different slag-based calcium silicates by 0.5 M HCl and 2 % citric acid at 30 °C were strongly correlated with the increment of the plant-available Si in both alluvial deposit- and red soil-derived paddy soils fertilized by slag-based calcium silicates. Considering that plant roots release organic acids to the rhizosphere, Wang et al. (1995) believe that citric acid is recommended as a more suitable extractant to estimate plant-available Si in fertilizers. It was reported that, however, Si extraction with 0.5 M HCl at 30 °C for 1 h was not an appropriate procedure to predict Si uptake by rice as the amount of Si extracted from the Si fertilizer was not related to the Si uptake by plants (Takahashi 1981; Kato and Owa 1997; Snyder 2001).

2.5.2 Acetate Buffer Method

The 1.0 M acetate buffer at pH 4 method proposed by Imaizumi and Yoshidai (1958) has also been recommended for assessing Si availability in slags (NIAES 1987). However, this method has been found unsuitable for gauging plant-available Si in soils fertilized by slag-based calcium silicate (Takahashi 1981; Takahashi and Nonaka 1986; Nonaka and Takahashi 1990).

2.5.3 Acidic Cation Exchange Method

Kato and Owa (1997) also demonstrated that the acid extraction method commonly used in Japan was unsuitable for gauging available Si content in slag-based calcium silicate fertilizers. They developed a procedure using a weakly acidic cation exchange H^+-resin (Table 2.11).

This method is reported to provide the best indicator of plant-available Si in the fertilizer, and the Si extracted is correlated well with both the indirect chemical

Table 2.11 The procedure of acidic cation exchange H-resin method

Extractant	Procedure
Acidic cation exchange H-resin	(1) Put 0.2 g air-dried and sieved (<2 mm) slag into a 500-mL plastic extraction bottle
	(2) Add 0.5 g of weekly acidic cation exchange H-resin (e.g. Amberlite IRC-50) into the bottle
	(3) Add 400 mL distilled water
	(4) Immediately shake the bottle for a while by hand
	(5) Shake on a reciprocating shaker at 100 rpm at 25°C for 96 h
	(6) Filter and collect the extract into plastic containers
	(7) Measure Si in the supernatant colorimetrically at 650 nm

According to Kato and Owa (1997), also see Snyder (2001)

Table 2.12 The procedure of column method

Extractant	Procedure
Acidic cation exchange H-resin	(1) Put 3 g of silicon source and 5.0 g medium-density polyethylene into a 20-mL plastic syringe slag and mix
	(2) Use glass wool above and below the mix to ensure that the mixture is retained in the centre of the syringe
	(3) Insert a stopper fitted with a glass tuber in place of the syringe plunger
	(4) Use a peristaltic pump to pass 0.1 M TRIS buffer (pH 7) upwards through the syringe (the 'column') at a rate of 1 mL min^{-1}
	(5) Collect the water solutions passing through the column in each of 2 successive 24-h periods for analysis of Si
	(6) Use a similar analysis of finely ground wollastonite with each analysis of candidate Si sources to serve as a reference

According to Snyder (2001)

extraction results and also soil and plant Si and yield in the pot experiments (Kato and Owa 1997). Pereira et al. (2003) also tested this 'resin' method and found it to give the highest correlation between Si extracted from 12 different sources of Si materials and Si uptake by rice.

2.5.4 *'Column' Method*

Snyder (2001) and his colleagues also developed a 'column' method to compare mineral Si sources on the basis of plant availability of Si, maintenance of neutral solution pH, low Ca concentration near the Si source and sufficiently low dissolved Si concentration to minimize polymerization (Table 2.12). However, Si laboratory analysis to identify promising mineral Si sources is required to correlate with both the crop responses and crop Si uptake through greenhouse and field trials to ultimately assess the Si-supplying capacity of the sources.

2.5.5 *Sodium Carbonate–Ammonium Nitrate Method*

More recently, Sebastian et al. (2013) have developed a 5-day method for determining the soluble Si concentrations in nonliquid fertilizer products using a $Na_2CO_3+NH_4NO_3$ extractant followed by visible spectroscopy with heteropoly blue analysis at 660 nm. This 5-day Na_2CO_3–NH_4NO_3 soluble Si extraction method has recently been approved by the Association of American Plant Food Control Officials (AAPFCO) for testing nonliquid Si fertilizer products (Sebastian et al. 2013) (Table 2.13).

Table 2.13 The procedure of 5-day Na_2CO_3–NH_4NO_3 soluble Si extraction method

Extractant	Procedure
0.047 M sodium carbonate and ammonium nitrate solutions	(1) Grind test sample: grind fertilizer material to pass a 300-μm sieve
	(2) Weigh out a 0.2 g test portion (±0.005 g at most), transfer to a 250-mL tarred plastic flask, weigh again after transfer, and record test portion weight
	(3) Add 100 ml each of sodium carbonate and ammonium nitrate solutions (0.047 M Na_2CO_3 + 0.10 M NH_4NO_3) using a plastic graduated cylinder
	(4) Cap flask tightly and shake solution at 140 rmp (table unit) at ambient temperature for 1 h
	(5) Remove from shaker and let stand undisturbed for 5 days
	(6) Prepare a spiked talc sample by extracting talc using steps (1–5) above. Before step 4, add 3 mL 500 mg L^{-1} Si spike solution to the talc test sample. This talc-spiked test sample is processed and used for the matrix spike recovery test to verify that soluble (spike) rather than insoluble Si (talc) is extracted and reported using this method
	(7) Make a duplicate of at least one of the unknown test samples
	(8) At the end of 5 days, transfer 2 mL (4 mL for materials expected to be < 3 % Si) of resting extraction sample (step 5 above) to a 200-mL polypropylene volumetric flask and dilute to 200 mL with distillate water. Stopper flask and mix by inverting 10 times and pipette 20 mL of diluted test solution into a plastic test tube
	(9) Prepare Si calibration standards, blank (0 mg Si L^{-1}) and standards 1–4 (0.25, 0.50, 1.0 and 2.0 mg Si L^{-1})
	(10) Add 2 mL 0.42 M ammonium molybdate solution containing 1.84 M H_2SO_4 and mix well for 10 s using a touch agitator; wait 10 min and then add 2 ml 1.33 M tartaric acid solution. Stopper test tube and mix well for 10 s using a touch agitator. Wait 5 min and then add 2- mL 0.017 M ascorbic acid solution. Stopper test tube and mix well for 10 s using a touch agitator
	(11) Allow test sample, blank and standards to stand for 1 h for colour development. Colour gradation from blue to purple should be seen with increasing Si concentration
	(12) Measure the absorbance at 660 nm

According to Sebastian et al. (2013)

References

Acquaye DM, Tinsley J. Soluble silica in soils. In: Hallsworth EG, Crawford DV, editors. Experimental pedology (Easter School in Agricultural Science, University of Nottingham). London: Butterworths; 1965. p. 127–48.

Ayres AS. Calcium silicate slag as a growth stimulant for sugarcane on low-silicon soils. Soil Sci. 1966;101:216–27.

Barbosa MP, Snyder GH, Elliott CL, Datnoff LE. Evaluation of soil test procedures for determining rice-available silicon. Commun Soil Sci Plant Anal. 2001;32:1779–92.

Berthelsen S. An assessment of the silicon status of soils in north Queensland, and the impact of sub-optimal plant available soil silicon on sugarcane production systems. MSc thesis, Tropical plant Sciences within the School of Tropical Biology, James Cook University, Townsville; 2000.

Berthelsen S, Noble AD, Kingston G, Hurney A, Rudd A, Garside A. Improving yield and ccs in sugarcane through the application of silicon based amendments. Final Report, Sugar Research and Development Corporation Project CLW009; 2003.

Berthelsen S, Korndörfer GH. Methods for Si analysis in plant, soil and fertilizers. Proceedings of the third International Conference on Silicon in Agriculture; 2005 Oct 22–26; Uberlandia, Brazil: Universidade Federal de Uberlandia; 2005. p 85–91.

Bishop RT. Aluminium and silica relationships in growth failure areas. Proceedings of the South African Sugar Technologists Association; 1967. p. 190–6.

Buck GB, Korndorfer G, Datnoff LE. Extractors for estimating plant available silicon from potential silicon fertilizer sources. J Plant Nutr. 2011;34:272–82.

Conley DJ. An interlaboratory comparison for the measurement of biogenic silica in sediments. Mar Chem. 1998;63:39–48.

Davy H. The elements of agricultural chemistry. Hartford: Hudson and Co.; 1819.

DeMaster DJ. The supply and accumulation of silica in the marine environments. Geochim Cosmochim Acta. 1981;45:1715–32.

Eaton AD, Clesceri LS, Greenberg AE. Standard methods for the examination of water and wastewater. 19th ed. Washington, DC: American Public Health Association; 1995.

Elawad SH, Gascho GJ, Street JJ. Response of sugarcane to silicate source and rate. I. Growth and yield. Agric J. 1982;74:481–4.

Elgawhary SM, Lindsay LW. Solubility of silica in soils. Soil Sci Soc Am Proc. 1972;36:439–42.

Elliott CL, Snyder GH. Autoclave-induced digestion for the colorimetric determination of silicon in rice straw. J Agric Food Chem. 1991;39:1118–9.

Follett EAC, McHardy WJ, Mitchell BD, Smith BFL. Chemical dissolution techniques in the study of soil clays: Part I. Clay Miner. 1965;6:23–34.

Foster MD. The determination of free silica and free alumina in montmorillonites. Geochim Cosmochim Acta. 1953;3:143–54.

Fox RL, Silva JA, Younge OR, Plucknett DL, Sherman GD. Soil and plant silicon and silicate response by sugarcane. Soil Sci Soc Am Proc. 1967;31:775–9.

Fox RL, Silva JA. Symptoms of plant nutrition: silicon, an agronomically essential nutrient for sugarcane. Illustrated Concepts in Tropical Agriculture. Dept. of Agronomy and Soil Sci., College of Tropical Agriculture and Human Resources, University of Hawaii; 1978.

Gascho GJ. Silicon Sources for Agriculture. In: Datnoff LE, Snyder GH, Korndorfer GH, editors. Silicon in agriculture. Amsterdam: Elsevier Science B.V.; 2001. p. 197–207.

Gillman GP, Bell LC. Soil solution studies on weathered soils from tropical North Queensland. Aust J Soil Res. 1978;16:67–77.

Guntzer F, Keller C, Meunier JD. Determination of the silicon concentration in plant material using Tiron extraction. New Phytol. 2010;188:902–6.

Hallmark CT, Wilding LP, Smeck NE. Silicon. In: Page AL, Miller RH, Keeney DR, editors. Methods of soil analysis. Part 2: Chemical and microbiological properties, Agronomy monograph no. 9. 2nd ed. Madison: The America Society of Agronomy and Soil Science Society of America; 1982. p. 263–73.

Hashimoto J, Jackson ML. Rapid dissolution of allophane and kaolinite-halloysite after dehydration. Proceedings of the 7th National Conference on Clays and Clay Minerals; 1958 Oct 20–23; Washington, DC; 1960. p. 102–13.

Haynes RJ, Belyaeva ON, Kingston G. Evaluation of industrial wastes as sources of fertilizer silicon using chemical extractions and plant uptake. J Plant Nutr Soil Sci. 2013;176:238–48.

Haysom MBC, Chapman LS. Some aspects of the calcium silicate trials at Mackay. Proc Qld Soc Sugar Cane Technol. 1975;42:117–22.

Haysom MB, Kingston G. Soil analysis for predicting sugarcane yield response to silicon. In: Datnoff LE, Snyder GH, Korndorfer GH, editors. Silicon in agriculture. Amsterdam: Elsevier Science; 2001. p. 373.

He DY, Zang HL, Zhang XP. The effect of silicate slags on rice in paddy soils derived from red earth. Acta Pedol Sin. 1980;17:355–63 (In Chinese with English abstract).

Hong XJ, Yang J, Yang XH. The effect of different extractants on silicon content extracted from paddy soils. Commun Agric Sci Technol. 2011;11:31–3 (In Chinese).

Hurney AP. A progress report on the calcium silicate investigations. Proc Qld Soc Sugar Cane Technol. 1973;40:109–13.

Iler RK. The chemistry of silica. New York: Wiley; 1979.

Imaizumi K, Yoshidai S. Edaphological studies on the silicon supplying power of paddy soils. Bull Natl Inst Agric Sci Tokyo. 1958;B8:261–304.

Jones RL, Dreher GB. Chapter 22. In: Bigham JM, editor. Methods of soil analysis, Part 3: Chemical methods. Soil science society of America. Book series No. 5. Madison: The Soil Science Society of America and America Society of Agronomy; 1996.

Kanamugire A, Meyer JH, Haynes RJ, Naidoo G, Keeping MG. An assessment of soil extraction methods for predicting the silicon requirement of sugarcane. Proc South Afr Sugar Technol Assoc. 2006;80:287–90.

Kato N, Owa N. Dissolution of slag fertilizers in a paddy soil and Si uptake by rice plants. Soil Sci Plant Nutr. 1997;43:329–41.

Kato N, Sumida H. Evaluation of silicon availability in paddy soils by an extraction using a pH 6.2 phosphate buffer solution. Annual report of the Tohoku Agricultural Experiment Station for 1999; 2000. p. 73–5.

Kawaguchi K. Tropical paddy soils. Jpn Agric Res Quart. 1966;1:7–11.

Khalid RA, Silva JA, Fox RL. Residual effects of calcium silicate in tropical soils. I. Fate of silicon during five years of cropping. Soil Sci Soc Am Proc. 1978;42:89–94.

Kilmer VJ. Silicon. In: Black CA, editor. Methods of soil analysis, Part 2: Chemical and microbiological properties, Agronomy series, No. 9. Madidon: America Society Agronomy; 1965. p. 959–62.

Koning E, Epping E, Van Raaphorst W. Determining biogenic silica in marine samples by tracking silicate and aluminium concentrations in alkaline leaching solutions. Aquat Geochem. 2002;8:37–67.

Korndörfer GH, Coelho NM, Mizutani CT, Snyder GH. Evaluation of soil extractants for silicon availability in upland rice. In: Datnoff LE, Snyder GH, Korndorfer GH, editors. Silicon in agriculture. Amsterdam: Elsevier Science; 2001. p. 376.

Korndörfer GH, Pereira HS. Silicon testing, silicon fertilizer manufacturing techniques and standards. Proceedings of the 5th International Conference on Silicon in Agriculture; 2011 Sept 11–19; Beijing, China; 2011. p. 89–98.

Kraska JE, Breitenbeck GA. Simple, robust method for quantifying silicon in plant tissue. Commun Soil Sci Plant Anal. 2010;41:2075–85.

Li CH, Zhou W, Rong XN. Standardization of analytical method for available SiO_2 in silicon-containing fertilizer. Phosphate Compd Fertil. 2004;19:60–1 (In Chinese with English abstract).

Lian S. Silica fertilization of rice. In: The fertility of paddy soils and fertilizer applications for rice. Taipei: Food Fertilizer Technology Center; 1976. p. 197–220.

Liang YC, Ma TS, Li FJ, Feng YJ. Silicon availability and response of rice and wheat to silicon in calcareous soils. Commun Soil Sci Plant Anal. 1994;25:2285–97.

Lindsay WL. Chemical equilibrium in soil. New York: Wiley; 1979.

Liu MD. Methods for assessing silicon-supplying capacity in paddy soils and efficacy of silicon fertilizers on rice. PhD dissertation, Shenyang Agricultural University; 2002.

Ma JF, Takahashi E. Soil, fertilizer, and plant silicon research in Japan. Amsterdam: Elsevier; 2002.

Ma TS, Qian ZR, Zhang ZF, Shu JM, Wang HY. Study on the paddy soils V. Preliminary study on the availability of silica in paddy soil of Jiangsu hilly area and the effect of slag silica fertilizer. J Nanjing Agric Univ. 1985;4:67–73 (In Chinese with English abstract).

Maxwell W. Lavas and soils of the Hawaiian islands. Honolulu: Hawaiian Sugar Planters' Association; 1898. p. 189.

Medina-Gonzales OA, Fox RL, Bosshart RP. Solubility and availability sugarcane (Saccharum Spp.) of two silicate materials. Fertil Res. 1988;16:3–13.

Menzies NW, Bell LC. Evaluation of the influence of sample preparation and extraction technique on soil solution composition. Aust J Soil Res. 1988;26:451–64.

Meyer ML, Bloom PR. Lithium metaborate fusion for silicon, calcium, magnesium, and potassium analysis of wild rice. Plant Soil. 1993;153:281–5.

Mortlock RA, Froelich PN. A simple method for the rapid determination of biogenic opal in pelagic marine sediments. Deep-Sea Res. 1989;36:1415–26.

Nayar PK, Misra AK, Patnaik S. Rapid microdetermination of silicon in rice plant. Plant Soil. 1975;42:491–4.

Nayer PK, Misra AK, Patnaik S. Evaluation of silica-supplying power of soils for growing rice. Plant Soil. 1977;47:487–94.

NIAES. Official methods of analysis of fertilizers, vol. 124. Tsukuba: National Institute of Agro-environmental Sciences, Foundation Nohrin Kohsaikai; 1987. p. 36–7.

Nonaka K, Takahashi K. A method of measuring available silicon in paddy soils. Jpn Agric Res Quart. 1988;22:91–5.

Nonaka K, Takahashi K. A method of assessing the need of silicate fertilizers in paddy soils. XIV International Congress of Soil Science; Kyoto, Japan 4; 1990. p. 513–514.

Park CS. Past and future advances in silicon research in the Republic of Korea. In: Datnoff LE, Snyder GH, Korndörfer GH, editors. Silicon in agriculture. Amsterdam: Elsevier; 2001. p. 359–71.

Pereira HS, Korndörfer GH, Moura WF, Correa GF. Extratores de silício disponível em escórias e fertilizantes. Revista Brasileira Ciencia do Solo. 2003;27:265–74 (In Portuguese).

Qin RC. Effect of silicon fertilizer on disease resistance and yield increase in rice. Zhejiang Agric Sci. 1979;1:12–5 (In Chinese).

Ramsey MH, Potts PJ, Webb PC, Watkins P, Watson JS, Coles BJ. An objective assessment of analytical method precision: comparison of ICP-AES and XRF for the analysis of silicate rocks. Chem Geol. 1995;124:1–19.

Reidinger S, Ramsey MH, Hartley SE. Rapid and accurate analyses of silicon and phosphorus in plants using a portable X-ray fluorescence spectrometer. New Phytol. 2012;195:699–706.

Robinson WO. Method and procedure of soil analysis used in the Division of Soil Chemistry and Physics. USDA Circ. 139. Washington, DC: USDA; 1930.

Robinson WO. The fusion analysis of soils. Soil Sci. 1945;59:7–11.

Saito K, Yamamoto A, Sa TM, Saigusa M. Rapid, micro-methods to estimate plant silicon content by dilute hydrofluoric acid extraction and spectrometric molybdenum method. I. Silicon in rice plants and molybdenum yellow method. Soil Sci Plant Nutr. 2005;51:29–36.

Sauer D, Saccone L, Conley DJ, Herrmann L, Sommer M. Review of methodologies for extracting plant-available and amorphous Si from soils and aquatic sediments. Biogeochemistry. 2006;80:89–108.

Savant NK, Snyder GH, Datnoff LE. Silicon management and sustainable rice production. Adv Agron. 1997;58:151–99.

Sebastian D, Rodrigues H, Kinsey C, Korndörfer G, Pereira H, Buck G, Datnoff L, Miranda S, Provance-Bowley M. A 5-day method for determination of soluble silicon concentrations in nonliquid fertilizer materials using a sodium carbonate-ammonium nitrate extractant followed by visible spectroscopy with heteropoly blue analysis: single-laboratory validation. J Assoc Off Anal Chem Int (AOAC Int). 2013;96:251–9.

Snyder GH. Methods for silicon analysis in plants, soils, and fertilizers. In: Datnoff LE, Snyder GH, Korndörfer GH, editors. Silicon in agriculture. Amsterdam: Elsevier; 2001. p. 185–96.

Struve GA. De silicia in plantis nonnullis. Berolini: Dissertaion; 1835.

Sumida H. Silicon supplying capacity of paddy soils and characteristics of silicon uptake by rice plants in cool regions in Japan. Bull Tohoku Natl Agric Exp Stn. 1992;85:1–46 (In Japanese with English summary).

Takahashi K. Effects of slags on the growth and the silicon uptake by rice plants and the available silicates in paddy soils. Bull Shikoku Agric Exp Stn. 1981;38:75–114 (In Japanese with English summary).

Takahashi K, Nonaka K. Available silicates in paddy soils. Part 2. Development of method of measuring available silicates and its application to soil silicate analysis. Bull Shikoku Agric Exp Stn. 1986;47:16–39 (In Japanese with English summary).

Takijima Y, Wijayaratna HMS, Seneviratne CJ. Nutrient deficiency and physiological disease of lowland rice in Ceylon. III. Effect of silicate fertilizers and dolomite for increasing rice yields. Soil Sci Plant Nutr. 1970;16:11–6.

Ueda K, Yamaoka M. Studies on degraded paddy soils. VII. Examination of a method for determining available silicate in soil. Jpn J Soil Fertil. 1959;30:393–6 (In Japanese).

Wang C, Schuppli PA. Determining ammonium oxalate-extractable Si in soils. Can J Soil Sci. 1986;66:751–5.

Wang FH, Long RJ, Hu DJ. Preliminary study on the method of determining available silicon in Si Ca fertilizer. J Huazhong Agric Univ. 1995;14:537–41 (In Chinese with English abstract).

Weaver RM, Syers JK, Jackson ML. Determination of silica in citrate-bicarbonate-dithionite extracts of soils. Soil Sci Soc Am Proc. 1968;32:497–501.

Wickramasinghe DB (1994) The solubility of rice straw silica and its use as a silicon source in paddy cultivation. PhD thesis, Department of Soil Science, University of Reading.

Wong Y, Cheong Y, Halais P. Needs of sugarcane for silicon when growing in highly weathered latosols. Explan Agric. 1970;6:99–106.

Yoshida S, Forno DA, Cock JH, Gomez KA. Laboratory manual for physiological studies of rice. 3rd ed. Los Baños: International Rice Research Institute; 1976. p. 17–22.

Young RS. Chemical analysis in extractive metallurgy. London: Griffin; 1971. p. 302–4.

Zang HL, Zhang XP, He DY. On the silicon supplying capacity of paddy soils in South China. Acta Pedol Sin. 1982;19:131–40 (In Chinese with English abstract).

Zhang XP, Zang HL. Methods for determining soil available silicon. Soils. 1982;5:188–92 (In Chinese).

Chapter 3
Silicon Biogeochemistry and Bioavailability in Soil

Abstract Whether silicon (Si) deficiency occurs in plants depends largely on plant-available Si concentration in soils but not on total Si content. Si bioavailability in soils is related closely to soil Si biogeochemistry including biogeochemical cycling of Si, forms and solubility of Si in soils and solubility of Si in soils. This chapter deals in detail with soil Si biogeochemistry and bioavailability, solid forms of Si in soils, Si in natural waters, soluble and plant-available Si in soils, and Si-supplying power.

Keywords Plant-available silicon • Silicon in natural water • Silicon-supplying power • Soil silicon chemistry

3.1 Silicon in Soil

Silicon (Si), atomic number 14, molecular weight 28.0855, has 4 outer valance electrons and oxidation states of +2, +4, and −4 with its melting point of 1,410 °C and boiling point of 2,355 °C (Gascho 2001). Interestingly, in the periodic table of elements, Si is surrounded by such near neighbours as boron (B), carbon (C), nitrogen (N), oxygen (O), phosphorus (P) and sulphur (S) (all recognized as 'essential elements'), and aluminium (Al), gallium (Ga), germanium (Ge) and arsenic (As), which, together with Si itself, are all recognized as 'nonessential and/or beneficial elements' or even toxic elements.

Si is the second most abundant element after oxygen in the Earth's crust and in soil. The average content of elemental Si in the lithosphere is approximately 28 %. Total Si content in soil ranges normally from 25 to 35 % with an average of 30 %, depending greatly upon soil types. However, in some highly weathered soils such as latosols or latosolic red soils in the tropics where desilification and fersialitization processes are extremely active, Si content can be as low as less than 1 %. Si content and its availability in soil depend greatly upon soil-forming processes and consequently soil types (Fig. 3.1). Si is mainly present in soils including various categories of aluminosilicates and quartz (SiO_2), which may account for up to 75–95 % of soil inorganic constituents. Except organic soils (histosols), most mineral soils are composed of sands (largely SiO_2), various crystalline forms of primary (e.g. olivine,

Y. Liang et al., *Silicon in Agriculture*, DOI 10.1007/978-94-017-9978-2_3

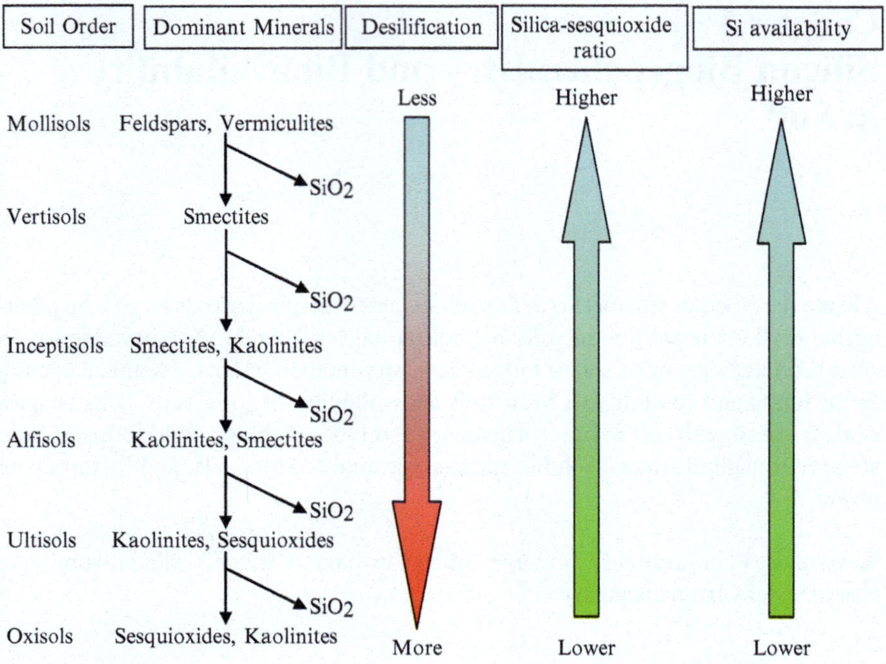

Fig. 3.1 Simplified acid weathering sequence in soil in relation to Si availability (Modified from Friesen et al. 1994; Savant et al. 1997)

augite, hornblende, quartz, feldspars-orthoclase, plagioclase, albite and mica) and secondary silicate minerals (clay minerals like illite, vermiculite, montmorillonite, chlorite and kaolinite) and amorphous (noncrystalline) Si-containing minerals such as allophone and opal. These forms of silicate compounds are only sparingly soluble and usually biogeochemically inert. Monosilicic acid or orthosilicic acid (H_4SiO_4) is the only form of water-soluble Si, while poly-silicic acid (polymerized silicic acid) is only partially water soluble in soil. Water-soluble Si is present in soil solution and can be adsorbed on the surfaces of inorganic, organic and organic–inorganic colloids in soils such as soil clays, soil organic matter and organic–inorganic complex.

3.1.1 Solid Forms of Silicon

Solid forms of Si can be divided into two categories: one is various forms of silicates bound to Al and/or other elements; and the other is quartz (SiO_2). The commonly present primary silicate minerals are crystalline silicates and can be divided into the following five categories (Hurlbut and Klein 1985; Deer et al. 1992):

1. Nesosilicates: Nesosilicates or orthosilicates, which have isolated (insular) negatively charged silicon–oxygen tetrahedra, $[SiO_4]^{4-}$, are connected only by interstitial cations such as iron (Fe), magnesium (Mg), etc. The representative minerals are forsterite (Mg_2SiO_4) and fayalite (Fe_2SiO_4), which belong to the olivine group.
2. Cyclosilicates or ring silicates: Cyclosilicates, which have linked silicon–oxygen tetrahedra with $(T_xO_{3x})^{2x-}$ or a ratio of 1:3, exist usually as 3-member $(T_3O_9)^{6-}$ and 6-member $(T_6O_{18})^{12-}$ rings, where T stands for a tetrahedrally coordinated cation. The minerals typically representing this category are benitoite $[BaTi(Si_3O_9)]$ (3-member ring) and axinite $[(Ca,Fe,Mn)_3Al_2(BO_3)(Si_4O_{12})OH]$ (6-member ring).
3. Inosilicates: Inosilicates or chain silicates consist of both single-chain and double-chain silicates. The single-chain silicates have interlocking chains of tetrahedra with a silicon–oxygen ratio of 1:3 (SiO_3), while the double-chain silicates have interlocking chains of tetrahedra with a silicon–oxygen ratio of 4:11 (Si_4O_{11}). The typical minerals representing the single-chain and double-chain inosilicates are augite $[(Ca,Na)(Mg,Fe,Al)(Si,Al)_2O_6]$ and hornblende $[Ca_2(Mg,Fe)_5(OH)_2Si_8O_{22}]$, respectively.
4. Tectosilicates: Tectosilicates are characterized by a three-dimensional framework of silicate tetrahedra with a silicon–oxygen ratio of 1:2 (SiO_2). This category accounts nearly for 75 % of the Earth's crust. Except for the quartz group, the minerals representing this category are aluminosilicates. The commonly present tectosilicates belong to the feldspar family including alkali feldspars (potassium feldspars) and plagioclase feldspars. The typical minerals of alkali feldspars (potassium feldspars) are microcline ($KAlSi_3O_8$), orthoclase ($KAlSi_3O_8$) and anorthoclase $[(Na,K)AlSi_3O_8]$, while albite ($NaAlSi_3O_8$) and anorthite ($CaAl_2Si_2O_8$) are the representing minerals of plagioclase feldspars. The quartz group is characterized by a three-dimensional framework of silicon–oxygen tetrahedra, including quartz, tridymite and cristobalite, all corners being shared except those protruding at the outer surfaces. This category of mineral is extremely resistant to weathering.
5. Phyllosilicates: Phyllosilicates or sheet silicates form parallel sheets of silicate tetrahedra with a silicon–oxygen ratio of 2:5 (Si_2O_5). Mica group is the typical primary minerals representing phyllosilicates including biotite $[K(Mg,Fe)_3(AlSi_3)O_{10}(OH)_2]$, muscovite $[KAl_2(AlSi_3)O_{10}(OH)_2]$, phlogopite $[KMg_3(AlSi_3)O_{10}(OH)_2]$, lepidolite $[K(Li,Al)_{2-3}(AlSi_3)O_{10}(OH)_2]$, margarite $[CaAl_2(Al_2Si_2)O_{10}(OH)_2]$ and chlorite $[(Mg,Fe)_3(Si,Al)_4O_{10}(OH)_2 \cdot (Mg,Fe)_3(OH)_6]$.

All these crystalline aluminosilicates or SiO_2 are highly weather resistant, with the resistance depending largely upon their structure. The stability of these silicates increases in the order of olivine (nesosilicates) < augite (single-chain inosilicates) < hornblende (double-chain inosilicates) < biotite (phyllosilicates) < potassium-feldspar (tectosilicates) < muscovite (phyllosilicates) < quartz (tectosilicates).

The noncrystalline silicates, which are commonly present in soil, are various layered clay minerals. The formation of these clay minerals by neogenesis implies that all polyhedra of primary minerals, especially of olivine, augite, hornblende and feldspars, are liberated by hydrolysis and then reassembled to clays either inside or adjacent to the parent crystal or after movement to a different horizon site (Jenny 1980). In soils, the commonly present clay minerals (secondary silicates) are phyllosilicates which are characterized by a structure of silicate tetrahedrons arranged in sheets. The typical phyllosilicates in soils are illites, vermiculites, montmorillonites, chlorites, kaolinites and halloysites. The principal clays that might result from the weathering of mica can be summarized as follows (Jenny 1980):

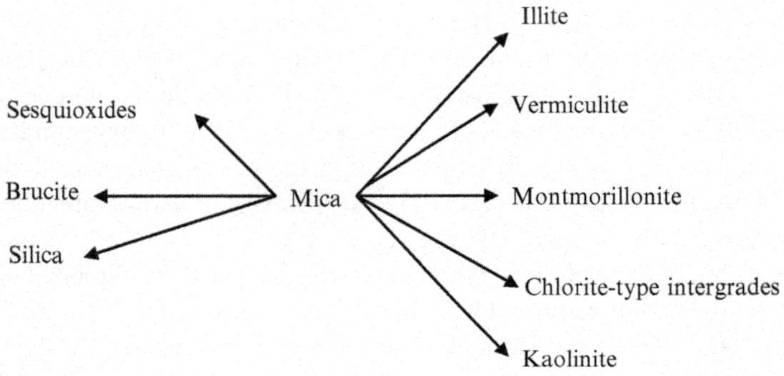

The left arrows point to the oxide clay minerals resulting from severe lattice disruption during both rock weathering and soil-forming processes, while the right arrows result in the commonly present silicate clay minerals derived from weathering of mica in soils (Jenny 1980). Clay genesis is one of the most important indicators of soil formation from parent materials or parent rocks, which is a function of the soil-forming factors. For instance, the effect of the climatic factors on clay formation processes can be illustrated by the transformations of biotite (Tardy et al. 1973): in arid climates it is typically converted to montmorillonite and in temperate environments to vermiculite, while intensive weathering in the humid tropics leads to the formation of kaolinite; on the other hand, weathering of the plagioclase feldspar progresses directly to kaolinite in all climatic conditions.

The commonly present soil amorphous aluminosilicate, which is particularly abundant in soils derived from volcanic ash in humid regions, is allophane (Al_2O_3,$SiO_2 \cdot nH_2O$), with its SiO_2/Al_2O_3 ratio and CEC being highly variable. According to Jackson (1958), allophane can be divided into two groups, unstable and stable, depending largely upon its dissolution characteristics by dilute acid (e.g. HCl) or alkali (e.g. 2 % Na_2CO_3). The amorphous SiO_2 is composed of both silica glass, which is scarce in soils, and opal, which is commonly present. The bioavailability of amorphous opal ($SiO_2 \cdot nH_2O$) to higher plants is higher than that of allophane or laterite ($Fe_2O_3 \cdot 2SiO_2 \cdot nH_2O$).

The solid forms of silicates and Si compounds may undergo weathering with their weathering rate related to not only the mineral itself and its specific surface area but also the environmental factors such as temperature, water, pH, etc.

3.1.2 Solubility of Silicon

As is well known, Si, which is surrounded by four oxygen atoms, remains tetrahedral as SiO_4 in most silicate minerals. In general, quartz is considered to be the most stable SiO_2 mineral at normal temperatures and pressures. Different forms of silica have their solubility relationships (see Table 3.1 and Fig. 3.2). The solubility of silica minerals ranges from $10^{-2.74}$ M (amorphous Si) to 10^{-4} M (quartz) (Lindsay 1979). According to Jenny (1980), the solubility of quartz dissolved in water in amounts of SiO_2 ranges from 5 to 20 mg L^{-1}. Hydration, as $SiO_2 + 2H_2O$, produces $Si(OH)_4$ molecules. The solubility of amorphous silica increases with an increase in temperature, approaching 0 at temperature < 0 °C (Iler 1979). Amorphous silica precipitates in a structure of a porous gel with its solubility in water ranging generally from 100 to 150 SiO_2 mg L^{-1} at pH 7 at room temperature (Jenny 1980). However, amorphous silica of such high solubility is thermodynamically unstable such that the silica gel is expected to form quartz. Because $Si(OH)_4$ can be strongly specifically adsorbed on the surfaces of sesquioxide clays, soluble Si concentration is less than 10 mg kg^{-1} in many highly weathered acid soils in the tropical areas (Fox et al. 1967).

Table 3.1 (Reactions 8 to 12) and Fig. 3.3 show the relationships between the dissociation of H_4SiO_4 and the polymerization of silicate species in solution under the condition that H_4SiO_4 in soil solution is controlled by SiO_2 (soil) (Lindsay 1979). The major silicate species as shown in Fig. 3.4 are the forms of silicate compounds that can be expected in soil solution. Clearly, the ionic silicates contribute significantly to total silica in solution only at pH > 8.5, whereas H_4SiO_4 is the major form of Si species in solution at pH values ranging from 4.5 to 8.0 (Lindsay 1979).

Table 3.1 The major equilibrium reactions of SiO_2 minerals with water and dissociation of H_4SiO_4 (Modified from Lindsay 1979).

Reaction no.	Equilibrium reaction equation	Log K^o
Hydration of SiO_2 minerals		
1	SiO_2 (silica glass) + $2H_2O = H_4SiO_4^o$	−2.71
2	SiO_2 (amorphous) + $2H_2O = H_4SiO_4^o$	−2.74
3	SiO_2 (coesite) + $2H_2O = H_4SiO_4^o$	−3.05
4	SiO_2 (soil) + $2H_2O = H_4SiO_4^o$	−3.10
5	$\alpha - SiO_2$ (tridymite) + $2H_2O = H_4SiO_4^o$	−3.76
6	$\alpha - SiO_2$ (cristobalite) + $2H_2O = H_4SiO_4^o$	−3.94
7	$\alpha - SiO_2$ (quartz) + $2H_2O = H_4SiO_4^o$	−4.00
Dissociation of silicic acid		
8	$H_4SiO_4^o = H_3SiO_4^- + H^+$	−9.71
9	$H_4SiO_4^o = H_2SiO_4^{2-} + 2H^+$	−22.98
10	$H_4SiO_4^o = HSiO_4^{3-} + 3H^+$	−32.85
11	$H_4SiO_4^o = SiO_4^{4-} + 4H^+$	−45.95
12	$4H_4SiO_4^o = H_6Si_4O_2^{2-} + 2H^+ + 4H_2O$	−13.32

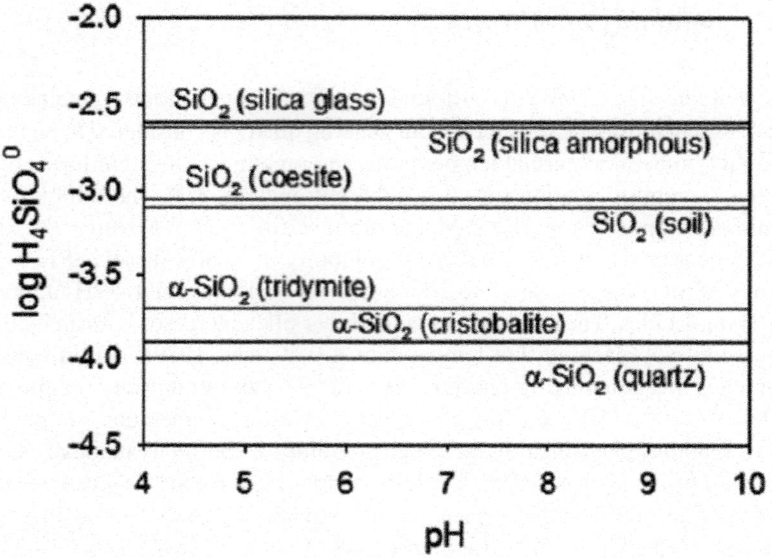

Fig. 3.2 The activity of H_4SiO_4 maintained by various forms of silica (Redrawn by Miroslav Nikolic from Lindsay 1979)

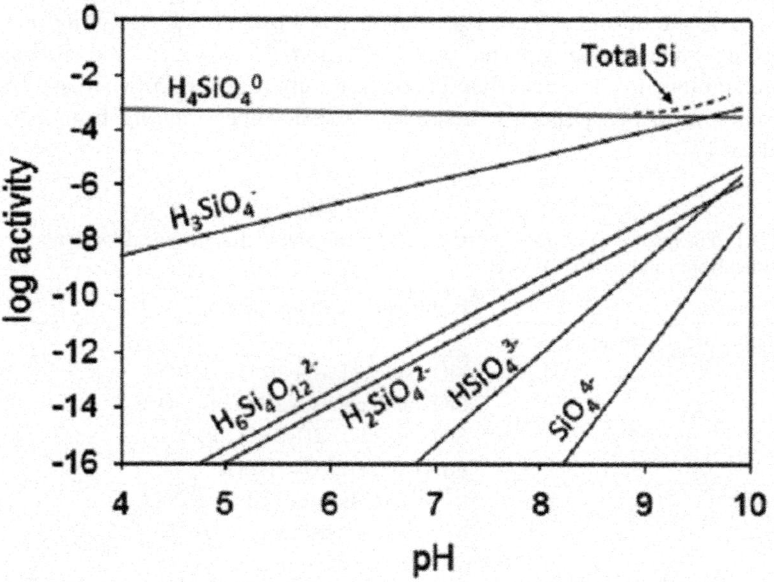

Fig. 3.3 Silicate species in equilibrium with SiO_2 (soil) represented by $10^{-3.10}$ M H_4SiO_4 (Redrawn by Miroslav Nikolic from Lindsay 1979)

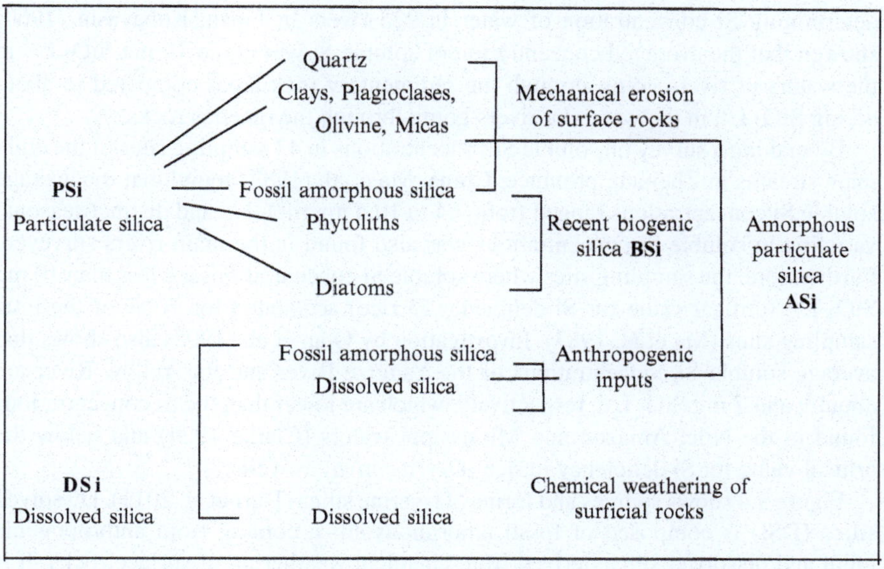

Fig. 3.4 Sources and forms of riverine Si. Amorphous Si (ASi); particulate Si (PSi); biogenic particulate Si (BSi) (Redrawn by Yongchao Liang from Dürr et al. 2011)

3.2 Silicon in Natural Waters

Si in natural waters such as rivers and lakes is primarily derived from the weathering of silicates and aluminosilicates in the bedrock and soils of an area (Berner and Berner 1996). The concentrations of Si in natural waters are much lower than 120 mg SiO_2 L^{-1} (Martin 1970) and in soil solutions decrease with an increase in pH up to about pH 8 (Jones and Handreck 1967). Si in natural waters, which accounts for a considerable proportion of Si absorbed by plants, is greatly influenced by Si in soil solution. The low concentration of silica in natural waters has been attributed largely to both adsorption of silica by Fe and Al hydroxides (Jones and Handreck 1967) and recombination of silica with aluminium silicates (Mackenzie et al. 1967).

Rainwater normally contains little Si, with Si accretion and input from rainwater being < 1 kg ha^{-1} $year^{-1}$ (Alexandre et al. 1997). The concentration of Si in irrigation water varies with the source of water used. If a form of rainwater harvesting is employed, then the water will contain little Si, whereas if the water comes from a wadi or river, the concentration of Si will be possibly higher (Imaizumi and Yoshidai 1958; Meybeck 1987; Bluth and Kump 1994; White and Blum 1995).

By calculating the averaged component of the global rivers, the US Geological Survey pioneer, Clarke (1924), showed that silica in the rivers accounted for approximately 12 % of the total dissolved solids and the dissolved Si concentration of Nile, Amazon and Mississippi waters was 17, 11 and 11 mg SiO_2 L^{-1}, respectively. By determining Si concentration of water in 225 rivers in Japan, Kobayashi (1960) showed that the averaged concentration of soluble Si was about 10 mg SiO_2 L^{-1} in the waters of rivers going through the sedimentary rock areas compared to about 45 mg SiO_2 L^{-1} in the waters of rivers going through the igneous rocks.

By making a survey on soluble Si concentrations in 43 sampling sites of the eight main streams in Zhejiang province, China, Ma et al. (1987) found that the average soluble Si concentrations ranged from 7.4 to 10.6 mg SiO_2 L^{-1} and distinct seasonal variation in soluble Si concentrations was also found in the main rivers surveyed. Furthermore, the sampling sites where soluble Si concentrations are less than 10 mg SiO_2 L^{-1} (critical value for Si deficiency in rice) accounted for 70 % of the total sampling sites (Ma et al. 1987). Investigation by Qian et al. (1995) also shows that average soluble Si concentrations of the Yangtze River and the Yellow River are about 9 and 7 mg SiO_2 L^{-1}, respectively, which are lower than the Si concentrations found in the Nile, Amazon and Mississippi waters (Clarke 1924) and below the critical value for Si deficiency in rice (*Oryza sativa*) as well.

Figure 3.4 shows sources and forms of riverine silica (Dürr et al. 2011). Dissolved silica (DSi) is composed of fossil amorphous silica coming from anthropogenic input and dissolved silica derived from chemical weathering of surface rocks. It is reported that the averaged global dissolved Si (DSi) values range from 8.3 to 13.1 mg SiO_2 L^{-1} with small variation having taken place over the last 9 decades (Dürr et al. 2011). African rivers have the highest Si concentrations (12.6 mg SiO_2 L^{-1} on average) followed by Australian rivers (11.8 mg SiO_2 L^{-1}) and Asian rivers (10.0 mg SiO_2 L^{-1}), while European rivers have the lowest silica levels on average: 5.6 mg SiO_2 L^{-1} vs. 9.5 mg SiO_2 L^{-1} for the global exorheic (externally drained water bodies) average. In addition, European rivers have the lowest averaged dissolved silica influx (13.5 Mt SiO_2 $year^{-1}$), whereas Asian rivers have the highest averaged dissolved silica influx (129.4 Mt SiO_2 $year^{-1}$) followed by South American rivers (106.5 Mt SiO_2 $year^{-1}$) (Dürr et al. 2011).

Liu et al. (2008) reported the concentrations of silicic acid and particulate biogenic Si (PBSi; μmol L^{-1}) in major streams entering into Jiaozhou Bay, China. The concentrations of silicic acid varied considerably among different streams (e.g. the Licunhe and Daguhe) and also by a factor of 1–12 between dry and flood seasons. They ranged from 10.5 to 370.8 μmol L^{-1} from March to August 2002 and from April to August 2004 with the Licunhe and Moshuihe having much higher concentrations and Daguhe, Yanghe and Baishahe having much lower concentrations.

Liu et al. (2008) also calculated the Si budget in Jiaozhou Bay, showing that among all input pathways for biologically available Si, the exchange flux between Jiaozhou Bay and the Yellow Sea is the major contribution, accounting for 72 %, followed by riverine input (21 %) and then wastewater discharge (6 %).

By measuring dissolved silica (DSi) concentrations at the Pawcatuck river-mouth in Westerly, Rhode Island, over 70 times, Fulweiler and Nixon (2005) found that DSi concentrations varied greatly during the annual cycle, with a minimum of about 90 μM Si in late spring and early summer and a maximum of about 200 μM Si during winter. Moreover, DSi concentrations are lower during spring than they are during fall at equivalent water temperatures. For example, the lowest monthly average DSi concentration and the highest occurred at approximately 12 °C (108 μM Si in April and 201 μM Si in October). A distinct seasonal variation in Si concentration was also reported in the water of the Mill Stream (downstream site), Dorset, England, between June 1991 and June 1994 (House et al. 2001) with Si concentration ranging from 25 to 180 μM. The minimum Si concentrations occurred in the spring, and the variations in concentration of Si during the periods between the spring minima were relatively small and generally ranged from 100 to 150 μM (House et al. 2001).

According to Neal et al. (2005), the Si concentration in the eastern UK rivers considerably varied, from 1.2 to 7.5 mg Si L^{-1}; the lowest averages occurred in the rural areas of Scotland and northern England, while higher concentrations were found in the industrial areas of the Humber region and in the agricultural regions of southeastern England. These trends of increased Si concentrations in rivers along the north–south gradient are consistent with the underlying north–south climatic gradient of increasing temperatures. Neal et al. (2005) also reported that dissolved Si concentrations for waters were significantly affected by temperature, pH and the type of crystalline Si with which equilibrium was established.

Husnain et al. (2008) reported that in Citarum River, Indonesia, the average Si concentration ranged from 22 to 37 mg SiO_2 L^{-1} in the upper stream and from 13 to 25 mg SiO_2 L^{-1} in the middle and lower streams, while in the Kaligarang River, the average Si concentration ranged from 34 to 54 mg SiO_2 L^{-1} in the upper stream and from 23 to 39 mg SiO_2 L^{-1} in the middle and lower streams, showing the difference was affected by the geological condition of the watersheds. Kawaguchi and Kyuma (1977) summarized that the average soluble Si concentrations in rivers on Java Island are 42.3, 30.2, 18.2 and 28.6 mg SiO_2 L^{-1}, respectively, with the geological substrate being volcanic ash, marl, lime and acid tuff loam. Ma and Takahashi (2002) also stated that the Si content was three- to fourfold higher in volcanic ash than in granite and sedimentary rocks. According to Kawaguchi and Kyuma (1977), Si concentrations in irrigation water ranged from 10.1 to 23.7 mg SiO_2 L^{-1} in Thailand, from 5.3 to 16.3 mg SiO_2 L^{-1} in West Malaysia and from 7.3 to 21.8 mg SiO_2 L^{-1} in Sri Lanka.

3.3 Soluble and Available Silicon in Soils

3.3.1 Soluble Silicon in Soils

It has been well documented that monosilicic acid (H_4SiO_4) is the only form of Si present in soil solutions (Jones and Handreck 1967; Epstein 1994). Thermodynamically, the solubility of silicate minerals in terms of H_4SiO_4 ranges from $10^{-2.74}$ M (amorphous Si) to 10^{-4} M (quartz) with the solubility of soil Si corresponding to $10^{-3.10}$ M (Lindsay 1979). Nevertheless, the measured concentrations of monosilicic acid (H_4SiO_4) in soil solutions were only 0.1–0.6 mM (Drees et al.

1989; Epstein 1994), which is much less than that in saturated monosilicic acid solution and is mainly controlled by the pH-dependent adsorption–desorption processes on sesquioxides (Jones and Handreck 1967; Mckeague and Cline 1963).

3.3.2 Available Silicon in Soils

Available Si in soils refers to the amount of Si that can be taken up by plants during the growing season and is usually considered an index of Si-supplying power or capacity in soil. Generally, Si is absorbed and transported by plants in the form of monosilicic acid. However, in monosilicic acid-saturated soil solution, H_4SiO_4 easily polymerizes into polymeric $Si(OH)_4$ which is in a dynamic equilibrium with noncrystalline (amorphous) and crystalline silicates, exchangeable silicates and sesquioxides. Thus, available Si in soils include monosilicic acid in soil solution and parts of silicate components that can be easily converted into monosilicic acids such as polymerized silicic acid, exchangeable silicates and part of colloidal silicates. At pH 2–9, especially at physiological pH values, Si in soil is mainly present as monosilicic acid and conversion of monosilicic acid into ionic silicates is possible only at pH >9. The main factors influencing soil Si availability or Si-supplying power include types of soil and parent material, historical land-use change, soil pH, soil texture, soil Eh, organic matter, temperature and accompanying ions (Kawaguchi and Kyuma 1977; He and Li 1995; Cai et al. 1997; Sumida 2002; Husnain et al. 2008; Struyf et al. 2010a, b).

3.3.2.1 Types of Soil and Parent Material

Soil availability and Si-supplying power vary with soil types, depending mainly on the type of parent materials, weathering and eluviation and illuviation. Soils derived from granite, quartz porphyry and peat are prone to Si deficiency, while those developed from basalt and volcanic ash are Si sufficient. The paddy soils which are subjected to intensive weathering and eluviation are prone to Si deficiency, whereas those which are subjected to slight weathering and eluviation are generally rich in plant-available Si.

He (1993) determined available Si content in 64 paddy soil samples collected from Hunan province, South China. The results showed that the available Si content in paddy soils ranged from 26 to 256 mg SiO_2 kg^{-1}. It is estimated that 51.5 % of the total area of soils is Si deficient. The paddy soils developed from plate shale, red sandstone, alluvial deposits, granite and acidic purple sandstone and those derived from quaternary red earth with lower pH and lighter soil texture were deficient in available Si, while those paddy soils derived from quaternary red earth with higher pH and heavier soil texture and those derived from lacustrine deposits contained high amount of available Si. By determining soil-available Si content of 410 paddy soil samples collected from Jiangsu province, China, Ma et al. (1993) found that 30 %

of the areas of paddy soils were Si deficient with the available Si content of less than 100 mg SiO_2 kg^{-1} and 50 % of those contained available Si of less than 150 mg SiO_2 kg^{-1}. As reported by Li et al. (1999), there are three categories of soils in terms of Si availability. The first is the soils derived from granite, quartzite and alluvial deposits. These soils, because of their sandy texture and strong leaching loss, had lowest averaged available Si contents, ranging from 33.3 to 43.3 mg SiO_2 kg^{-1} and 94 to 100 % of the tested soil samples are Si deficient or severely Si deficient. The second is the soils developed from red sandstone, pelite, lacustrine deposits and quaternary red earth. These soils, mainly due to desilification and fersialitization, had lower average available Si content, ranging from 52.9 to 66.7 mg SiO_2 kg^{-1} and 80 to 85 % of these soils are deficient or severely deficient in Si. The third is the soils derived from purple rock, limestone and Xiashu loess. These soils had higher available Si content due to their clayey soil texture, ranging from 98 to 125.8 mg SiO_2 kg^{-1} and only 33 to 60 % of these soils are Si deficient.

Cai et al. (1997) collected 179 upland and paddy rice soil samples in Fujian province, Southeast China to analyse their Si-supplying power. The results showed that the available Si content extracted by 1.0 M acetate buffer (pH 4.0) ranged from 6 to 450 mg SiO_2 kg^{-1} with an average value of 80 mg SiO_2 kg^{-1}. More importantly, according to the critical value for Si deficiency (105 mg SiO_2 kg^{-1}), 80 % of the upland soil samples and 84 % of paddy rice soil samples tested were deficient (50–100 mg SiO_2 kg^{-1}) or severely deficient in Si (< 50 mg SiO_2 kg^{-1}).

In Anhui province, out of 251 soil samples extracted by 1.0 M acetate buffer (pH 4.0), Si deficiency was established in about 54 %, with 28 % of the samples having severe Si deficiency (Zheng 1998). The available Si content was the highest in vertisols, followed by alluvial soils, limestone soils, yellow-brown soils, paddy soils, red soils and purple soils.

He and Li (1995) also reported that in Hubei province, soil-available Si content was the highest in paddy soils derived from limestone, followed by those from red-purplish sandy shale, alluvial and lacustrine deposits, quaternary red earth, granitic gneiss and sandy shale, and approximately 50 % of the 50 samples tested were Si deficient with available Si content below 100 mg SiO_2 kg^{-1}.

By analysing available Si content of 1818 surface soil samples collected from Shandong province in North China, Quan et al. (1999) reported that the available Si content ranged between 15.4 and 779.5 mg SiO_2 kg^{-1} with an average of 234 mg SiO_2 kg^{-1} and approximately 17.21 % of the soil samples tested were found to be Si deficient (<100 mg SiO_2 kg^{-1}) and 35.68 % of the soil samples were found to be potentially Si deficient (100–200 mg SiO_2 kg^{-1}). The available Si content varied with different types of soil with the highest observed in cinnamon soils (Typic Haplustalfs), followed by lime concretion black soils (Vertic Haplustalfs), saline soils (Typic Halaquepts), alluvial soils (Typic Fluvaquents), brown soils (Typic Paleustalfs) and litho soils (Lithic Ustochrepts). Investigating into 159 soil samples representing 17 major soil types distributed in Shaanxi province of Northwest China showed that available Si content extracted by 0.025 M citric acid ranged from 75 to 980 mg SiO_2 kg^{-1} with an average value of 391 mg SiO_2 kg^{-1}, suggesting that the soil Si-supplying capacity is higher (Dai et al. 2004). The available Si content varied with soil type in a descending order of paddy soils > brown soils > limestone soils > cinnamon soils.

Compared with lowland paddy soils, upland soils (i.e. ultisol and oxisol soil orders) which are often leached, acidic and highly weathered in the humid tropical areas of Southeast Asia including Indonesia, Laos, Myanmar, Thailand, Vietnam and neighbouring areas, and West Africa and South America such as Nigeria and Colombia are more prone to be Si deficient because the Si concentration and content in irrigation water, soil and crop were found to be 80 and 90 % lower in the uplands than in the lowlands (Winslow et al. 1997 and references therein).

3.3.2.2 Land-Use Pattern

Intense biogeochemical cycling of silica occurs in soils (Blecker et al. 2006; Gérard et al. 2008). It is reported that the annual ecosystem biogeochemical cycling of Si exceeds the annual export from continents to the ocean by two orders of magnitude (Conley 2002; Struyf et al. 2010a). It has been recently reported that land use is the most important controlling factor on baseflow Si mobilization in a temperate European river basin (Scheldt basin), with historical soil disturbance and sustained cultivation (> 250 years) of formerly forested areas leading to a twofold to threefold decrease in baseflow delivery of Si from the land surface to the aquatic continuum (Struyf et al. 2010b, Fig. 3.5). It seems to suggest that such human cultivation as land use can significantly alter the biogeochemical silica cycle, thus affecting terrestrial silica mobilization and the availability of Si for the growth of terrestrial plants and oceanic phytoplankton blooms (Struyf et al. 2010b).

3.3.2.3 Soil pH

The concentration of monosilicic acid is strongly dependent upon soil pH. The lowest concentration is observed at pH 8–9, below or above which the concentration of monosilicic acid increases significantly. Si concentration in soil solution may rise sharply when pH value decreases from 7 to 2 (Beckwith and Reeve 1963). Numerous studies show that soil-available Si content is closely positively correlated with soil pH values (Beckwith and Reeve 1963; Kawaguchi and Kyuma 1977; Zang 1987; He 1993; Ma et al. 1993; Wan et al. 1993; Liang et al. 1994; Shen et al. 1994; He and Li 1995; Zhang et al. 1996, 2003; Cai et al. 1997; He and Wang 1998; Zheng 1998; Li et al. 1999; Qin et al. 2012). Soil-available Si content in acid soils increased with increasing pH, organic matter and clay content (Lian 1976; He et al. 1980; Qin et al. 2012).

In contrast to South China, available Si content as extracted by acetate buffer solution in paddy soils located in the north of Yangtze River is higher (Wu et al. 1987; Ma et al. 1993; Liang et al. 1994; Zhang et al. 2003). Liang et al. (1994) showed that soil-available Si content as extracted by sodium acetate buffer was also increased with increase of pH, silt and clay content. The soils derived from river

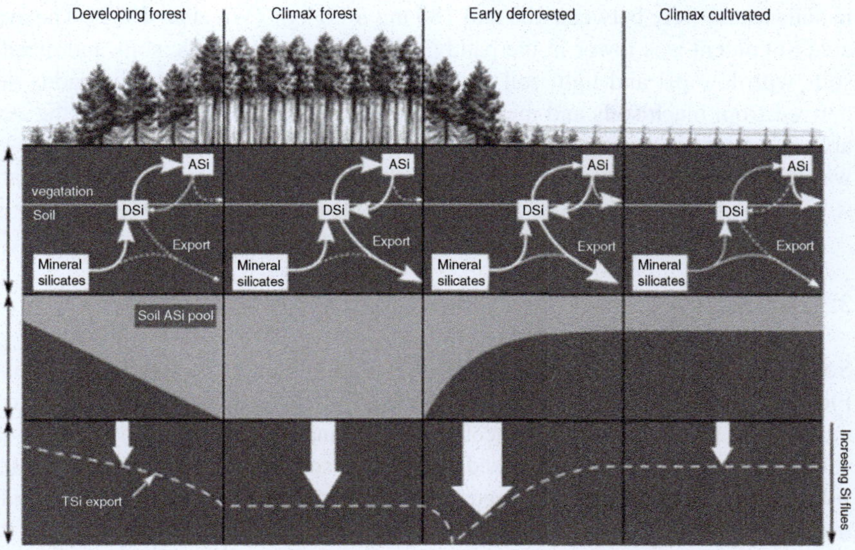

Fig. 3.5 New conceptual model for changes in Si cycling with long-term soil disturbance (From Struyf et al. 2010b). (**a**) Hypothesized Si cycling in developing forest, climax forest, early deforested areas and equilibrium cultured areas, the associated soil ASi stock (**b**) and the resultant magnitude of TSi export (**c**). *Boxes* represent stocks of Si. *Arrows* represent fluxes: the thickness of *arrows* is representative for flux size. *Dashed* arrows represent irrelevant fluxes. In (**b**), the *light grey* area represents the size of the soil ASi pool. In (**c**), the sizes of the *arrows* represent relative TSi fluxes. The *dashed line* represents the hypothesized evolution of the magnitude of the TSi fluxes: the closer the *dashed line* is to the figure *bottom*, the higher the TSi export (as also indicated by the *arrow* to the *right below*). *ASi* amorphous Si, *DSi* dissolved TSi, ASi + Dsi

alluvial and marine deposits distributed widely through the north of Yangtze River to the North China Plain to Northeast China are calcareous and alkaline. The acetate buffer-extractable Si contents in these soils ranged from 150 to 350 mg SiO_2 kg^{-1}, but the availability to plants was still lower, and rice plants responded positively to Si fertilizers (Ma et al. 1993; Liang et al. 1994; Ma et al. 1994; Zhang et al. 2003). The main reasons are that part of carbonate-bound silicates extracted by the acetate buffer solution at pH 4.0 are not plant available and the acetate buffer method overestimates the Si-supplying power of these calcareous paddy soils (Liang et al. 1994; Ma et al. 1994; Zhang et al. 2003). For these soils, there is a need either to develop a suitable extraction method or to modify the critical value of available Si content above which no positive rice yield response can be expected under field conditions (Liang et al. 1994).

Zhang et al. (2003) reported that the available Si content in Liaoning province ranged from 60 to 630 mg SiO_2 kg^{-1} with the highest levels observed in the paddy soils derived from both calcareous alluvial deposits in the North Liaoning, China, and saline marine deposits in the South Liaoning and the lowest available Si content observed in the paddy soils derived from acid alluvial deposits with light soil texture in the Central and East Liaoning. In Heilongjiang province, the available Si contents in soils tested were between 128 and 180 mg SiO_2 kg^{-1} (Wu et al. 1987). The available Si content was lower in the paddy soils derived from albic soils and meadow soils with low pH and light soil texture, but was much higher in the paddy soils derived from black soils and meadow black soils. In East Jilin province, the available Si content in soils tested ranged from 118 to 262 mg SiO_2 kg^{-1} and the lowest was found in cold-water paddy soils, followed by the paddy soils derived from albic soils, alluvial soils, black soils and meadow soils (Zhang et al. 1994).

3.3.2.4 Soil Eh

Soil Eh is one of the most important factors that influence the solubility of soil Si. Flooding results in soil reduction, lowering soil Eh and normally leading to an increase in soil-available Si concentration. Ponnamperuma (1965) reported that submergence of soil led to a marked decrease in soil Eh and concurrent increase in solubility of soil Si with submergence time. In one soil with pH of 4.8 and organic matter of 4.4 %, the concentration of Si increased from 24 to 41 mg kg in less than 50 days after submergence. However, it was also reported that Si concentration in the soil solution slightly increased after flooding and then decreased gradually and, after several months of submergence, may be lower than at the beginning (Liang et al. 1992). Liang et al. (1992) reported that the concentration of Si in soil solution ranged from 5.2 to 10.5 mg SiO_2 L^{-1} in a waterlogging paddy soil (pH 6.5) derived from Xiashu loess and a de-gleyed paddy soil (pH 7.8) developed from bog soil in Jiangsu province in China during a 60-day-period incubation. The concentration of soil-available Si in the waterlogging paddy soil doubled after submergence for 5 days and then reached a peak at 40 days and dropped slightly at 60 days. The soil-available Si content was found to have a logarithmic dependence on the days of submergence. However, for the de-gleyed paddy soil (pH 7.8) developed from bog soil, the soil-available Si concentration had a different dynamics: after 5 days of submergence, it increased only slightly, then had a peak on the 10th day, followed by a slow decrease on the 20th day, and continued to drop afterwards to a level before submergence. The distinct difference in flooding effect on soil-available Si concentration in these two paddy soils tested, especially during the early period of submergence (at 5 days), may be attributed to the differences in redox conditions; the de-gleyed paddy soil (pH 7.8) was initially more reduced than the waterlogged paddy soils (pH 6.5) (Liang et al. 1992).

The increase in Si concentration after flooding may be due to the release of silica following (a) reduction of hydrous ferric oxides which adsorb silica and (b) action of CO_2 on aluminosilicates (Imaizumi and Yoshidai 1958; Nayer et al. 1977; Ponnamperuma 1978; Liang et al. 1992). The subsequent decrease may be ascribed to recombination of Si with aluminosilicates following the decrease in the pressure of CO_2 (Ponnamperuma 1978). However, Wei et al. (1997) reported that Eh effect on Si availability depended upon soil types. For example, the available Si content increased progressively with soil reduction in a clay-textured purple soil but decreased in two sand-textured purple soils.

3.3.2.5 Soil Texture

Many studies have shown that soils with light or sandy texture are usually deficient in available Si and thus have low Si-supplying power, while those with heavy or clayey texture are Si sufficient (Kawaguchi and Kyuma 1977; He 1993; Ma et al. 1993; Wan et al. 1993; Liang et al. 1994; He and Li 1995; Zhang et al. 1996, 2003; Cai et al. 1997; Zheng 1998; Li et al. 1999). Soil-available Si content is positively correlated with clay content in soils (Wan et al. 1993; Zhang et al. 1996; Dai et al. 2004) as soil clay minerals with high specific surface have a high capacity to adsorb silicates. Some researchers reported that soil-available Si content was positively correlated with physical clay (< 0.01 mm) fraction but not with smaller clay (< 0.002 mm) in soils (Shen et al. 1994; Yu et al. 1998). He and Wang (1998) reported that the effect of soil particle size on soil-available Si content was dependent on soil acidity. In acid soils (pH bellow 6.5), pH and clay content were significantly positively related to the soil-available Si content, whereas in the soils with pH above 6.5, only pH and silt and sand fractions were negatively correlated with soil-available Si content.

3.3.2.6 Organic Matter

So far, contrasting results on the effects of soil organic matter on Si availability have been reported. Most of the authors agree that soil-available Si content is positively correlated with soil organic matter content (Lian 1976; Shen et al. 1994; Yu et al. 1998; Qin et al. 2012), while others believe that little or even negative relationship existed between soil-available Si content and organic matter content (Zang 1987; Wan et al. 1993). These contrasting results may be ascribed to the differences of soil types investigated.

3.3.2.7 Adsorption–Desorption Balance

The concentrations of plant-available Si fractions in soil were shown to be governed by processes of Si adsorption on reactive soil materials (primarily sesquioxides) and of Si desorption in the form of soluble Si in soil solution (Mckeague and Cline 1963; Jones and Handreck 1967). The properties of soil adsorption complex (its sorption and desorption characteristics) were consequently shown to largely affect the plant-available Si fraction (Elgawhary and Lindsay 1972; Wei et al. 1997; Yu and Li 1999; Yang et al. 2008, 2010, 2012).

The sorption–desorption characteristics of soils are dependent largely upon soil type and amendment with soluble or amorphous Si. The desorption of Si from a suspension of a calcareous silt loam soil was shown to be different in treatments with and without Si addition (Elgawhary and Lindsay 1972). In the soil suspension without Si addition, the concentration of Si was 18 mg Si L^{-1} at 1 day, 22 mg Si L^{-1} at 10 days and 25 mg Si L^{-1} at 50 days of incubation, while the soil suspension initially treated with 100 mg L^{-1} of soluble Si contained 27 mg Si L^{-1} at 1 day and 25 mg Si L^{-1} after 10 days and remained at 25 mg Si L^{-1} during the subsequent 50-day reaction period (Elgawhary and Lindsay 1972). This suggests that such a calcareous soil maintains an equilibrium level of Si of approximately 25 mg Si L^{-1}. The addition of 10 % amorphous Si led to a steady soil suspension Si concentration at 50 mg Si L^{-1} during the 30th to 50th day of incubation, whereas inclusion of 100 mg Si L^{-1} of soluble Si with the amorphous Si treatment showed little effect. These results seem to suggest that the solid phase of Si in this soil maintained Si levels below the solubility of amorphous Si (51 mg L^{-1}) but above that of quartz (2.8 mg L^{-1}) (Elgawhary and Lindsay 1972). By contrast, in an acidic sandy loam humid soil, the same authors found different Si adsorption–desorption behaviour. Without Si addition, the dynamics of Si in a soil suspension showed an increased trend during the 50-day experimental period and approached 19 mg Si L^{-1} after 50 days, while the addition of 100 mg L^{-1} of soluble Si caused a drop from 32 mg L^{-1} to a constant level of 19 mg L^{-1} between days 30 and 50. The amorphous Si treatment approached equilibrium at about 46 mg L^{-1}. Based on the experimental results of Elgawhary and Lindsay (1972) who measured the solubility of Si in two soils in which equilibrium was approached from both undersaturation and supersaturation, the SiO_2 (soil) concentration was $10^{-3.10}$ M and was intermediate between quartz and amorphous silica, corresponding to Reaction 4 of Table 3.1. In highly weathered tropical acid soils such as oxisols, SiO_2 may be severely leached out of the soil profiles and become depleted due to desilification and fersialitization during weathering and soil-forming processes. Consequently, sesquioxides rather than silicate clay minerals are the dominant residual secondary minerals. In such soils, the solubility of H_4SiO_4 is below that of quartz (10^{-4} M).

Using five paddy soils derived from lake and alluvial deposits and one latosol as testing materials, Yu and Li (1999) found that (1) the isothermal sorption of Si fitted well into Langmuir, Freundlich and Temkin equation and the conventional Langmuir equation seemed to be the best one to describe the sorption characteristics with high significant correlation coefficients; (2) the parameters calculated in the Langmuir equations, namely, the adsorption maximum (Xm) and maximum buffering capacity (MBC), were good parameters to assess Si sorption characteristics; (3) Xm and MBC were closely related to contents of soil clay (<0. 01 mm) and sesquioxides; and (4) approximately only 26–53 % of the adsorbed Si could be desorbed by water. However, Yang et al. (2010) reported recently that neither Langmuir equation nor Freundlich equation could be used to describe Si adsorption on meadow soils used for vegetable production in Northeast China. Rather, both the linear equation ($y = bx - a$) and Temkin equation could be used to accurately describe Si adsorption characteristics of these soils and the linear equation was more suitable for describing Si adsorption characteristics. Furthermore, significant positive linear correlations and quadratic equation existed between the regression parameters (a and b) and soil pH or organic matter. So the isothermal adsorption equation could be used to evaluate Si-supplying capability in the vegetable soils tested. It seems that soil sorption and desorption characteristics are largely dependent upon soil type, pH and clay type.

It was reported that exponential equations could be used to express the Si sorption kinetics of soils incorporated with blast furnace slag, coal fly ash and diamond slag; the accumulative releasing amount of Si at 127 days after incubation increased exponentially with time and was much higher in slag- or coal fly ash-treated soils than in control soil, suggesting that the soil Si-supplying capacities were greatly improved by addition of slag materials (Yang et al. 2008). In 20 meadow soils used for vegetable production, Yang et al. (2012) recently investigated the Si release kinetics and found that the relationship between the cumulative amount of released Si (Y) and the incubation time (X) could be expressed in the following exponential equation: $Y = kX^m$, where k and m are constants, k represents the initial Si release amount and is linearly related to soil pH, organic matter and free iron oxide contents, and m represents initial Si release rate and is also linearly correlated with soil pH, organic matter and free iron oxide contents.

3.4 Silicon-Supplying Power and Silicon Deficiency

It is critically important to understand Si-supplying power or capacity in soils in order to assess whether the amount of plant-available Si in soils is sufficient for optimized crop production. In general, Si-supplying power can be estimated by testing Si status in soils, plants and irrigation water. Generally, Si content in plant tissues is the most relevant value to directly indicate whether Si deficiency occurs or not, but soil testing can help predict Si nutritional status in plants and prevent Si deficiency by applying Si fertilizers before planting. Si level in irrigation water is also one of the useful parameters for prognosis of crop Si deficiency. In practice, it is essential to obtain information on Si status in soils, plants and irrigation water in combination with crop species and varieties to be planted for integrated crop Si nutrient management.

3.4.1 Soil Testing

Analysis of soil-available Si is one of the most commonly used approaches to determine whether Si deficiency will occur or not in a given soil. However, soil-available Si is determined by different methods involving different extracts (see Chap. 2 for more details). The most commonly used extractant is sodium acetate buffer method (pH 4.0) proposed by Imaizumi and Yoshidai (1958) as there exists a critical value or threshold value below which positive rice yield response to Si fertilization can be expected. The threshold value for soil-available Si content varies from one country to another, ranging from 105 mg SiO_2 kg^{-1} in Japan (Imaizumi and Yoshidai 1958; Lian 1976), 100 mg SiO_2 kg^{-1} in Korea (Lian 1976), 40 mg SiO_2 kg^{-1} in Chinese Taiwan (Lian 1976) and 95–100 mg SiO_2 kg^{-1} in China (He et al. 1980; Zhang et al. 2003). According to this criterion, soils low or deficient in available Si content are mainly scattered over the highly weathered acid soils in tropical and subtropical areas. In Japan, many studies were done on Si-supplying capacity and Si fertilizer application to rice crop in the 1950s (Imaizumi and Yoshidai 1958; Ma and Takahashi 2002). In Korea, investigations were carried out including testing and diagnosis of Si availability in Korean paddy soil samples taken randomly from 365 sites, modification of the acetate buffer method proposed by Imaizumi and Yoshidai (1958) and large-scale field trials on optimized application rates of various types of calcium silicates in rice crop during the 1960s (Park 2001). In South and Southeast China, Si deficiency has been frequently reported since 1970 (Qin 1979; He et al. 1980; Ma et al. 1985; Fan et al. 1989) with most of the work done in the 1990s (Yi and Zhang 1991; He 1993; Wan et al. 1993; Ma et al. 1994; Shen et al. 1994; He and Li 1995; Ge and Li 1996; Cai et al. 1997; Yu et al. 1998; Zheng 1998; He and Wang 1998; Li et al. 1999).

However, the suitability of this criterion for assessing Si deficiency, established half a century ago, is increasingly challenged by mounting evidence from field trials that has clearly demonstrated rice yield responses to Si fertilizers in soils with levels of available Si much higher than the universally recognized critical value (100–130 mg SiO_2 kg^{-1}). Si deficiency, diagnosed on the basis of the extraction with sodium acetate buffer, is reported more frequently than before in East China, Central China, Northeast China and even Northwest China (Wu et al. 1987; Ma et al. 1993, 1994; Liang et al. 1994; Zhang et al. 1994; Mao and Wang 2002).

Liang et al. (1994) reported that the sodium acetate buffer-extractable Si content of calcareous paddy soils in Nantong of North Jiangsu province ranged from 152 to 388 mg SiO_2 kg^{-1} and still rice and wheat yields responded positively to the application of Si fertilizer to these soils. Therefore, it seems that the acetate buffer probably overestimates the Si-supplying power in calcareous soils under field conditions as this buffer can extract some Si bound to carbonates which is not plant available (Liang et al. 1994; Ma et al. 1994). Similar findings of positive rice yield responses to Si fertilizers were reported in calcareous and alkaline paddy soils in Liaoning and Heilongjiang provinces of Northeast China (Li et al. 2002; Zhang et al. 2003; Luo 2004; Liu et al. 2006). For example, Li et al. (2002) investigated the Si-supplying power of the paddy soils with pH ranging from 5.2 to 8.3 during an 8-week-incubation experiment. The results indicated that the alkaline soil (pH 8.3) had the lowest Si-supplying power and the lowest water-soluble Si levels, although its acetate buffer-extractable Si content was the highest. The application of sodium metasilicate to this alkaline paddy soil increased rice yield by as much as 197 %. A pot experiment using an alkaline soil with pH of 9.24 and acetate buffer-extractable SiO_2 content of 438.5 mg SiO_2 kg^{-1} showed that application of sodium metasilicate and slag silicate fertilizer increased rice yield by 7.8 and 10.6 %, respectively (Liu et al. 2006). These results suggest that acetate buffer method is not suitable for assessing the Si-supplying power of alkaline soils, a result consistent with the reports on calcareous soils by Liang et al. (1994) and Ma et al. (1994). Liang et al. (1994) proposed that either the threshold or critical value below which positive rice yield response can be expected should be revised (raised) or a more suitable new soil testing method (probably extractant) should be developed for the high-pH calcareous soils. According to Sumida (1992), the critical value of the acetate buffer-extractable soil-available Si content for rice is 300 mg SiO_2 kg^{-1}. Based on seven field trials on slag silicate fertilizer in paddy soils derived from quaternary red earth, red sandstone and alluvial deposits in Dangyang City of Hubei province, Xu et al. (1993) proposed that the critical value of soil-available Si content for hybrid rice growth was 200 mg SiO_2 kg^{-1}. Kawaguchi and Kyuma (1977) reported that the soil-available Si content in tropical Asia ranged from 104 to 629 mg SiO_2 kg^{-1} and the available Si content was reported to be the highest in soils of Indonesia among Asian countries (Kawaguchi and Kyuma 1977). However, recent studies indicate that 12 sawah sites out of 16 in Citarum and 9 sawah out of 15 were low or deficient in available Si contents for rice growth (Husnain et al. 2008).

To summarize, soil testing for available Si using sodium acetate buffer method provides an estimate of Si-supplying power in acid soils. First, the threshold value of soil-available Si content set for acid soils generally works for rice although it differs from one country to another. Second, more and more convincing evidence shows that rice responds positively to Si fertilization in soils where available Si content is higher than the threshold value, suggesting that the threshold value should be revised (raised). Third, the sodium acetate buffer method seems to overestimate the Si-supplying power in calcareous and/or saline (alkaline) soils; thus, a new threshold value is needed for assessing whether Si is sufficient or deficient in calcareous and/or saline (alkaline) soils. Fourth, although there are many methods (mainly extractants) proposed to test soil-available Si content (see Chap. 2), no threshold values are available for the majorities of the extractants used except for the sodium acetate buffer method. Additional lab and field studies are needed to work out a universally recognized threshold value for Si extractants. Additionally, threshold values should be established for other crops of agricultural importance such as sugarcane, wheat, barley, maize, sorghum and vegetable and fruit crops.

3.4.2 *Plant Testing*

Soil-available Si status can affect plant growth, yield and quality and can be expressed in plant vegetative growth performance and yield and Si content in plant tissues. Although Si-deficient rice usually has some deficiency symptoms such as soft and droopy leaves, weeping willow-like appearance as described by Lewin and Reimann (1969), chemical analysis of Si content in rice straw is a reliable measure of Si status in plants. Therefore, plant Si testing is needed together with soil testing to help understand the Si nutritional status in plants. Several protocols are currently available for routine analysis of plant Si content (for details, see Chap. 2). On the basis of a large number of field trials conducted in Asian countries to correlate Si content of rice straw with rice yield response to Si fertilization, the threshold or critical Si content in straw for zero yield response to slag application was proposed. The threshold value or critical value varies from one country to another. For example, if SiO_2 content in the straw of mature rice is less than 13 % in Japan and Korea,

10 % in South China, and 11 % for both early and late rice in Chinese Taiwan, Si deficiency might occur and Si fertilization should be recommended (Lian 1976; Zang et al. 1982; Savant et al. 1997). It was reported that *japonica* rice with straw SiO_2 content of less than 11 % responded positively to slag-based calcium silicates in some cooler ecoregions (Savant et al. 1997), whereas in tropical ecoregions such as Sri Lanka and India, *indica* rice with straw SiO_2 content of less than 8 % might respond to Si applications (Nair and Aiyer 1968; Takijima et al. 1970). Snyder et al. (1986) reported that more than 6.4 % SiO_2 in rice straw was needed for good rice yield on organic soils (histosols) of the Everglades Agricultural Area in Florida. Xu et al. (1993) reported that rice with flag-leaf SiO_2 content of less than 14 % might respond positively to application of slag (averaged yield increase by 11 %). The rice yield (Y) was correlated linearly with flag-leaf SiO_2 content (X) in the following regression equation: $Y = 243.4 + 13.19X$.

3.4.3 *Irrigation Water Testing*

The average dissolved silica concentration (run-off weighted average) for the exorheic parts of Africa, Europe, North America, South America, Asia and Australia is 12.6, 5.6, 8.0, 9.0, 10.0 and 11.8 mg SiO_2 L^{-1}, respectively, with an average of 9.5 mg SiO_2 L^{-1} (Dürr et al. 2011). Si concentration for eastern UK rivers and tributaries ranges from 0 to 34.7 mg SiO_2 L^{-1} (Neal et al. 2005 and references therein), depending on many geological and soil factors. The Si concentration of surface waters across the UK ranges from 0 to 40.7 mg SiO_2 L^{-1} (Neal et al. 2005), which is well within that found for many rivers globally (Meybeck 1980; Berner and Berner 1996). As rice is known to be a typical Si-hyperaccumulating plant species, Si supply from irrigation water plays an important role in Si nutritional status in rice. After having analysed 400 irrigation water, soil and rice samples, Imaizumi and Yoshida (1958) pointed out that the amount of Si coming from irrigation water and from soils accounted approximately for 27 % and 73 % of the total amount of Si taken up by rice, respectively, showing that Si supply from irrigation water is of great significance to rice growth and development. Total Si content in rice straw is significantly correlated with the Si content in irrigation water. Generally, the SiO_2 concentration in irrigation water less than 10 mg L^{-1} implies a high probability of Si deficiency in rice (Liang et al. 1993). Mao (1986) found that the premature senility in rice grown in paddy soils derived from red sandstone could be attributed to both low plant-available Si content in soil and low Si concentration in irrigation water (11.8–21.3 mg SiO_2 L^{-1}). It was reported by Wang et al. (1999) that the application of slag Si fertilizers to paddy soils with an average plant-available Si content of 307 mg SiO_2 kg^{-1} still increased rice yield by 3–9 %, because the average Si concentration in irrigation water was only 4.8 mg SiO_2 L^{-1} (ranging from 1.2 to 11.8 mg SiO_2 L^{-1}), suggesting the importance of Si in irrigation water for sustainable rice production.

References

Alexandre A, Meunier JD, Colin F, Koud JM. Plant impact on the biogeochemical cycle of silicon and related weathering processes. Geochim Cosmochim Acta. 1997;61:677–82.

Beckwith RS, Reeve R. Studies on soluble silica in soils. I. The sorption of silicic acid by soils and minerals. Aust J Soil Res. 1963;1:157–68.

Berner EK, Berner RA. Global environment: Water, air and geochemical cycles. Upper Saddle River: Prentice Hall; 1996. 376 pp.

Blecker SW, McCulley RL, Chadwick OA, Kelly EF. Biologic cycling of silica across a grassland bioclimosequence. Global Biogeochem Cycles. 2006;20, GB3023. doi:10.1029/2006GB002690.

Bluth GJS, Kump LR. Lithologic and climatologic controls of river chemistry. Geochimica et Cosmochimica Acta. 1994;58:2341–59.

Cai AY, Xue ZZ, Peng JG, Lu HD, Lin YJ. Studies of available silica content in the soils of Fujian province and the prerequisite for its variation. J Fujian Acad Agric Sci. 1997;12(4):47–51 (In Chinese with English abstract).

Clarke FW (1924) Data of geochemistry (5th ed., No. 770). U.S. G.S. Survey Bull., 841 pp.

Conley DJ. Terrestrial ecosystems and the global biogeochemical silica cycle. Global Biogeochem Cycles. 2002;16:1121.

Dai GL, Duanmu HS, Wang Z, Zhao YB, Zhu G, Zhou J. Study on characteristics of available silicon content in Shaanxi province. J Soil Water Conserv. 2004;18:51–3 (In Chinese with English abstract).

Deer WA, Howie RA, Zussman J. An introduction to the rock-forming minerals. 2nd ed. London: Longman; 1992.

Drees LR, Wilding LP, Smeck NE, Senkayi AL. Silica in soils: quartz, and disordered silica polymorphs. In: Dixon JB, Weed SB, editors. Minerals in soil environments. 2nd ed. Madison: Soil Science Society of America; 1989. p. 471–552.

Dürr HH, Meybeck M, Hartmann J, Laruelle GG, Roubeix V. Global spatial distribution of natural riverine silica inputs to the coastal zone. Biogeosciences. 2011;8:597–620.

Elgawhary SM, Lindsay LW. Solubility of silica in soils. Soil Sci Soc Am Proc. 1972;36:439–42.

Epstein E. The anomaly of silicon in plant biology. Proc Natl Acad Sci U S A. 1994;91:11–7.

Fan YC, Tao QX, Zhang MH. Silicon nutrition of rice and yield effect of silicon fertilizer in Jiangxi province. In: Chemical fertilizer application in China. Beijing: Beijing Science and Technology Press; 1989. p. 434–5 (In Chinese).

Fox RL, Silva JA, Younge OR, Plucknett DL, Sherman GD. Soil and plant silicon and silicate response by sugarcane. Soil Sci Soc Am Proc. 1967;31:775–9.

Friesen DK, Sanz JI, Correa FJ, Winslow MD, Okada K, Datnoff LE, Snyder GH. Silicon deficiency of upland rice on highly weathered Savanna soils in Colombia. I. Evidence of a major yield constraint. IX Conferencia Internaciaonal de arroz para a America Latina e para o Caribe, V Reuniao Nacional de Pesquisa de Arroz; 21–25 Marco; Castro's Park Hotel. Goiania, Brasil; 1994.

Fulweiler RW, Nixon SW. Terrestrial vegetation and the seasonal cycle of dissolved silica in a southern New England coastal river. Biogeochemistry. 2005;74:115–30.

Gascho GJ. Silicon Sources for Agriculture. In: Datnoff LE, Snyder GH, Korndorfer GH, editors. Silicon in agriculture. Amsterdam: Elsevier Science B.V.; 2001. p. 197–207.

Ge XL, Li SL. General introduction to silicon nutrient status and silicon fertilizer application in croplands of Zhejiang province. Zhejiang Agric Sci. 1996;2:76–7 (In Chinese).

Gérard F, Mayer KU, Hodson MJ, Ranger J. Modelling the biogeochemical cycle of silicon in soils: application to a temperate forest ecosystem. Geochim Cosmochim Acta. 2008;72:741–58.

He DY, Zang HL, Zhang XP. The effect of silicate slags on rice in paddy soils derived from red earth. Acta Pedol Sin. 1980;17:355–63 (In Chinese with English abstract).

He DY. Form, content and availability of silicon in main farmland of Hunan province and research on furnace slag fertilizer. Res Agric Modern. 1993;14:43–7 (In Chinese).

He LY, Li XL. Content and distribution of available silicon in paddy soils of Hubei province. J Huazhong Agric Univ. 1995;14:363–8 (In Chinese with English abstract).

He LY, Wang ZL. Study on relationships between soil particle size or pH and soil available silicon content. Soils. 1998;5:243–6 (In Chinese).

House WA, Leach DV, Armitage PD. Study of dissolved silicon and nitrate dynamics in a freshwater stream. Water Res. 2001;35:2749–57.

Hurlbut CS, Klein C. Manual of mineralogy. 20th ed. New York: Wiley; 1985.

Husnain, Wakatsuki T, Setyorini D, Hermansah SK, Masunaga T. Silica availability in soils and river water in two watersheds on Java Island, Indonesia. Soil Sci Plant Nutr. 2008;54:916–27.

Iler RK. The Chemistry of Silica. New York: Wiley; 1979.

Imaizumi K, Yoshidai S. Edaphological studies on the silicon supplying power of paddy soils. Bull Natl Inst Agric Sci Tokyo. 1958;B8:261–304.

Jackson ML. Soil chemical analysis. Englewood Cliffs: Prentice-Hall; 1958.

Jenny H. The soil resource: origin and behavior, Ecological studies 37. New York: Springer; 1980. p. 95–112.

Jones LHP, Handreck KA. Silica in soils, plants and animals. Adv Agron. 1967;19:107–49.

Kawaguchi K, Kyuma K. Paddy soils in tropical Asia, their material nature and fertility. Honolulu: University Press of Hawaii; 1977.

Kobayashi J. A chemical study of the average quality and characteristics of river waters of Japan. Berichte Ohara Institute für Landwirtschaft Biologie. 1960;11:313–57.

Lewin J, Reimann BEF. Silicon and plant growth. Annu Rev Plant Physiol. 1969;20:289–304.

Li ZZ, Tao QX, Liu GR, Zhang HZ, Liu YR. Investigation of available silicon content in cultivated soil in Jiangxi province. Acta Agric Jiangxi. 1999;11:1–9 (In Chinese with English abstract).

Li J, Zhang YL, Liu MD, Yu N, Huang Y, Yang LJ. Study on silicon-supplying capacity and efficiency of siliceous fertilizer in paddy soils in Liaoning province. Chin J Soil Sci. 2002;33:142–4 (In Chinese with English abstract).

Lian S. Silica fertilization of rice. In: The fertility of paddy soils and fertilizer applications for rice. Taipei: Food Fertilizer Technology Center; 1976. p. 197–220.

Liang YC, Chen XH, Zhang YC, Ma TS. Effects of flooding and incorporation of organic materials on soil available silicon. Soils. 1992;5:244–7 (In Chinese).

Liang YC, Zhang YC, Ma TS. Silicon nutrition of higher plants. Prog Soil Sci. 1993;21:7–14 (In Chinese).

Liang YC, Ma TS, Li FJ, Feng YJ. Silicon availability and response of rice and wheat to silicon in calcareous soils. Commun Soil Sci Plant Anal. 1994;25:2285–97.

Lindsay WL. Chemical equilibrium in soil. New York: Wiley; 1979.

Liu MD, Zhang YL, Meng XF, Li J, Yang D, Wang YJ. Fertilizer response of silicon on rice in alkaline paddy soil. Chin J Soil Sci. 2006;37:554–9 (In Chinese with English abstract).

Liu SM, Ye XW, Zhang J, Zhang GS, Wu Y. The silicon balance in Jiaozhou Bay, North China. J Mar Syst. 2008;74:639–48.

Luo JB. Silicon effect on rice grown on soda solonchak paddy soils of Heilongjiang province. Inner Mongolia Agric Sci Technol. 2004;4:31–3 (In Chinese).

Ma TS, Qian ZR, Zhang ZF, Shu JM, Wang HY. Study on the paddy soils V. Preliminary study on the availability of silica in paddy soil of Jiangsu hilly area and the effect of slag silica fertilizer. J Nanjing Agric Univ. 1985;4:67–73 (In Chinese with English abstract).

Ma GR, He NZ, Shi WY. A survey on the concentration of sulphur, potassium and silicon and the pH value of the major water streams in Zhejiang province. Zhejiang Agric Sci. 1987;1:14–7 (In Chinese).

Ma TS, Li FJ, Liang YC, Chen XH. Silicon nutrient status and yield increase effect of silicon fertilizers. Chin J Soil Sci. 1993;24:262–3 (In Chinese).

Ma TS, Feng YJ, Liang YC, Hua W, Sun B, Wang QS, Huang SH, Zhang BK. Silicon supplying power and silicon fertilizer application in areas along Yangtze River. Soils. 1994;26:154–6 (In Chinese).

Ma JF, Takahashi E. Soil, fertilizer, and plant silicon research in Japan. Amsterdam: Elsevier; 2002. p. 1–190.

Mackenzie FT, Garrels RM, Bricker OP, Bickley F. Silica in sea water-control by silica minerals. Science. 1967;155:1404–5.

Mao DM. Analysis of causes for premature senility of rice grown on paddy soil derived from red sandstone and its prevention and control. Zhejiang Agric Sci. 1986;5:213–5 (In Chinese).

Mao GF, Wang CB. Available silicon status in croplands of Shanghai district and effectiveness of silicon fertilizers in rice. Soils. 2002;5:280–73 (In Chinese).

Martin DF. Marine Chemistry, vol. 2. New York: Dekker; 1970.

McKeague JA, Cline MG. Silica in soils. Adv Agron. 1963;15:339–96.

Meybeck M. Global chemical weathering of surficial rocks estimated from river dissolved loads. Am J Sci. 1987;287:401–28.

Meybeck M. Pathways of major elements from land to ocean through rivers. In: Martin JM, Burton JD, Eisma D, editors. Proceedings of the review and workshop on river inputs to ocean-systems; 1980. p. 18–30.

Nair PK, Aiyer RS. Status of available silica in the rice soils of Kerala State (India). II. Silicon uptake by different varieties of rice in relation to available silica contributed by soil and irrigation water. Agric Res J Kerala. 1968;6:88–94.

Nayer PK, Misra AK, Patnaik S. Evaluation of silica-supplying power of soils for growing rice. Plant Soil. 1977;47:487–94.

Neal C, Neal M, Reynolds B, Maberly SC, May L, Ferrier RC, Smitha J, Parker JE. Silicon concentrations in UK surface waters. J Hydrol. 2005;304:75–93.

Park CS. Past and future advances in silicon research in the Republic of Korea. In: Datnoff LE, Snyder GH, Korndörfer GH, editors. Silicon in agriculture. Amsterdam: Elsevier; 2001. p. 359–71.

Ponnamperuma FN. Dynamic aspects of flooded soils and the nutrition of the rice plant. In: The mineral nutrition of the rice plant. Baltimore: John Hopkins; 1965. p. 295–328.

Ponnamperuma FN. Soils and rice. Philippine: Los Banos; 1978. p. 427.

Qian FJ, Cai DL, Deng T. Silicon concentration in irrigation water and its distribution in Henan province. Area Res Dev. 1995;14:66–8 (In Chinese).

Qin RC. Effect of silicon fertilizer on disease resistance and yield increase in rice. Zhejiang Agric Sci. 1979;1:12–5 (In Chinese).

Qin FJ, Wang F, Lu H, Cen TX, Wang B, Han HX, Zhuang YQ, Zhang H. Study on available silicon contents in cultivated land and its influencing factors in Ningbo City. Acta Agric Zhejiangensis. 2012;24:263–7 (In Chinese with English abstract).

Quan WJ, Zou Q, Du CX. Available silicon content in soil and its distribution in Shandong province. Shandong Agric Sci. 1999;5:11–4 (In Chinese with English abstract).

Savant NK, Snyder GH, Datnoff LE. Silicon management and sustainable rice production. Adv Agron. 1997;58:151–99.

Shen YZ, Pan WQ, Xu JB, Fu CC. Dynamic changes of soil available silicon over last decade in Yangzhou City and application of silicon fertilizers to rice. Soil Fertil. 1994;5:23–6 (In Chinese).

Snyder GH, Jones DB, Gascho GJ. Silicon fertilization of rice on Everglades Histosols. Soil Sci Soc Am J. 1986;50:1259–63.

Struyf E, Smis A, Van Damme S, Meire P, Conley DJ. The global biogeochemical silicon cycle. Silicon. 2010a;1:207–13.

Struyf E, Smis A, Van Damme S, Garnier J, Govers G, Van Wesemael B, Conley DJ, Batelaan O, Frot E, Clymans W, Vandevenne F, Lancelot C, Goos P, Meire P. Historical land use change has lowered terrestrial silica mobilization. Nature Commun. 2010b. doi:10.1038/ncomms1128.

Sumida H. Silicon supplying capacity of paddy soils and characteristics of silicon uptake by rice plants in cool regions in Japan. Bull Tohoku Natl Agric Exp Station. 1992;85:1–46 (In Japanese with English summary).

Sumida H. Plant-available silicon in paddy soils. Proceedings of the Second Silicon in Agriculture Conference; 2002 August 22–26; Tsuruoka, Yamagata, Japan; 2002. pp. 43–49.

Takijima Y, Wijayaratna HMS, Seneviratne CJ. Nutrient deficiency and physiological disease of lowland rice in Ceylon. III. Effect of silicate fertilizers and dolomite for increasing rice yields. Soil Sci Plant Nutr. 1970;16:11–6.

Tardy Y, Bocquier G, Paquet H, Millot G. Formation of clay from granite and its distribution in relation to climate and topography. Geoderma. 1973;10:271–84.

Wan XS, He DY, Liao XL. Silicon forms of soils in Hunan province in relation to soil properties. Soils. 1993;25:146–51 (In Chinese).

Wang CB, Mao GF, Jiang ZT. Study on rice silicon nutrition status in Shanghai region and effects of silicon fertilizer. Acta Agric Shanghai. 1999;15:65–9 (In Chinese with English abstract).

Wei CF, Yang JH, Gao M, Xie DT, Li QZ, Li HL, Li J, Xiang TC. Study on availability of silicon in paddy soils from purple soil. J Plant Nutr Fertil. 1997;3:229–36 (In Chinese with English abstract).

White AF, Blum AE. Effects of climate on chemical weathering in watersheds. Geochimica et Cosmochimica Acta. 1995;66:1891–904.

Winslow MD, Okada K, Correa-Victoria F. Silicon deficiency and the adaptation of tropical rice ecotypes. Plant Soil. 1997;188:239–48.

Wu Y, Zhao XC, Li SF. Study on silicon fertilizer effect on rice in varying types of soils in Heilongjiang province. Heilongjiang Agric Sci. 1987;5:8–12 (In Chinese).

Xu SL, He SK, Gao GW, Zhou HL, Xiong BX, Shen CQ. Efficacy of calcium silicate fertilizer in hybrid rice. Chin J Soil Sci. 1993;24:264–6 (In Chinese).

Yang D, Zhang YL, Liu MD, Yu N. Silicon liberation characteristics of soil and its affecting factors after applying slag mucks. II Relationship between silicon liberation and pH. Chin J Soil Sci. 2008;39:726–9 (In Chinese with English abstract).

Yang D, Zhang YL, Liu MD, Wang CX, An J. Evaluation of silicon supplying capability in greenhouse soil with adsorption isotherm. J Shenyang Agric Univ. 2010;41:540–4 (In Chinese with English abstract).

Yang D, Zhang YL, Liu MD, Wang CX. Characteristics of silicon releasing kinetics in greenhouse soil. Chin J Soil Sci. 2012;43:42–6 (In Chinese with English abstract).

Yi XR, Zhang M. Available silicon content in paddy soils and perspectives of silicon fertilizer application in East and Central Guizhou province. Tillage Cultiv. 1991;5:51–3 (In Chinese).

Yu QY, Li XL. Study on adsorption and desorption of silicon in soils. J Anhui Agrotech Teachers Coll. 1999;13:1–6 (In Chinese with English abstract).

Yu QY, Li XL, Zhang YL. Available silicon distribution and affecting factors in paddy soils of Anhui province. J Anhui Agrotech Teachers Coll. 1998;12:5–9 (In Chinese with English abstract).

Zang HL. A preliminary study on changes in soil available silicon content. Soils. 1987;19:123–5 (In Chinese).

Zang HL, Zhang XP, He DY. On the silicon supplying capacity of paddy soils in South China. Acta Pedol Sin. 1982;19:131–40 (In Chinese with English abstract).

Zhang W, Wang WD, Zhan YL, Li DY, Zhang DL. Available silicon status and effectiveness of silicon fertilizer in paddy soils of East Jilin province. Chin J Soil Sci. 1994;25:37–9 (In Chinese with English abstract).

Zhang XM, Zhang ZY, Yin KS, Li GL, Qiu ZX, Liu YQ. Study on the available silicon content and its relationship with soil physico-chemical properties. J Heilongjiang August First Land Reclam Univ. 1996;8:42–5 (In Chinese with English abstract).

Zhang YL, Li J, Liu MD, Huang Y. A preliminary study of soil silicon status of paddy soils in Liaoning province. Chin J Soil Sci. 2003;34:543–7 (In Chinese with English abstract).

Zheng L. Available silicon content and its distribution in cultivated soils in Anhui province. Chin J Soil Sci. 1998;29:126–8 (In Chinese with English abstract).

Chapter 4
Silicon Uptake and Transport in Plants: Physiological and Molecular Aspects

Abstract Two different types of silicon (Si) transporter, influx and efflux, are involved in the Si-transport process of higher plants. The transporters responsible for Si uptake by roots (Lsi1 and Lsi2) have been identified in several higher plant species including rice (*Oryza sativa*), barley (*Hordeum vulgare*), maize (*Zea mays*), wheat (*Triticum aestivum*), pumpkin (*Cucurbita moschata*) and the primitive vascular plant, *Equisetum arvense* (horsetail). An influx transporter, Lsi6, involved in xylem unloading of Si has also been identified in graminaceous species, including rice, barley and corn. In this chapter, an overview will be provided to highlight the most recent breakthroughs in Si uptake and transport by plants.

Keywords Silicon accumulation • Efflux • Influx • Silicon transporters • Xylem loading of silicon

4.1 General

Although all terrestrial plants contain some silicon (Si) in their tissues, the concentration of Si in shoots varies greatly among plant species (from 0.1 to 10 % Si on a dry weight basis), showing an extremely uneven distribution within the plant kingdom (Fig. 4.1; Epstein 1999; Ma et al. 2001b; Richmond and Sussman 2003). For instance, Si is heavily accumulated in Bryophyta, Lycopsida and Equisetopsida in Pteridophyta, but less so in Filicopsida in Pteridophyta, Gymnospermae and most Angiospermae (Ma et al. 2001b; Ma and Takahashi 2002). In higher plants (Angiospermae), only few taxa show high (>4 % Si; Cyperaceae, Poaceae and Balsaminaceae) to moderate (2–4 % Si; Cucurbitales, Urticales and Commelinaceae) Si accumulation (Ma and Takahashi 2002; Hodson et al. 2005).

There is also a genotypic variation in the shoot Si concentration, although the variation within species is usually much lower than the variation among species. Grown under the same conditions, rice (*Oryza sativa*) contains 39 mg Si g^{-1} in the shoots, but chickpea (*Cicer arietinum*) contains only 3.0 mg Si g^{-1} (Broadley et al. 2011). On the other hand, the Si concentration in barley (*Hordeum vulgare*) grain showed a smaller variation, ranging from 1.2 to 3.8 mg Si g^{-1} (Ma et al. 2003). In sugarcane (*Saccharum officinarum*), the Si concentration in the shoots varied with the cultivars, ranging from 6.4 to 10.2 mg Si g^{-1} (Deren 2001). In rice (*Oryza sativa*),

Y. Liang et al., *Silicon in Agriculture*, DOI 10.1007/978-94-017-9978-2_4

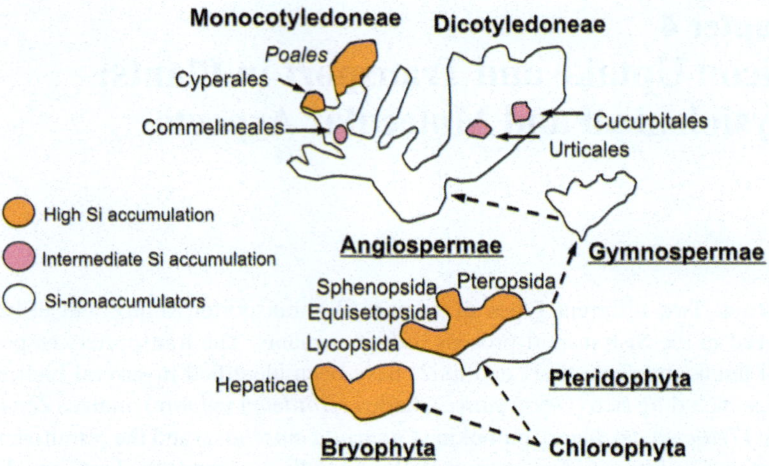

Fig. 4.1 Distribution of Si accumulators within the plant kingdom (Modified from Epstein 1999)

japonica cultivars usually showed a higher Si concentration than *indica* cultivars (Ma and Yamaji 2008 and references therein).

In general, the difference in Si accumulation of different plant species has been attributed to the specific ability of roots to take up Si (for review see Ma and Yamaji 2006). Plants take up Si in the form of the monomeric, uncharged molecule of orthosilicic acid (H_4SiO_4), as the only molecular species likely to cross the root plasma membrane at physiological pH (Raven 2001). Takahashi et al. (1990) proposed three possible types of Si uptake for higher plants assessed in relation to water uptake: active (faster/higher Si uptake than water uptake), passive (similar to water uptake), and even rejective (slower/lower than water uptake). Plants with an active Si uptake mechanism thus cause a significant depletion of Si concentration in the uptake solution, whereas for plants taking up Si passively Si concentration remains unchanged. Correspondingly, for plants with a tendency to exclude Si from their tissues, an increase in Si concentration in the uptake solution occurs over time.

Passive transport of uncharged monosilicic acid by diffusion across the lipid component of the plasma membrane and/or by facilitated diffusion via proteinaceous channels is an energy-independent component of Si uptake, which is present in all plant species regardless of their ability to accumulate Si (Raven 2003). However, a metabolically active, energy-dependent component of Si transport has been identified in various Si-accumulating species (both high accumulators and intermediate plants) including rice, barley, maize (*Zea mays*), wheat (*Triticum aestivum*), banana (*Musa* sp.), and cucumber (*Cucumis sativus*) (e.g. Tamai and Ma 2003; Henriet et al. 2006; Liang et al. 2005; Liang et al. 2006; Nikolic et al. 2007). Both passive diffusion of silicic acid and transporter-mediated uptake are involved in the radial root transport of Si with the transporter-mediated Si uptake being an energy-dependent process, because metabolic inhibitors and low temperatures inhibit Si transport (e.g. Mitani and Ma 2005; Liang et al. 2006; Nikolic et al. 2007).

In principle, two different types of Si transporter, influx and efflux, are operating in the Si-transport process. The transporters responsible for Si uptake by roots (Lsi1 and Lsi2) have been identified in several plant species (e.g. rice, barley, corn, wheat and pumpkin; see Sects. 4.2.1, 4.2.2, and 4.3.1 in this Chapter), and most recently in the primitive vascular plant *Equisetum arvense* (horsetail; Grégoire et al. 2012). Following Si uptake into the root symplast by the influx transporter, Lsi1, and the release of Si from the symplast into the apoplast (xylem loading) by efflux transporter, Lsi2, Si is translocated upwards to the shoots through the xylem vessels via the transpiration stream. Si in the xylem sap is presented in the form of monosilicic acid (Casey et al. 2003; Mitani et al. 2005). In the shoot, silicic acid is further concentrated through loss of water (transpiration) and polymerized to amorphous silica [$(SiO_2)_n \times nH_2O$; known as opal or phytolith] being finally deposited in specific cells. In certain species, xylem unloading of Si is also an active, transporter-mediated process (Ma et al. 2011). So far, an influx transporter, Lsi6, involved in xylem unloading of Si has been identified in graminaceous species, including rice, barley and maize (see Sect. 4.2.1 and 4.2.2 in this Chapter).

4.2 Monocots

4.2.1 Rice

Rice is able to accumulate Si to up to 10 % of shoot dry weight, a concentration considerably higher than that of macronutrients such as nitrogen (N), phosphorus (P) and potassium (K) (Ma and Takahashi 2002). Hence, a large amount of Si is required for high and sustainable rice production (Savant et al. 1997). Physiological studies show that uptake of Si by rice roots is a metabolically active, transporter-mediated process that is much faster than that of water transport and is not affected by the transpiration (e.g. Tamai and Ma 2003; Mitani et al. 2005; Liang et al. 2006; Nikolic et al. 2007). The capacity of roots to take up Si is much higher in rice than in other graminaceous species such as barley, corn, wheat, rye and sorghum (*Sorghum bicolor*) (Tamai and Ma 2003). The SIT gene family encoding Si transporters has been, for the first time, identified in the marine diatom *Cylindrotheca fusiformis*, an organism that requires Si as an essential element (Hildebrand et al. 1997). Rice *Lsi1* is the first gene encoding a Si transporter identified in higher plants (Ma et al. 2006). This gene was cloned using a rice mutant (*lsi1; low silicon 1*) that is defective in active Si uptake (Ma et al. 2002).

Rice Lsi1 (OsLsi1) and its homologs, which belong to the noduline-26 major intrinsic protein 3 (NIP3) subfamily of aquaporins, show influx activity for silicic acid in *Xenopus laevis* oocytes (Ma and Yamaji 2008). They are responsible for Si transport from the soil solution into the root cells (Fig. 4.2). The predicted amino acid sequence has six transmembrane domains and two Asn-Pro-Ala (NPA) motifs that are well conserved in typical aquaporins (Ma and Yamaji 2006). *OsLsi1* is constitutively expressed in the roots, but its expression appears downregulated by Si

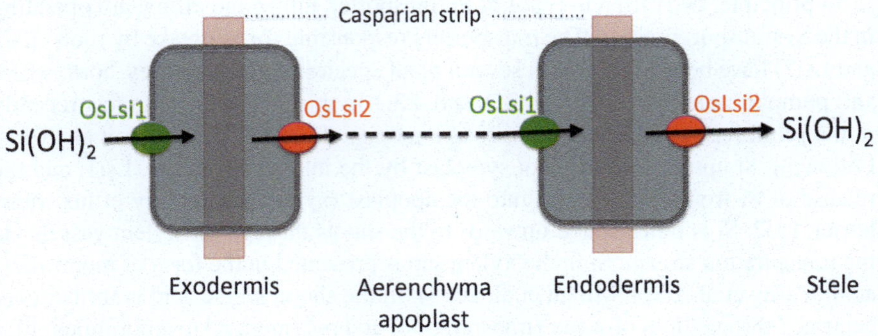

Fig. 4.2 Schematic presentation of Si transport in rice roots (Modified from Ma and Yamaji 2008)

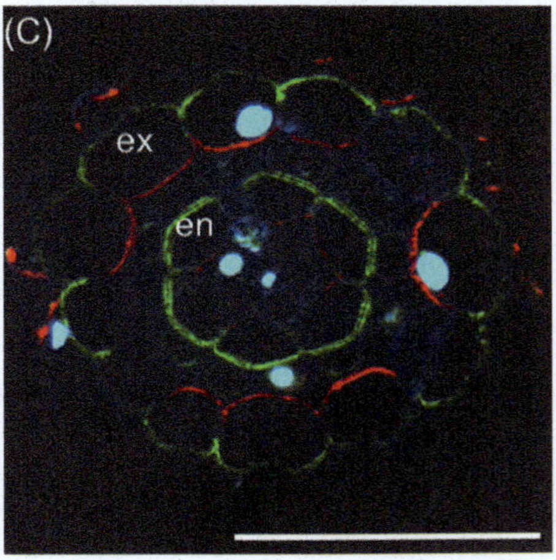

Fig. 4.3 Localization of Si transporters in lateral roots of rice (From Yamaji and Ma 2011). Lsi1 (*green*); Lsi2 (*red*); nuclei (*cyan*); exodermis (*ex*); endodermis (*en*)

(Ma et al. 2006). Within the root, the expression of *OsLsi1* is lower in the apical (up to 10 mm from the root tips) than in the basal root regions (above 10 mm from the root tips), suggesting that Si uptake occurs in the mature root zone rather than in the root tips (Yamaji and Ma 2007). In the basal zone, the Lsi1 protein is localized in the seminal, crown and lateral roots, but not in root hairs (Ma et al. 2006; Yamaji and Ma 2007). Previous physiological studies by Ma's group have shown that lateral roots contribute to Si uptake in rice whereas root hairs do not (Ma et al. 2001a, 2002). The Lsi1 transport protein is localized to the plasma membrane of both exodermis and endodermis, where the Casparian strips prevent apoplastic transport into the root stele (Fig. 4.3), and it shows polar localization at the distal side of both exodermal and endodermal cells (Yamaji and Ma 2007).

Rice *Lsi2* (*OsLsi2*) is also the first Si efflux transporter gene identified in higher plants, and was cloned using a novel rice mutant (*lsi2; low silicon 2*) defective in Si uptake (Ma et al. 2007a). The gene is predicted to encode a membrane protein with

11 transmembrane domains and is responsible for Si transport out of the root cells towards the stele (Fig. 4.2). The Lsi2 protein belongs to a putative anion-channel transporter without any similarity with the Si influx transporter Lsi1 (Yamaji and Ma 2011). In *Xenopus* oocytes Lsi2 exclusively showed efflux activity for silicic acid without any influx transport activity. Unlike OsLsi1, the efflux activity of OsLsi2 is inhibited by low temperature and protonophores (e.g. DNP, CCCP and FCCP), and the transport activity of OsLsi2 decreases at higher external pH values (Ma et al. 2007a). Therefore, Lsi2 is an active efflux transporter driven by the proton gradient, which can transport silicic acid against the concentration gradient. As such, a $Si(OH)_4/H^+$ antiport and involvement of H^+-ATPase are most likely responsible for energizing this secondary active transport (Ma et al. 2011; Fig. 4.4).

The expression pattern and tissue and cellular localization of OsLsi2 are the same as those of OsLsi1 (Yamaji and Ma 2011). Furthermore, there is little accumulation of OsLsi2 transcripts in the root tips and higher accumulation in the mature root zones (Ma et al. 2007a). Like Lsi1, Lsi2 is also localized to the plasma membranes of both exodermal and endodermal root cells but, in contrast to Lsi1, it is localized at their proximal side (Fig. 4.2; e.g. Yamaji and Ma 2011). The expression of both Lsi1 and Lsi2 was lower in a rice variety with low Si uptake capacity than in a variety with high Si uptake capacity, however with no difference in the cellular localization of these two transporters (Ma et al. 2007b). Therefore, the genotypic difference in the Si accumulation may result from the difference in the expression of Si transporter genes in rice roots.

Both Lsi1 and Lsi2 are also involved in arsenite [As(III)] uptake (Ma et al. 2008; Zhao et al. 2010). For instance, mutation in Si influx transporter OsLsi1 significantly decreased As uptake, while in the *lsi2*-rice mutants defective in the Si efflux transporter (OsLsi2), As(III) loading into the xylem and its accumulation in shoots and grain greatly decreased. In field-grown rice, mutation in Lsi2 had a much

Fig. 4.4 Schematic presentation of active efflux of Si via Lsi2

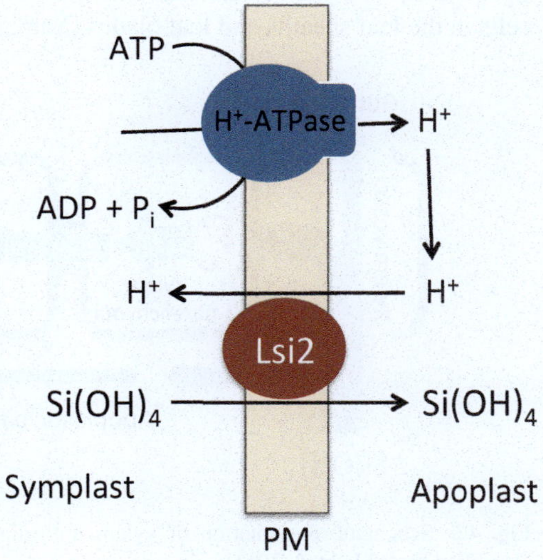

stronger impact on As accumulation in shoots and grains than that in Lsi1 (Ma et al. 2008). Arsenite transport in rice roots therefore shares the same highly efficient pathway as silicic acid, which explains why rice is efficient in As accumulation. One possible explanation is that As(III) and $Si(OH)_4$ have a similar size and both are mostly undissociated at physiological pH (Ma 2010). At a more practical level, ensuring sufficient Si availability in the soil is likely to suppress As accumulation in rice (Ma et al. 2008).

Once transported via Lsi1 and Lsi2 into the stele, Si is then translocated to the shoot through the xylem by the transpiration stream. More than 90 % of Si taken up by roots is translocated to the shoots (Ma and Takahashi 2002). In rice, the Si concentration in the xylem sap can be as high as 20 mM, severalfold exceeding the Si concentration in the external solution (Nikolic et al. 2007). Interestingly, such very high concentrations are probably only present transiently, because in vitro silicic acid polymerizes when its concentration exceeds 2 mM (Mitani et al. 2005). Also, relatively large amounts of Si are deposited in the cell walls of xylem vessels (Balasta et al. 1989) where it may prevent compression of the vessels when the transpiration rates are high (Raven 1983).

Finally, Si is deposited into the epidermis cell walls as a hydrated amorphous polymer (opal) forming silica–cuticle double layers and also deposited in specific shoot cells. There are two types of silicified cells in rice leaf blades, viz. silica cells, and silica bodies (phytoliths) or silicified motor cells (Ma and Takahashi 2002; Prychid et al. 2004). Therefore, Si must be transported out of the xylem before being deposited in shoots. A Si transporter, Lsi6, responsible for the transport of silicic acid from the xylem into xylem parenchyma cells (xylem unloading), has recently been identified in rice (Yamaji et al. 2008). This suggests that Lsi6 influences the subsequent Si distribution in rice shoots (Fig. 4.5). Lsi6 is an influx Si transporter and a homologue of Lsi1 (Yamaji et al. 2008). However, in contrast to Lsi1, Lsi6 is also expressed in the leaf sheaths and leaf blades in addition to the root tips (Yamaji et al. 2008). Lsi6 is localized in the adaxial side of xylem parenchyma cells in the leaf sheaths and leaf blades (Yamaji et al. 2008). Knockout of Lsi6 does

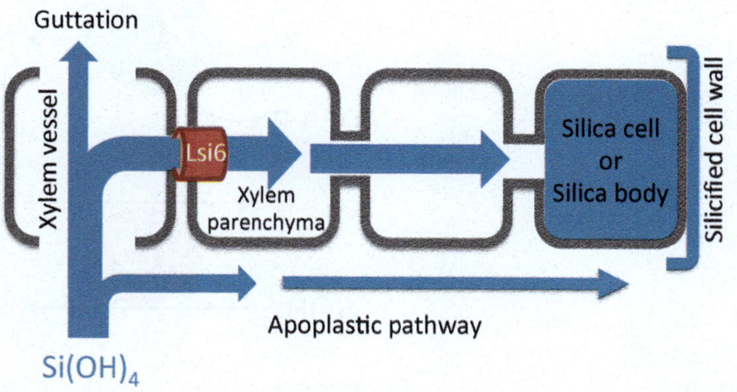

Fig. 4.5 Schematic presentation of xylem unloading of Si by Lsi6 and deposition in shoot (Modified from Ma et al. 2011)

not affect the root uptake of Si, but causes an increased Si deposition in the silicified epidermal cells of leaf blades and sheaths and also causes increased excretion of Si in the guttation fluid (Ma et al. 2011).

High Si accumulation especially in the husk is required for high productivity of rice. Preferential distribution of Si at the node, where the vascular systems are highly developed, is a key step for Si accumulation to the husk at the reproduction stage (Ma et al. 2011). At the first node below the panicles, there are two different vascular bundles (enlarged and diffuse vascular bundles). Therefore, inter-vascular transfer of Si between two different vascular bundles is required to deliver Si taken up by the roots to the panicle, the final sink of Si (Ma et al. 2011). Lsi6 is also highly expressed in the first node below the panicles where it is mainly localized at the proximal side of xylem transfer cells located at the outer boundary region of the enlarged large vascular bundles (Yamaji and Ma 2009). Knockout of Lsi6 resulted in decreased Si accumulation in the panicles, but increased accumulation in the flag leaf (Yamaji and Ma 2009). Therefore, Lsi6 appears to be a transporter involved in transfer of silicic acid from the enlarged large vascular bundles coming from the roots to the xylem transfer cells and then moved symplastically across the parenchyma cell bridge before being reloaded to the diffuse vascular bundles connected to the panicles. Silicic acid is unloaded via Lsi6. Beside Lsi6 responsible for unloading Si from transpirational flow in the enlarged vascular bundles, two other efflux Si transporters, Lsi2 and Lsi3, are also highly expressed in the first node, which would indicate that they are responsible for reloading Si to diffuse vascular bundles (Yamaji et al. 2011).

4.2.2 Other Monocots

In general, most graminaceous plants (e.g. ryegrass, wheat, triticale, sorghum, rye, corn and barley) are known as Si-accumulating species, although Si uptake by their roots is lower than that of rice (e.g. Jarvis 1987; Rafi and Epstein 1999; Tamai and Ma 2003; Liang et al. 2006; Nikolic et al. 2007). In a short-term experiment, Nikolic et al. (2007) demonstrated that both the Si concentration in the xylem sap and the relative proportion of root-to-shoot translocated Si was lower in barley (6 mM and 68 %) than in rice (21 mM and 97 %). Previous uptake studies indicated that the Si uptake by wheat and barley followed a typical Michaelis–Menten kinetics, and uptake was inhibited by DNP, suggesting that these species actively take up Si (Rains et al. 2006; Nikolic et al. 2007). For instance, K_m value for barley was similar to that found in rice (0.35 mM and 0.38 mM, respectively), but with a significantly lower V_{max} as compared to rice (Nikolic et al. 2007). Furthermore, both root uptake and xylem loading of Si in corn were strongly suppressed by metabolic inhibitors or low temperature, regardless of the external Si concentrations, which indicate the importance of an active component in both transport processes (Liang et al. 2006). The coexistence of an active mechanism of Si uptake has also been suggested in other monocots such as banana (Henriet et al. 2006).

Recently, Lsi1 and Lsi2 transporters have also been identified in barley and corn (Chiba et al. 2009; Mitani et al. 2009a, b). More recently, TaLsi1 a wheat Si transporter gene has been cloned (Montpetit et al. 2012). Barley HvLsi1, corn ZmLsi1 and wheat TaLsi1 showed a high level of homology with rice OsLsi1 (Ma et al. 2011; Montpetit et al. 2012). Both HvLsi1 and ZmLsi1 show Si influx transport activity as OsLsi1, however, with different localization and expression patterns than observed in rice (Ma et al. 2011). Although TaLsi1 has yet to be characterized to the same extent as OsLsi1, it is likely that TaLsi1 also functions as an influx Si transporter in wheat (Montpetit et al. 2012). HvLsi1 and ZmLsi1 are localized in the epidermal, hypodermal and cortical cells, and their expression levels are not affected by Si supply (Chiba et al. 2009; Mitani et al. 2009b). Homologues of rice OsLsi2, barley HvLsi2 and corn ZmLsi2 also share more than 80 % similarity at the amino acid level with OsLsi2 (Mitani et al. 2009a). Like OsLsi2, HvLsi2 and ZmLsi2 also function as efflux transporters. The expression of both HvLsi2 and ZmLsi2 is much higher in the mature root zone (>10 mm) than in root tips (Mitani et al. 2009a). Both HvLsi2 and ZmLsi2 are downregulated by Si supply and show a similar expression pattern as OsLsi2 (Mitani et al. 2009a). In contrast to rice OsLsi2, both HvLsi2 and ZmLsi2 are localized only to the endodermis of either barley or corn roots and do not show polar localization (Ma et al. 2011).

In summary, both influx and efflux transporters of Si in rice (OsLs1 and OsLs2, respectively) are localized polarly to the same cells of both the exodermis and endodermis (Ma et al. 2006; Yamaji and Ma 2011). However, in barley and corn, influx (HvLsi1/ZmLsi1) and efflux (HvLsi2/ZmLsi2) transporters are localized to different cells: HvLsi1/ZmLsi1 is polarly localized to epidermal, hypodermal and cortical cells, while HvLsi2/ZmLsi2 is nonpolarly localized exclusively in the endodermis (Chiba et al. 2009; Mitani et al. 2009a, b). These differences result in differential pathways of Si from the external solution to the xylem vessel in barley/corn and rice (Ma et al. 2011). In barley and corn, Si can be taken up from the external solution (soil solution) by HvLsi1/ZmLsi1 localized at the distal side of epidermal and cortical cells and then transported symplastically to the endodermis, where it is released by an active Si efflux transporter (HvLsi2/ZmLsi2) to the stele (Fig. 4.6). On the other hand, in rice, Si is taken up from the external solution by OsLsi at the distal

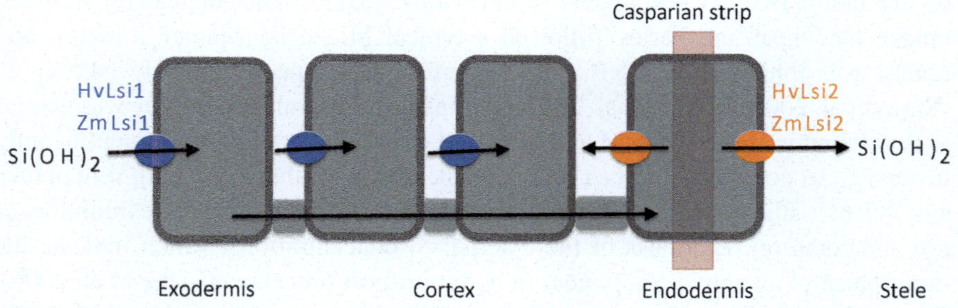

Fig. 4.6 Schematic presentation of Si transport in barley and maize roots (Modified from Mitani et al. 2009b)

side and released into the apoplast of aerenchyma cells by OsLsi2 at the proximal side of exodermal cells (see Fig. 4.2); then, Si is transported into the stele by both OsLsi1 and OsLsi2 from endodermal cells (Ma et al. 2011).

In barley, HvLsi2 was also found in parenchyma cells adjacent to the transfer cells with opposite polarity of HvLsi6, suggesting that the coupling of HvLsi6 and HvLsi2 is involved in the inter-vascular transfer of Si at the nodes (Yamaji et al. 2012). In addition to the xylem parenchyma cells, HvLsi6 was also detected in the outer parenchyma cells surrounding the phloem area. Si translocated via the enlarged vascular bundles is unloaded to the transfer cells by HvLsi6, followed by HvLsi2 to reload Si to the diffuse vascular bundles, which are connected to the panicles (Yamaji et al. 2012).

In corn, ZmLsi6, a homologous to rice influx Si transporter OsLsi6, is localized in the xylem parenchyma cells that are adjacent to the vessels in both leaf sheaths and leaf blades (Mitani et al. 2009b). Like OsLsi6, ZmLsi6 exhibits polar localization on the side facing towards the vessel and mainly functions as a Si transporter for xylem unloading (Mitani et al. 2009b).

4.3 Dicots

4.3.1 Cucumber

Cucumber is known as a species that accumulates Si in the shoots to a much greater extent than other dicots (Wiese et al. 2007). The concentration of Si in cucumber leaves is in the range of 1.8–2.9 % (Miyake and Takahashi 1983a, b; Wiese et al. 2007), and the concentration of Si in the xylem sap is severalfold higher than in the external solutions, regardless of the external Si supply (Liang et al. 2005). In a short-term experiment, Nikolic et al. (2007) demonstrated that more than half of the Si, taken up by roots, was translocated to the shoots in cucumber. Liang et al. (2005) were the first to suggest that Si uptake in cucumber was a concentration-independent and metabolically active process, strongly inhibited by low temperature and metabolic inhibitors. The kinetic uptake studies indicate that Si uptake by cucumber follows a typical Michaelis–Menten curve, but with a significantly lower V_{max} value compared to rice (Mitani et al. 2005; Nikolic et al. 2007).

Pumpkin *Lsi1* (*CmLsi1*) is the first gene encoding an influx Si transporter identified in dicots (Mitani et al. 2011). Unlike rice OsLsi1 transporter, pumpkin CmLsi1 is localized at all root cells, similar to that in barley and corn (Ma et al. 2006; Chiba et al. 2009; Mitani et al. 2009b). This difference has been attributed to the anatomical difference of the roots in paddy and field species (Chiba et al. 2009; Mitani et al. 2009b). Indeed, paddy rice roots develop aerenchyma, wherein almost all cortical cells are destructed, a phenomenon not occurring under normal environmental conditions with field species, including barley, maize and pumpkin. Also, unlike Lsi1 from rice, barley and maize, CmLsi1 did not show polar localization (Mitani et al. 2011).

The surface of cucumber fruits is usually coated with white and fine powders (bloom) primarily composed of SiO_2. However, bloomless cucumber, produced by grafting cucumber on some specific pumpkin cultivars, became more popular in Japan nowadays (Mitani-Ueno et al. 2011). It is reported that the difference in Si accumulation between bloom and bloomless rootstocks of pumpkin cultivars is attributed to different mechanisms of Si uptake by the roots (Mitani et al. 2011). Heterogeneous expression in both *Xenopus* oocytes and *lsi1*-rice mutant defective in Si uptake shows that the influx transporter from the bloom pumpkin rootstock (CmLsi1B^+) transports Si, whereas that from the bloomless rootstock (CmLsi1B^-) does not. Further analysis shows that the transporters from the two rootstocks differ in two amino acid residues (Mitani et al. 2011). A mutation (proline to leucine) at position 242 in bloomless rootstocks results in the loss of Si-transport activity. The Si transporter from the bloom pumpkin rootstock (CmLsi1B^+) is localized at the plasma membrane of all root cells, whereas the one from the bloomless rootstock (CmLsi1B^-) is localized at the endoplasmic reticulum (Mitani et al. 2011).

Two Si efflux transporters (CmLsi2-1 and CmLsi2-2) have also been isolated from two pumpkin cultivars used for the rootstocks of either bloom or bloomless cucumber (Mitani-Ueno et al. 2011). Both CmLsi2-1 and CmLsi2-2 show an efflux transport activity for Si and are expressed in both roots and shoots. Unlike the Si influx transporter CmLsi1, there is no difference in the sequence of both CmLsi2 transporters between the two rootstock cultivars (Mitani-Ueno et al. 2011). Hence, a mutation in CmLsi1, but not in CmLsi2-1 and CmLsi2-2, is responsible for lower Si uptake in the bloomless phenotype.

4.3.2 Other Dicots

In contrast to Cucurbitaceae, little information is available in the literature about Si-transport properties in other dicots. Recently, putative influx Si transporter genes *GmNIP2-1* and *GmNIP2-2* belonging to NIP2 subfamily of aquaporins have been identified, characterized and cloned from soybean (*Glycine max*) (Deshmukh et al. 2013). Expression of both genes was detected in shoot and root tissues, and likewise both *GmNIP2-1* and *GmNIP2-2* were downregulated by increasing supply of Si (Deshmukh et al. 2013). Physiological studies by Liang et al. (2006) showed that the Si uptake was inhibited by low temperature and metabolic inhibitors in the sunflower and wax guard, particularly at the lower external Si concentrations. This might indicate that both active and passive components of Si uptake machinery coexist in the roots of sunflower and wax gourd, similar to that in cucumber (Liang et al. 2005), with their relative contribution much depending upon the external Si concentrations. However, measured Si uptake in faba bean (*Vicia faba*) was significantly lower than the calculated uptake, suggesting a rejective type of Si uptake (Liang et al. 2005). In contrast to Si deposition in leaves, little is known about the compartmentation of Si in the roots, particularly in dicots (Lux et al. 2003). The high proportion of symplastic Si fraction in Si accumulators probably facilitates the transfer of Si into the xylem, and thus translocation to the shoots, as suggested by Sangster and Hodson. However, in tomato (*Lycopersicon esculentum*) (Si excluder) roots, only a very small part (less than 10 %) of the total root Si was present in the symplast (Heine et al. 2005).

The Si concentration in xylem sap of both faba bean and tomato was lower than that of the external solutions, regardless of the external Si concentration (Liang et al. 2005; Heine et al. 2005; Mitani and Ma 2005; Nikolic et al. 2007). The relative proportion of root-to-shoot translocated Si in tomato was about 30 %, which is much lower than in cucumber and rice, especially (54 % and 97 %, respectively; Nikolic et al. 2007). Therefore, the uptake system in tomato and faba bean differs greatly from other plant species that accumulate more Si in their shoots. Furthermore, tomato exhibited no tendency towards saturation kinetics in the total Si uptake, suggesting a lack of the active component of Si uptake in the cortex of tomato (Nikolic et al. 2007). In principle, plants lacking Si transporters will not be able to accumulate Si efficiently. For instance, *Arabidopsis thaliana* is known as a low Si accumulator and, accordingly, has no close homologues of Lsi1 or Lsi2 in its genome (Ma et al. 2006, 2007a; Deshmukh et al. 2013).

The so-called non-accumulators actually exclude silicic acid from their roots, because they contain less Si in the shoots than would be expected if there was a nonselective passive influx of Si with water (Liang et al. 2005, 2006). In the uptake study with radioactive ^{68}Ge (a tracer for Si), application of DNP caused an increase of uptake and root-to-shoot translocation of Si in tomato, suggesting that the restriction of Si from symplast to apoplast in the cortex and endodermis requires a metabolically derived energy (Nikolic et al. 2007). The existence of an active, transporter-mediated Si exclusion from the root cells might explain the rejective type of Si uptake in Si excluders (postulated by Takahashi et al. 1990), which is also demonstrated by the increasing concentration of Si in the uptake solution (Mitani et al. 2005). However, the physiological meaning of the existence of such metabolically energized export of Si out of the roots to prevent Si shoot accumulation is not understood.

References

Balasta MLFC, Perez CM, Juliano BO, Villareal CP, Lott JNA, Roxas DB. Effect of silica level on some properties of *Oryza sativa* straw and hull. Can J Bot. 1989;67:2356–63.

Broadley M, Brown P, Cakmak I, Ma JF, Rengel Z, Zhao F. Beneficial elements. In: Marschner P, editor. Marschner's mineral nutrition of higher plants. 3rd ed. Amsterdam: Elsevier; 2011. p. 257–61.

Casey WH, Kinrade SD, Kinght TG, Rains DW, Epstein E. Aqueous silicate complexes in wheat, *Triticum aestivum* L. Plant Cell Environ. 2003;27:51–4.

Chiba Y, Yamaji N, Mitani N, Ma J. HvLsi1 is a silicon influx transporter in barley. Plant J. 2009;57:810–8.

Deren CW. Plant genotype, silicon concentration, and silicon-related responses. In: Datnoff LE, Snyder GH, Korndorfer GH, editors. Silicon in agriculture. Amsterdam: Elsevier Science; 2001. p. 149–58.

Deshmukh RK, Vivancos J, Guérin V, Sonah H, Labbé C, Belzile F, Bélanger RR. Identification and functional characterization of silicon transporters in soybean using comparative genomics of major intrinsic proteins in Arabidopsis and rice. Plant Mol Biol. 2013;83:303–15.

Epstein E. Silicon. Annu Rev Plant Physiol Plant Mol Biol. 1999;50:641–64.

Grégoire C, Rémus-Borel W, Vivancos J, Labbé C, Belzile F, Bélanger RR. Discovery of a multigene family of aquaporin silicon transporters in the primitive plant *Equisetum arvense*. Plant J. 2012;72:320–30.

Heine G, George Tikum GI, Horst WJ. Silicon nutrition of tomato and bitter gourd with special emphasis on silicon distribution in root fractions. J Plant Nutr Soil Sci. 2005;168:600–6.

Henriet C, Draye X, Oppitz I, Swennen R, Delvaux B. Effects, distribution and uptake of silicon in banana (Musa spp.) under controlled conditions. Plant Soil. 2006;287:359–74.

Hildebrand M, Volcani BE, Gossmann W, Schroeder J. A gene family of silicon transporters. Nature. 1997;385:688–9.

Hodson MJ, White PJ, Mead A, Broadley MR. Phylogenetic variation in the silicon composition of plants. Ann Bot. 2005;96:1027–46.

Jarvis SC. The uptake and transport of silicon by perennial ryegrass and wheat. Plant Soil. 1987;97:429–37.

Liang YC, Si J, Römheld V. Silicon uptake and transport is an active process in *Cucumis sativus*. New Phytol. 2005;167:797–804.

Liang YC, Hua HX, Zhu YG, Zhang J, Cheng CM, Römheld V. Importance of plant species and external silicon concentration to active silicon uptake and transport. New Phytol. 2006;172:63–72.

Lux A, Luxová M, Abe J, Tanimoto E, Taiichiro H, Shinobu I. The dynamics of silicon deposition in the sorghum root endodermis. New Phythol. 2003;158:437–41.

Ma JF. Si transporters in higher plant. In: Jhon PT, Bienert PG, editors. MIPs and their role in the exchange of materials. Texas: Landes Bioscience; 2010. p. 99–109.

Ma JF, Takahashi E. Soil, fertilizer, and plant silicon research in Japan. Amsterdam: Elsevier; 2002.

Ma JF, Yamaji N. Silicon uptake and accumulation in higher plants. Trends Plant Sci. 2006;11:392–7.

Ma JF, Yamaji N. Functions and transport of silicon in plants. A review. Cell Mol Life Sci. 2008;65:3049–57.

Ma JF, Goto S, Tamai K, Ichii M. Role of root hairs and lateral roots in silicon uptake by rice. Plant Physiol. 2001a;127:1773–80.

Ma JF, Miyake Y, Takahashi E. Silicon as a beneficial element for crop plants. In: Datnoff LE, Snyder GH, Korndörfer GH, editors. Silicon in agriculture. Amsterdam: Elsevier; 2001b. p. 17–39.

Ma JF, Tamai K, Ichii M, Wu K. A rice mutant defective in active Si uptake. Plant Physiol. 2002;130:2111–7.

Ma JF, Higashitani A, Sato K, Tateda K. Genotypic variation in Si content of barley grain. Plant Soil. 2003;249:383–7.

Ma JF, Tamai K, Yamaji N, Mitani N, Konishi S, Katsuhara M, Ishiguro M, Murata Y, Yano M. A silicon transporter in rice. Nature. 2006;440:688–91.

Ma JF, Yamaji N, Mitani M, Tamai K, Konishi S, Fujiwara T, Katsuhara M, Yano M. An efflux transporter of silicon in rice. Nature. 2007a;448:209–12.

Ma JF, Yamaji N, Tamai K, Mitani N. Genotypic difference in Si uptake and expression of Si transporter genes in rice. Plant Physiol. 2007b;145:919–24.

Ma JF, Yamaji N, Mitani N, Xu XY, Su YH, McGrath S, Zhao FJ. Transporters of arsenite in rice and their role in arsenic accumulation in rice grain. Proc Nat Acad Sci U S A. 2008;105:9931–5.

Ma JF, Yamaji N, Mitani-Ueno N. Transport of silicon from roots to panicles in plants. Jpn Acad, Ser B. 2011;87:377–85.

Mitani N, Ma JF. Uptake system of silicon in different plant species. J Exp Bot. 2005;56:1255–61.

Mitani N, Ma JF, Iwashita T. Identification of silicon form in the xylem of rice (*Oryza sativa* L.). Plant Cell Physiol. 2005;46:279–83.

Mitani N, Chiba Y, Yamaji N, Ma JF. Identification and characterization of maize and barley Lsi-2-like silicon efflux transporters reveals a distinct silicon uptake system from that in rice. Plant Cell. 2009a;21:2133–42.

Mitani N, Yamaji N, Ma JF. Identification of maize silicon influx transporters. Plant Cell Physiol. 2009b;50:5–12.

Mitani N, Yamaji N, Ago Y, Iwasaki K, Ma JF. Isolation and functional characterization of an influx silicon transporter in two pumpkin cultivars contrasting in Si accumulation. Plant J. 2011;66:231–40.

Mitani-Ueno N, Yamaji N, Ma JF. Silicon efflux transporters isolated from two pumpkin cultivars contrasting in Si uptake. Plant Signal Behav. 2011;6:991–4.

Miyake Y, Takahashi E. Effect of silicon on the growth of solution-cultured cucumber plant. Soil Sci Plant Nutr. 1983a;29:71–83.

Miyake Y, Takahashi E. Effect of silicon on the growth of cucumber plant in soil culture. Soil Sci Plant Nutr. 1983b;29:463–71.

Montpetit J, Vivancos J, Mitani-Ueno N, Yamaji N, Remus-Borel W, Belzile F, Ma JF, Belanger RR. Cloning, functional characterization and heterologous expression of TaLsi1, a wheat silicon transporter gene. Plant Mol Biol. 2012;79:35–46.

Nikolic M, Nikolic N, Liang Y, Kirkby EA, Römheld V. Germanium-68 as an adequate tracer for silicon transport in plants. Characterization of silicon uptake in different crop species. Plant Physiol. 2007;143:495–503.

Prychid CJ, Rudall PJ, Gregory M. Systematics and biology of silica bodies in monocotyledons. Bot Rev. 2004;69:377–440.

Rafi MM, Epstein E. Silicon absorption by wheat (*Triticum aestivum* L.). Plant Soil. 1999;211:223–30.

Rains DW, Epstein E, Zasoki RJ, Aslam M. Active silicon uptake in wheat. Plant Soil. 2006;280:223–8.

Raven JA. The transport and function of Si in plants. Biol Rev. 1983;58:179–207.

Raven JA. Silicon transport at the cell and tissue level. In: Datnoff LE, Snyder GH, Korndörfer GH, editors. Silicon in agriculture. Amsterdam: Elsevier; 2001. p. 41–55.

Raven JA. Cycling silicon–the role of accumulation in plants. New Phytol. 2003;158:419–21.

Richmond KE, Sussman M. Got silicon? The non-essential beneficial plant nutrient. Curr Opin Plant Biol. 2003;6:268–72.

Savant NK, Snyder GH, Datnoff LE. Silicon management and sustainable rice production. Adv Agron. 1997;58:151–99.

Takahashi E, Ma JF, Miyake Y. The possibility of silicon as an essential element for higher plants. Comm Agric Food Chem. 1990;2:99–122.

Tamai K, Ma JF. Characterization of silicon uptake by rice roots. New Phytol. 2003;158:431–6.

Wiese H, Nikolic M, Römheld V. Silicon in plant nutrition. In: Sattelmacher B, Horst WJ, editors. The apoplast of higher plants: compartment of storage. Transport and reactions. Heidelberg: Springer; 2007. p. 33–47.

Yamaji N, Ma JF. Spatial distribution and temporal variation of the rice silicon transporter Lsi1. Plant Physiol. 2007;143:1306–13.

Yamaji N, Ma JF. A transporter at the node responsible for intervascular transfer of silicon in rice. Plant Cell. 2009;21:2878–83.

Yamaji N, Ma JF. Further characterization of a rice Si efflux transporter, Lsi2. Soil Sci Plant Nutr. 2011;57:259–564.

Yamaji N, Mitatni N, Ma JF. A Transporter regulating silicon distribution in rice shoots. Plant Cell. 2008;20:1381–9.

Yamaji N, Mitani-Ueno N, Ma JF. Transporters involved in preferential distribution of Si to the panicles at the node in rice. Proceedings The 5th International Conference on Silicon in Agriculture; 2011 September 13–18; Beijing, China; 2011. p. 210.
Yamaji N, Chiba Y, Mitani-Ueno N, Ma JF. Functional characterization of a silicon transporter gene implicated in silicon distribution in barley. Plant Physiol. 2012;160:1491–7.
Zhao FJ, Ago Y, Mitani N, Li RY, Su YH, Yamaji N, McGrath SP, Ma JF. The role of the rice aquaporin Lsi1 in arsenite efflux from roots. New Phytol. 2010;186:392–9.

Chapter 5
Silicon-Mediated Tolerance to Metal Toxicity

Abstract Silicon (Si) has been well documented to mitigate phytotoxicity of metals including manganese (Mn), iron (Fe), aluminium (Al), cadmium (Cd), arsenic (As), chromium (Cr), lead (Pb), copper (Cu) and zinc (Zn). However, the mechanisms explaining this phenomenon are still not clear. In general, there are two mechanisms, i.e. an external (*ex planta*) and an internal (*in planta*) one, proposed for explaining why and how Si can regulate plant resistance and/or tolerance to metal toxicity. Among the interactions of Si with metals, the possible roles of Si in enhancing tolerance to Mn and Al have been most extensively investigated, while over the last decade more work has been done on the Si-mediated alleviation of toxicity of the most important heavy metals, i.e. Cd, As, Cr and Pb. Nevertheless, direct evidence is still lacking that can clearly dissect the mechanisms involved in Si-mediated tolerance against metal toxicity, especially at the molecular level.

Keywords Aluminium • External mechanisms • Heavy metals • Internal mechanisms • Manganese • Silicon

5.1 Silicon-Alleviated Metal Toxicity Symptoms

5.1.1 Manganese and Iron

Manganese (Mn) is an important essential micronutrient for plant growth, but it easily becomes toxic above physiological levels. Mn toxicity occurs frequently in highly reduced paddy soils (waterlogged lowland soils) or in highly weathered acidic soils of tropical and subtropical areas. In general, the visual symptoms of Mn toxicity will vary with the plant species and plant sensitivity to excess Mn supply. Toxicity occurs at leaf Mn concentrations ranging from 200 to 5,300 mg kg^{-1} (Edwards and Asher 1982; Clarkson 1988), depending largely on the species; genotype; environmental conditions such as temperature, humidity, pH, light and source of nitrogen; nutritional interactions with calcium (Ca), magnesium (Mg), iron (Fe), phosphorus (P), molybdenum (Mo) and Si; and the action of mycorrhizae (El-Jaoual and Cox 1998). Symptoms of Mn toxicity are quite diverse among plant species, but brown spots on older leaves near the main and secondary veins surrounded by

Y. Liang et al., *Silicon in Agriculture*, DOI 10.1007/978-94-017-9978-2_5

chlorotic zones are typical symptoms of Mn toxicity (Morris and Pierre 1949; Foy et al. 1978, 1995; El-Jaoual and Cox 1998; Li et al. 2012a).

Mn toxicity symptoms are reported to be alleviated considerably by the addition of Si (Figs. 5.1 and 5.2). As early as 1957, Williams and Vlamis (1957a, b) showed that Si decreased Mn toxicity in barley (*Hordeum vulgare*) by causing Mn to be more evenly distributed instead of being concentrated in discrete necrotic spots, although the total Mn in the leaves was unaffected by Si treatment. Subsequently, alleviation of Mn toxicity by Si was reported in many crops (Table 5.1).

Okuda and Takahashi (1965) demonstrated significant alleviative effect of Si on Mn toxicity in rice plants and attributed this effect to Si-mediated inhibition of Mn uptake. It has been reported by Júnior et al. (2010) that dry matter weight of roots, sheaths, leaves and aerial parts of Mn-stressed rice (*Oryza sativa*) plants was significantly higher in plants treated with Si and the leaf erectness was significantly increased with addition of Si. Tavakkoli et al. (2011) reported that interactions of Si slag with phosphorus increased dry matter yield of rice grown on a highly weathered, acidic lateritic soil and probably mitigated Mn toxicity by decreasing Mn concentration. Li et al. (2012a) investigated the alleviative effect of Si on excess Mn in rice. Mn toxicity symptoms generally appeared as dark brown spots, necrotic lesions, chlorosis of leaf edges and tips and crinkle leaf. Clearly, Si supply could significantly alleviate Mn toxicity symptoms in both Mn-sensitive and Mn-tolerant cultivars (Fig. 5.2; Li et al. 2012a). Horst and Marschner (1978a) reported that without Si supply, bean growth depression and toxicity symptoms occurred at concentrations as low as 0.5 μM Mn in the nutrient solution, while the Si-treated plants could tolerate 5.10 μM Mn. They argued that the increase in Mn tolerance was not caused by a depressing effect of Si on the uptake or translocation of Mn but rather by an increase in Mn tolerance in leaf tissues. Iwasaki and Matsumura (1999) investigated Si-induced alleviation of Mn toxicity in two contrasting pumpkin (*Cucurbita*

Fig. 5.1 Si alleviates symptoms of Mn toxicity in cucumber (Photograph by Miroslav Nikolic)

Fig. 5.2 Toxicity symptoms of two rice cultivars grown in a nutrient solution containing normal (6.7 μM) or high Mn (2.0 mM) with or without Si (1.5 mM) supply (From Li et al. 2012a)

moschata) cultivars (cultivar Shintosa, a bloom type, and cultivar Superunryu, a bloomless type). The results showed that high Mn significantly depressed root and shoot biomass of both cultivars in the absence of Si, while in the presence of Si, high Mn only inhibited the growth of Superunryu cultivar, but not that of Shintosa which accumulated more Si than Superunryu. Interestingly, Mn uptake was similar in both cultivars. Iwasaki et al. (2002a, b) and Führs et al. (2009) showed that in cowpea receiving 50 μM Mn, dark brown spots were the first visible Mn toxicity symptoms, which completely disappeared in Si-fed plants throughout the experiment. However, distinct brown spots were noted in the Si-pretreated plants grown hydroponically with excess Mn, although the symptoms were less severe in the Si-pretreated plants than in the untreated ones (Iwasaki et al. 2002b). Rogalla and Römheld (2002) reported that brown spots on leaves, a typical symptom of Mn toxicity, appeared at over 50 μM Mn in Si-deprived plants, but at 500 μM and above in Si-fed plants (Table 5.2). Both shoot and root fresh weights of plants grown with Mn levels ranging from 50 to 1,000 μM were significantly increased by the addition of Si (Rogalla and Römheld 2002; Wiese et al. 2007). The experiments with cucumber (*Cucumis*

Table 5.1 Plant species reported to be positively responsive to mitigation of Mn toxicity by Si

Plant species	References
Barley	Williams and Vlamis (1957a, b), Vlamis and Williams (1967), and Horiguchi and Morita 1987
Maize	Doncheva et al. (2009)
Rice	Okuda and Takahashi (1962), Júnior et al. (2010), Tavakkoli et al. (2011), and Li et al. (2012a)
Sugarcane	Clements (1965) and Fox et al. (1967)
Wheat	Vlamis and Williams (1967)
Oats	Vlamis and Williams (1967)
Rye	Vlamis and Williams (1967)
Sorghum	Galvez et al. (1987)
Sudan grass	Bowen (1972)
Bean	Horst and Marschner (1978a)
Cowpea	Horst et al. (1999), Iwasaki et al. (2002a, b), and Führs et al. (2009)
Cucumber	Rogalla and Römheld (2002), Shi et al. (2005a), Wiese et al. (2007), Feng et al. (2009), Dragisic Maksimovic et al. (2007, 2012)
Pumpkin	Iwasaki and Matsumura (1999)

Table 5.2 Mn toxicity symptoms of *Cucumis sativus* plants grown in a nutrient solution with 1.8 mM (+Si) or without (–Si) Si

	Manganese concentration (μM)				
–Si concentration	0.5	5	50	500	1,000
–Si	–	–	+	++	+++
+Si	–	–	–	+	++

From Rogalla and Römheld (2002)
– none; +, marginal; ++, strong; +++, severe Mn toxicity symptoms

sativus) by Shi et al. (2005a) confirmed that plants grown hydroponically with excess Mn (600 μM Mn) showed typical visible symptoms of Mn toxicity on leaves where severe chlorosis developed on both older and younger leaves, while limited chlorosis occurred on leaves of plants treated with Si. All variables including shoot height, root length, shoot and root biomass and plant biomass were greatly suppressed by excess Mn, but were significantly reversed by Si supply (Shi et al. 2005a). A series of experiments with cucumber by Dragisic Maksimovic et al. (2007) and Dragisic Maksimovic et al. (2012) also showed that although the Mn concentration in the leaves of cucumber plants treated with 100 μM was 10–40 times higher than the optimum, no Mn toxicity symptoms appeared in Si-fed plants, compared to Si-deprived plants. Both the root and shoot biomass of plants grown hydroponically in presence of high Mn was improved by the presence of Si (Dragisic Maksimovic et al. 2007; Dragisic Maksimovic et al. 2012).

Fe is another important essential micronutrient for plant growth. Like Mn toxicity, Fe toxicity is primarily a problem for plants growing in highly reduced conditions (e.g. waterlogged lowland soils, poorly drained, compacted or poorly aerated soils) or in highly weathered acid soils of tropical and subtropical areas. One of the most commonly visible Fe toxicity symptoms is leaf necrotic spots; additional symptoms include dark green foliage, stunted top and root growth, as well as leaf bronzing, especially on rice, which causes dark brown to purple spots on the foliage. Silicon has been reported to significantly mitigate Fe toxicity symptoms in rice (Okuda and Takahashi 1965; Fu et al. 2012; Dufey et al. 2014). Okuda and Takahashi (1965) speculated that Si-mediated inhibition of Fe uptake explained the positive effect of Si against Al. It was also reported that application of slag at a rate of 2 t ha^{-1} to soils containing high exchangeable Fe and Mn led to a decrease in both Fe and Mn contents in rice plants without producing deficiency symptoms (IRRI 1965, 1966). More recent findings showed that a whole range of toxicity symptoms such as reduction of root and shoot biomass, primary root length and root number, aberrant root morphology and destruction of cytoplasmic organelles and nuclei in root epidermal cells were successfully mitigated by the addition of Si in an Fe-sensitive rice cultivar grown hydroponically with 100 µM Fe^{2+} (Fu et al. 2012). The inclusion of Si to rice plants grown in presence of a toxic level of Fe was found to significantly reduce Fe toxicity symptoms (Dufey et al. 2014).

5.1.2 Aluminium

Aluminium (Al) is not only a major constituent of most mineral soils but also a major constraint for plant growth at pH values ranging from 3.5 to 4.5. When in excess, Al severely inhibits root growth and nutrient uptake and limits crop production on highly weathered acid soils of tropical and subtropical areas worldwide. A typical symptom of Al toxicity is growth inhibition of the root system (Rorison 1960). As early as the 1960s, Peaslee and Frink (1969) demonstrated that the uptake of Al and Mn by tomato (*Lycopersicon esculentum*) could be depressed by the addition of silicic acid to the soil although no explanation was given for this result. Over the last 20 years, the possible impacts of Si on Al toxicity in plants have been extensively investigated. So far, positive effects of Si on Al-induced inhibition of root growth have been reported in many plants (crops) (Table 5.3). Clearly, less work has been done on Si and Al interactions since one decade ago.

Despite the previous reports, not all the experiments have convincingly shown the benefits of Si against Al toxicity. Some indicated that the alleviative effect was genotype and time dependent (Li et al. 1989; Cocker et al. 1998a, b; Kidd et al. 2001; Zsoldos et al. 2003). Time-dependent changes in Si-induced amelioration of Al toxicity in maize (*Zea mays*) were first demonstrated for the first time by Kidd et al. (2001). In a 48-h experiment, they showed that after an 8-h exposure to high Al, inhibition of root elongation in an Al-sensitive cultivar was significantly alleviated by Si pretreatment (root elongation rate was 55 % of the corresponding controls, compared to only 12 % when not pretreated with Si), whereas in an Al-resistant cultivar such effects of Si were not observed. In a long-term experiment (after 4 days), the effects of Si pretreatment on Al toxicity subsided (Zsoldos et al. 2003). Wang et al. (2004) found that clear alleviative effects of Si on Al-induced inhibition

Table 5.3 Interactions between Al and Si in experiments

Plant species	Authors and/or references
Barley	Hammond et al. (1995), Liang et al. (2001), and Morikawa and Saigusa 2002
Sorghum	Galvez et al. (1987), Galvez and Clark (1991), Hodson and Sangster (1993), and Li et al. (1996)
Rice	Rahman et al. (1998), Hara et al. (1999), and Singh et al. (2011)
Mung bean	Yang et al. (1999)
Maize	Ma et al. (1997), Corrales et al. (1997), Kidd et al. (2001), and Wang et al. (2004)
Teosinte	Barceló et al. (1993)
Wheat	Cocker et al. (1998a, b) and Zsoldos et al. (2003)
Cotton	Li et al. (1989)
Soybean	Baylis et al. (1994)
Melastoma malabathricum	Watanabe et al. (1997)
Holcus lanatus	Kidd and Proctor (2001)
Norway spruce	Ryder et al. (2003)
Stylosanthes	Zhang et al. (2009)

of root elongation (after 12 h) were noted when Si was supplied both as a pretreatment and during the exposure to excessive Al, but essentially no such effects were observed when Si was supplied only as a pretreatment. On the other hand, working with five varieties of *Stylosanthes*, Zhang et al. (2009) found that both Si pretreatment and continuous Si supply significantly overcame the inhibitory effect of Al on root elongation, and more importantly, the phenotypical difference in Al tolerance among the five genotypes tested disappeared after Si treatment.

Furthermore, the benefits of Si on Al toxicity were found to be dependent on the Si source (Morikawa and Saigusa 2002). Also, Si effects on Al detoxification in plants varied greatly with both Al and Si levels used (Li et al. 1989; Liang et al. 2001; Ma et al. 1997; Yang et al. 1999). Hammond et al. (1995) found that the alleviative effect of Si on Al-induced root elongation increased with Si concentrations.

Cocker et al. (1998b) examined the effects of Si on Al toxicity in both Al-tolerant and Al-sensitive wheat cultivars. Interestingly, Si was found to alleviate Al toxicity at low Al concentrations, but Si supply failed to exhibit significant alleviative effect in the Al-sensitive cultivar exposed to higher Al concentrations. It thus seems to suggest that the effect of Si on Al detoxification depends largely upon the plant species, Al and Si levels and the duration of plants exposed to Al.

5.1.3 Cadmium

Cadmium (Cd) is a nonessential toxic heavy metal that is highly mobile in the environment. Exposure to Cd can cause serious problems to microorganisms, plants and animals even at trace concentrations and can be highly toxic to human beings through its bioaccumulation in the food chain (Song et al. 2009). Once absorbed, transported and accumulated into plant tissues in excess amounts, Cd will disturb plant metabolisms in multiple ways. Many studies have indicated that excess Cd accumulated in plant tissues not only causes Fe, Mg and Ca deficiency (Welch 1995) and reduces chlorophyll content (Feng et al. 2010) but also inhibits plant growth and respiration; destroys the ultrastructure of plant cells such as the nucleus, chloroplast and mitochondria; and alters the activity and quantity of the key enzymes involved in various metabolic pathways (Nwugo and Huerta 2008; Cunha and Nascimento 2009; Song et al. 2009). Over the last decade, growing evidence has been presented in the literature showing that Si can significantly mitigate Cd toxicity symptoms and improve the growth and development of higher plants (Table 5.4).

It was reported that typical Cd toxicity symptoms such as chlorosis, leaf rolls and stunting appeared in the leaves of maize grown under Cd stress (Liang et al. 2005; see Fig. 5.3). The addition of Cd (20 and 40 mg Cd kg^{-1}) considerably reduced shoot

Table 5.4 The beneficial effects of Si on mitigation of Cd toxicity in plants

Plant species	References
Rice	Shi et al. (2005b), Nwugo and Huerta (2008a, b, 2011), Zhang et al. (2008), Gu et al. (2011), Tripathi et al. (2012a), Liu et al. (2013b), and Kim et al. (2014b)
Maize	Liang et al. (2005), Cunha et al. (2008), Cunha and Nascimento (2009), and Vaculík et al. (2012)
Wheat	Rizwan et al. (2012)
Pakchoi	Song et al. (2009)
Cucumber	Feng et al. (2010)
Strawberry	Treder and Cieslinski (2005)
Cotton	Farooq et al. (2013)
Peanut	Shi et al. (2010)
Solanum nigrum	Liu et al. (2013a)
Mangrove	Zhang et al. (2013a)
Mustard	Vatehova et al. (2012)

Fig. 5.3 Silicon-enhanced shoot growth and silicon-alleviated Cd toxicity symptoms of maize grown on an oxisol treated with 0 (CK), 20 and 40 mg Cd kg^{-1} with or without 400 mg Si kg^{-1} (Photograph by Yongchao Liang)

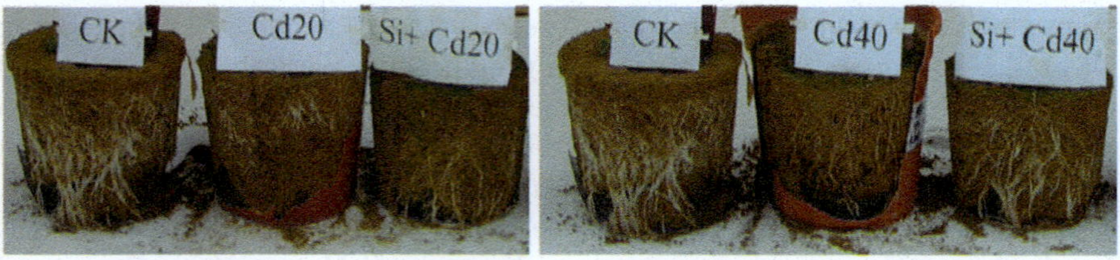

Fig. 5.4 Silicon-enhanced root growth and silicon-alleviated Cd toxicity symptoms of maize grown on an oxisol treated with 0 (CK), 20 and 40 mg Cd kg^{-1} with or without 400 mg Si kg^{-1} (Photograph by Yongchao Liang)

dry weight of maize on an oxisol (by 54 % and 25 % respectively; Liang et al. 2005), a condition that was significantly alleviated by the addition of Si (see Fig. 5.3). The shoot dry weight of plants exposed to Cd20 and Cd40 treatments was increased by addition of Si by 69 % and 119 %, respectively (Liang et al. 2005).

Root growth under Cd stress was also found to be improved significantly by addition of Si (see Fig. 5.3). The root tips of maize exposed to Cd were blackish (more apparently in the higher Cd treatment, Cd40), but they remained normally white coloured in the Cd plus Si treatment (see Fig. 5.4). Working on the genotypic differences between two contrasting pakchoi (*Brassica chinensis*) cultivars grown with 5 mg Cd L^{-1} plus 1.5 mM Si, Song et al. (2009) found that, in the Cd-sensitive cultivar, the inclusion of Si increased the shoot and root dry weight by 25 % and 40 %, respectively. They also found that the alleviative effect of Si on plant growth was more significant in the Cd-tolerant cultivar than in the Cd-sensitive cultivar, suggesting that Si and Cd interactions are genotype dependent.

Nwugo and Huerta (2008a, b) showed that treatment with 2.5 μM Cd significantly reduced the growth of rice seedlings, while inclusion of Si significantly increased all growth variables of Cd-stressed rice plants. They further showed that the inhibitory effects of Cd toxicity on photosynthesis, water use efficiency and light use efficiency were significantly mitigated by the addition of Si (Nwugo and Huerta 2008a, b). Feng et al. (2010) also reported that the commonly present Cd toxicity symptoms in cucumber, i.e. leaf yellowing and necrosis caused by excess Cd-induced nutrient imbalances, were markedly prevented by the application of Si (see Fig. 5.5); this was in accordance with the increased levels of chlorophyll a, chlorophyll b, total chlorophyll and carotenoid in Si-fed cucumber plants exposed to Cd stress. In addition, water use efficiency (WUE) was greatly decreased by exposure to Cd treatment for 10 or 15 days, but was significantly increased by the addition of Si in Cd-stressed plants (Feng et al. 2010).

5.1.4 *Arsenic*

Arsenic (As) is a nonessential highly toxic oxianionic metalloid that occurs naturally in all soils (Cullen and Reimer 1989). Human exposure to excess As will pose a major health risk, leading to a variety of health problems including skin conditions

Fig. 5.5 Effects of exogenous silicon on the phenotype of cucumber leaves and roots treated with Cd (From Feng et al. 2010)

and respiratory, pulmonary, cardiovascular and neurological diseases (Mandal and Suzuki 2002). As toxicity to humans has been an ever-increasingly important public concern because As contamination events are frequently reported throughout the world, particularly in some developing Southeast Asian countries such as Bangladesh, India and China (Dhar et al. 1997; Nordstrom 2002). As contamination may occur in soil and water through repeated application of As-containing pesticides and fertilizers, atmospheric deposition from the burning of fossil fuels, disposal of industrial and animal wastes and mining activities (Meharg et al. 2009). Contamination of soil and irrigation water with As may consequently lead to As accumulation in rice grain or straw (Abedin et al. 2002; Meharg et al. 2009; Finnegan and Chen 2012).

Upon exposure of plant roots to As, the metalloid inhibits root extension and proliferation. Once taken up and transported into the cell, arsenate (As^{V}) can be readily converted to arsenite (As^{III}), which is more toxic to living organisms. It has been demonstrated that both forms of As (As^{V} and As^{III}) can disrupt plant metabolism and severely inhibit both plant vegetative and reproductive growth and consequently lead to fertility and yield losses (Garg and Singla 2011; Finnegan and Chen 2012).

So far, limited information is available on Si and As interactions as only a few papers have been published on this topic (Guo et al. 2005; Bogdan and Schenk 2008; Shi 2008; Tripathi et al. 2013). Guo et al. (2005) investigated the effect of silicate on the growth and arsenate uptake by rice seedlings grown hydroponically and found that shoot dry weights increased with increasing external Si concentrations in the culture solution. The addition of Si to the culture solution at rates of 0.5, 1.0 and 2.8 mM Si increased shoot dry weight by 24, 37 and 33 %, respectively, but had little effect on root dry weight.

In a pot experiment on rice grown with three Si levels (0, 200 and 400 mg Si kg^{-1}) and three As levels (0, 25 and 50 mg As kg^{-1}) Shi (2008) showed that on the average, compared with Si-untreated treatments, the application of Si at 200 and 400 mg kg^{-1} increased shoot biomass of rice plants by 22 % and 32 %, respectively, and root biomass by 22 % and 34 %, respectively. However, the response of rice grain yield of As-stressed plants was shown to depend on the application rate of Si (Shi 2008). The application of Si increased the grain yield by 53 % at a lower rate (200 mg kg^{-1}), but reduced the grain yield at a higher rate (400 mg Si kg^{-1}) (Shi 2008). In the case of the rice *Lsi1* mutant defective in Si uptake, the application of Si (200 and 400 mg Si kg^{-1}) increased shoot biomass by 55 % and 27 %, respectively, and root biomass by 14 % and 7 %, respectively. However, the application of Si (200 and 400 mg Si kg^{-1}) decreased grain yield by 35 % and 74 %, respectively. The reason why grain yield for the rice *Lsi1* mutant was reduced by the Si application is still unclear. On the other hand, Bogdan and Schenk (2008) investigated the effect of Si on As accumulation in the grain of rice grown on six paddy soils in the greenhouse condition. They found a significant inhibitory effect of indigenous silicic acid in the soil solution on As uptake and accumulation, implying that soils with high plant-available Si contents result in low plant As accumulation and that Si application to soils may further decrease the As content of rice.

5.1.5 Chromium

Chromium (Cr) is another highly toxic oxianionic metalloid. Cr contamination is now receiving increased public attention because it has become a serious environmental problem as a consequence of anthropogenic activities including mining or industrial activities and repeated use of metal-enriched chemical fertilizers, pesticides, sewage sludge and wastewater irrigation in agriculture (Zayed and Terry 2003). The three species of Cr (Cr^{III}, Cr^{IV} and Cr^{VI}) are all highly phytotoxic, while Cr^{VI} is toxic to all living organisms. For plants, exposure to excessive Cr leads to inhibition of seed germination and plant growth, nutrient and water imbalance, degradation of photosynthetic pigments, inactivation of mitochondrial electron transport and inhibition of antioxidant defence enzyme activities (Dixit et al. 2002; Shanker et al. 2005; Panda 2007). Recently, Si has been reported to be effective in mitigating Cr toxicity in rice (Zeng et al. 2011; Tripathi et al. 2012b), barley (Ali et al. 2013) and pakchoi plants (Zhang et al. 2013b). In a hydroponic experiment with two Cr levels (0 and 100 μM), three Si levels (0, 1.25, and 2.5 mM) and two rice genotypes differing in grain Cr accumulation, Zeng et al. (2011) showed that seedling height and dry biomass of the two rice varieties were significantly reduced under Cr stress, but the reduction in seedling height was significantly alleviated by the inclusion of Si to the solution, and this alleviative effect was particularly distinct in the highest Si treatment. Tripathi et al. (2012b) also investigated the Cr^{IV} and Si interactions in rice and found that Cr treatment inhibited plant growth and decreased photosynthetic pigment and protein contents, which was accompanied by a significant increase in Cr accumulation and lipid peroxidation (as malondialdehyde; MDA), whereas the addition of Si to the Cr treatment alleviated Cr toxicity and promoted growth of rice by decreasing Cr accumulation, root-to-shoot Cr transport and MDA level.

In a pot experiment with pakchoi plants grown at three levels of Cr (50, 100 and 200 mg kg^{-1} Cr) and three levels of Si (0.5, 1.0 and 1.5 g kg^{-1} Si), Zhang et al. (2013b) demonstrated that Si supply increased plant growth but only at the lowest Cr level (50 mg Cr kg^{-1}). Increasing Si supply decreased shoot dry weight at both 100 and 200 mg Cr kg^{-1}. Ali et al. (2013) showed that application of Si mitigated Cr toxicity in barley plants as reflected by a significant increase in growth and photosynthetic parameters. They also demonstrated that the alleviative effect of Si on Cr toxicity was stronger at a higher Si level than at a lower Si level (2 mM Si vs. 1 mM Si).

5.1.6 Lead

Lead (Pb) is also a nonessential but highly toxic heavy metal that tends to accumulate in the environment because of its strong sorption to the minerals in soils and sediments. Pb contamination in soil, water and sediment arising from agricultural and industrial activities is a worldwide environmental problem. Exposure to excessive Pb can cause serious health risks and represents a genuine public concern. In plant tissues, excessive Pb affects membrane structures, inhibits photosynthesis and causes nutrient, water and phytohormone imbalance. Typical toxicity symptoms include stunted growth, chlorosis and root blackening.

In recent studies, Si has been reported to be effective in mitigating Pb toxicity in plants (Araujo et al. 2011; Li et al. 2012b; Bharwana et al. 2013). Araujo et al. (2011) showed that Si supply alleviated the inhibitory effect of Pb on the growth of maize. Bharwana et al. (2013) found that the addition of Pb (50 μM and 100 μM) to the culture solution caused a severe inhibition of cotton growth, which was fully alleviated by addition of 1 mM Si. Li et al. (2012b) reported that soil treatment with 800 mg Pb kg^{-1} decreased both shoot and root dry weight of banana seedlings, while the amendment of Si at 800 mg kg^{-1} (as sodium metasilicate) significantly enhanced banana biomass.

5.1.7 Zinc and Copper

Both zinc (Zn) and copper (Cu) are essential micronutrients for plant growth with numerous important metabolic functions (Marschner 1995). However, exposure to excessive Zn or Cu concentrations can cause severe phytotoxicity. Commonly visible symptoms of Zn toxicity include dark pigmented spots or blotches on older leaves, particularly on petioles and around the margins and veins of leaf blades

(O'Sullivan et al. 1997). Yellowing and wilting of sweet potato leaves are typical consequences of severe damage to the roots exposed to high Zn concentrations. Symptoms of severe Fe deficiency induced by Zn toxicity include a pale yellow to white interveinal chlorosis and eventual necrosis of leaf blades and growing points (O'Sullivan et al. 1997). Cu toxicity symptoms observed typically in moderate cases include pale green to white chlorotic mottles occurring on some mature leaves with even minor veins remaining green. However, in more severe cases, slow growth and poor root system development were generally accompanied by prominent chlorosis (O'Sullivan et al. 1997). So far, it has been demonstrated that the addition of Si can mitigate Zn toxicity in maize (Cunha et al. 2008; Cunha and Nascimento 2009; Kaya et al. 2009) and rice (Song et al. 2011; Gu et al. 2012). On the other hand, Masarovic et al. (2012) have shown that Si supply did not significantly alleviate Zn toxicity in young sorghum with an exception of Si-stimulated growth of primary seminal roots. Kaya et al. (2009) reported that significantly lower chlorophyll content, leaf relative water content (RWC) and biomass were noted in maize plants growing under excess Zn (0.5 mM) stress than in the control plants, whereas the addition of 1.0 mM Si significantly reduced those symptoms. As shown in Fig. 5.6, Song et al. (2011) showed that symptoms of Zn toxicity in rice were typically defined by a yellow colour on the lower leaves starting from the tips and spreading towards the base of the leaves. However, all symptoms of Zn toxicity were significantly alleviated by the addition of Si (Fig. 5.6). Gu et al. (2012) also showed that the addition of Si overcame growth inhibition of rice seedlings subjected to high Zn. Moreover, the alleviative effect of Si on Zn toxicity was dependent on the Si level used. Cunha et al. (2008) found that the root and shoot biomass of maize was increased by 80 % and 59 %, respectively, by applying 200 mg Si kg^{-1} to an acidic tropical soil contaminated with 100 mg Zn kg^{-1}.

Fig. 5.6 Toxicity symptoms on two rice cultivars grown hydroponically with either normal (0.15 μM) or high (2 mM) Zn with (1.5 mM) or without Si for 7 days (From Song et al. 2011)

Cu toxicity has been recently reported to be mitigated by the addition of Si in *Arabidopsis thaliana* (Li et al. 2008; Khandekar and Leisner 2011), *Zinnia elegans* (Frantz et al. 2011), *Erica andevalensis* (Rossini et al. 2011) and rice (Kim et al. 2014a). Collin et al. (2014) reported that bamboo grown hydroponically with 100 mM Cu induced suffered a significant growth inhibition with severe visual toxicity symptoms (i.e. chlorotic leaves and brown root coloration); the addition of 1.1 mM Si significantly resorbed those symptoms. Frantz et al. (2011) investigated Si-mediated tolerance to Cu toxicity in both a Si-accumulating [zinnia (*Zinnia elegans*)] and a Si-non-accumulating [snapdragon (*Antirrhinum majus*)] species based on visible stress indicators and dry weight analysis and showed that initially Si had a positive effect on zinnia but not on snapdragon. However, enzymatic assays and nutrient analysis of plant tissues showed that both species responded to supplemental Si, showing evidence of alleviated stress and balanced nutritional status more similar to healthy, control plants than plants exposed to high Cu stress. Rossini et al. (2011) examined the influence of Si added at 0.5 and 1.0 mM on responses to Cu excess in *Erica andevalensis* grown in nutrient solutions containing 1 or 500 μM Cu. Plants treated with 500 μM Cu showed differences in growth and shoot water content depending on Si supply, and the addition of 1 mM Si to high-Cu nutrient solutions significantly improved plant growth and reduced water loss, thus preventing plant death related to Cu excess. Li et al. (2008) investigated the role of Si in plant response to Cu stress in *Arabidopsis thaliana* and found that leaf chlorosis and shoot and root biomass reduction caused by excess Cu were significantly alleviated by Si amendments.

5.2 Mechanisms of Alleviation of Metal Toxicity by Silicon

The alleviative effect of Si on metal toxicity has been well documented, but the underpinning mechanisms are still unclear and controversial. Based on the available information in the literature, two major mechanisms, i.e. external (*ex planta*) and internal (*in planta*), have been proposed for explaining how Si can mitigate metal toxicity in plants. The *ex planta* mechanism was proposed prior to the *in planta* mechanism. From a chemical point of view, the *ex planta* mechanism is much simpler and easier to understand than the *in planta* mechanism. Nevertheless, the latter is more fascinating for plant physiologists. .

*5.2.1 External (*Ex Planta*) Mechanisms*

5.2.1.1 Effect of pH

External (*ex planta*) mechanisms include pH effects and reduced bioavailability of metals and are based partly on both solution chemistry of metals and interactions of metals with Si that are strongly affected by pH values in solution. Although Al and

Si are considered to form biologically unavailable complexes in solution, it is still unclear what concentrations of Al and Si and what pH ranges are required to form these complexes (Cocker et al. 1998a). The formation of aluminosilicate (AS) and its species, like Al per se, depends upon pH and the AS species that are insoluble at near neutral pH values (Cocker et al. 1998a).

The activity or bioavailability of metals in the root growth medium is strongly associated with the pH value, while inclusion of Si increases pH, especially when the basic or alkaline silicates such as sodium silicate, potassium silicate, calcium, magnesium silicate, etc., are used as Si sources. This *ex planta* mechanism was the first proposed to explain how Si can mitigate Al toxicity in plants (Hodson and Evans 1995; Hammond et al. 1995; Cocker et al. 1998b).

Li et al. (1996) proposed that the mechanism of alleviation of Al toxicity by Si could be attributed to pH effects but not the direct interactive effects of Si on Al in solutions. Their conclusions were drawn mainly from the following observations: (1) in solutions containing 296 μM Al with no Si added, the growth of sorghum seedlings was significantly decreased compared to the controls; (2) the root growth of seedlings grown hydroponically with 296 μM Al and varying Si levels was improved significantly with increasing levels of Si, but such effects were observed only when Al was added to the basal nutrient solution following the addition of Si without lowering the solution pH; (3) the addition of Si to the basal nutrient solution led to a pH increase up to or over 10.0, which facilitated precipitation of Al with Si and consequently decreased Al phytotoxicity. However, if the solution pH was lowered to 4.0 following the addition of Si, prior to the addition of Al, Al toxicity was maintained in presence of Si in solutions containing Al ranging from 534 to 1602 μM (Li et al. 1996). However, it was reported that, prior to the addition of the Al salt, even when the solution pH was adjusted to be within the acidic pH range to avoid precipitation of Al with Si, Al toxicity in wheat was still significantly alleviated by Si (Cocker et al. 1998b), suggesting that the formation of hydroxyaluminosilicate (HAS) was limited in bulk solutions with free Al levels in solutions unaffected by Si, but rather the alleviation mechanisms of Al toxicity by Si might involve the formation of HAS in the root apoplast (*in planta* mechanism).

It is generally recognized that the *ex planta* mechanisms similar to Si-alleviated Al toxicity applies to Si-mediated detoxification of Mn, Cd, Cr, Pb, Zn and Cu in soil/plant systems, i.e. reduction in metal availability via metal immobilization arising from an increase of pH (Galvez et al. 1987; Chen et al. 2000; Liang et al. 2005; Gu et al. 2011; Li et al. 2012b; Zhang et al. 2013a, b). This is easy to understand and also true when basic sodium metasilicate, slag or alkaline Si-containing materials such as biosolids and coal fly ash are incorporated into metal-contaminated soils as Si sources (Chen et al. 2000; Liang et al. 2005; Gu et al. 2011, Zhang et al. 2013b). Chen et al. (2000) reported that furnace slag was more effective in suppressing Cd uptake by rice and wheat than $CaCO_3$ or steel sludge. They speculated that the pH rise effect caused by the furnace slag used together with the plant-available Si released from the furnace slag contributed to the reduction of Cd uptake by plants (Chen et al. 2000). In a pot experiment with maize, Liang et al. (2005) showed that the application of sodium metasilicate at a higher rate of Si (400 mg Si kg^{-1}) to an acidic soil increased soil pH by 1.2 units, which contributed partly to the alleviation of Cd toxicity by Si. However, such Si-induced pH rise effects could not sufficiently explain the mitigation of Cd toxicity in maize. To further reveal the whole picture of Si effects on Cd detoxification in maize, Liang et al. (2005) ran another pot experiment in which less amount of silicate (50 mg Si kg^{-1}) was incorporated into the same Cd-contaminated soil tested to keep the soil pH unchanged. They reported that no significant differences were found either in soil pH or in plant-available Cd concentrations (e.g. $CaCl_2$-extractable fraction and water-soluble fraction of Cd) between the Si-treated soil and the Si-untreated soil. Nevertheless, Cd concentration in xylem sap was significantly lower in Si-treated plants than in Si-deprived plants, and Cd toxicity in maize was still significantly mitigated. These findings clearly suggest that the Si-induced pH rise effects are insufficient to explain the Si-mediated Cd detoxification, and both *ex planta* and *in planta* mechanisms must be involved. Gu et al. (2011) reported that the pH value of an acidic soil tested was increased by 1–2 units following the incorporation of coal fly ash and steel slag, leading to an over 60 % decrease in plant-available concentrations of heavy metals (Cd, Zn, Cu and Pb) in the soil tested. They further demonstrated significant negative correlations between heavy metal concentrations and pH values. Li et al. (2012b) reported that the addition of 800 mg Si kg^{-1} as sodium metasilicate to a Pb-contaminated soil significantly increased soil pH and carbonate- and residual-bound fractions of Pb, but decreased exchangeable fraction of Pb, thus reducing the plant availability of Pb. By contrast, addition of 100 mg Si kg^{-1} did not influence soil pH but still exhibited similar effects of Pb immobilization. All these findings seem to suggest that, apart from Si-induced pH effects, metal immobilization or inactivation in the growth medium by complexation of metals with silicates incorporated is also attributed to Si-mediated metal detoxification in plants (see Sect. 5.2.2 in this chapter).

5.2.1.2 Reduced Metal Bioavailability

It has been demonstrated that the availability of Al to living organisms in natural waters could be limited due to the fact that, chemically, Al and Si can form metastable species of AS in dilute acidic solutions (Birchall 1990; Exley and Birchall 1992, 1993; Chappell and Birchall 1988). Birchall et al. (1996) agreed that as the precursors of amorphous minerals such as proto-imogolite and allophanes, hydroxyaluminosilicate (HAS) species can form in dilute Al solutions containing high concentrations of silicic acid at pH ranging from 4.5 to 7.5. By using membrane filtration, ion exchange and dialysis, Exley and Birchall (1992) were able to identify HAS formation in solutions of low Al concentration (4 μM) and Si concentration (0–500 μM). Birchall (1992, 1993) and Exley and Birchall (1992) provided evidence for the formation of HAS and concluded that the reduction of Al bioavailability by Si could be attributed to the formation of HAS species in solutions at pH ≥4.0, especially in the rhizosphere where a boundary layer of near neutral pH may exist and is likely to be enhanced in the case of plants grown in hydroponics with nitrate as a sole nitrogen source (Hodson and Evans 1995). Taylor et al. (1997)

reported that the affinity of silica for Al was at least a million times stronger in an oligomeric form of silica (a soluble low-molecular-mass form) than in a monomeric form of silica, suggesting that this oligomeric form of silica may be responsible for the amelioration of Al toxicity observed, but these compounds have yet to be detected in the natural environment.

Barceló et al. (1993) showed that Si-mediated alleviative effects on Al detoxification in teosinte were attributed to HAS formation, leading to a significant reduction in concentration of monomeric Al species in the nutrient solution in the presence of Si. Ma et al. (1997) also concluded that the amelioration of Al toxicity in maize by Si was due to the formation of Al and Si complexes in solution rather than a physiological effect of Si on the plant, because they measured a decrease of the concentration of toxic Al^{3+} in the presence of Si.

More recently, this mechanism of Al immobilization or inactivation by Si has also been considered with respect to other heavy metals such as Cd, Cr, Pb, Zn and Cu (Liang et al. 2005; Yang et al. 2005; Gu et al. 2011; Gu et al. 2011; Zhang et al. 2013b). Liang et al. (2005) reported that both water- and $CaCl_2$-extractable fractions of Cd were significantly lowered by the application of Si at a rate of 400 mg kg^{-1}, but no effect was detected when only 50 mg Si kg^{-1} was applied. The same group (Yang et al. 2005) found that the ameliorative effect of Si on Cd toxicity in maize was a silicate-induced alteration of Cd forms in soil, resulting in a reduction of Cd bioavailability. They found that, at pH 5.9, Si application increased the carbonate-bound and the residual fractions of Cd, but decreased the exchangeable and Fe/Mn oxide-bound fractions of Cd. Although Si was applied as sodium metasilicate, these authors neutralized the induced pH change before application and showed that the pH effects of Si application on Cd immobilization in soil were negligible (Yang et al. 2005). Similar conclusions were drawn by Cunha et al. (2008) where the Cd detoxification in maize by silicates was attributed to Cd immobilization, but not to the increase of soil pH driven by the application of calcium silicate. They showed that Si altered the Cd and Zn fractions in soil, decreasing the most labile pools and increasing the more stable fractions such as those bound to organic matter and crystalline Fe oxides. More recently, Gu et al. (2011) showed that Cd, Zn, Cu and Pb concentrations estimated by the diffusive gradients in thin films technique were significantly decreased (by at least 84 %) in the steel-slag treatments and metal fluxes from the soil solid phase to solution were significantly reduced by steel-slag amendments as well. X-ray diffraction (XRD) analysis of steel slag spiked with heavy metals showed that the heavy metal precipitates (depositions) were mainly composed of insoluble Cd, Zn, Cu and Pb silicates, phosphates and hydroxides, showing the immobilization of heavy metals by steel-slag amendments (Gu et al. 2011). Zhang et al. (2013b) reported that shoot accumulation of Cr in pakchoi grown in a soil contaminated with Cr was significantly decreased by application of Na_2SiO_3. The alleviative effect of Si was mainly ascribed to the reduction of exchangeable-bound Cr fractions in the soil and the formation of precipitate-bound and organic matter-bound Cr.

It has been suggested that a critical concentration of Si at which the formation of HAS occurs is 100 μM (Exley and Birchall 1992). However, Cocker et al. (1998b) demonstrated that Al toxicity in Al-sensitive wheat grown at 1.5 μM Al could be significantly alleviated by the addition of 5 μM Si, which is well below the suggested critical Si requirement. They also demonstrated that Al toxicity symptoms were reduced in wheat at the lower end of the pH range ($pH \leq 5.0$) in which HAS formation is believed to occur, and Si addition did not reduce free Al concentration in solution.

Clearly, it is not conclusive yet whether co-deposition or precipitation of metals with Si in nutrient solutions or solid growth media always plays an important role in Si-mediated reduction of metal toxicity. However, information available from the literature so far shows that the results depend largely upon the experimental conditions employed. Obviously, although the *ex planta* mechanisms can explain in part why silica, particularly the basic silicates or Si-containing materials applied to the growth medium, can effectively alter metal toxicity in some cases, it is undoubtedly not the sole mechanism involved in Si-mediated metal detoxification in plants.

5.2.2 *Internal (*In Planta*) Mechanisms*

As early as 1940, it was found that the concentrations of mineral nutrients were higher in aerial parts of Si-deficient plants than in those of Si-sufficient plants (Wagner 1940). Later, more researchers confirmed this phenomenon (Yoshida et al. 1959; Okuda and Takahashi 1962; Tanaka and Park 1966). This role of Si in influencing accumulation of Mn and Fe seemed to be the most apparent in rice because rice plants normally have five to ten times higher leaf Mn concentrations compared to other grasses (Vlamis and Williams 1967). Barley plants grown hydroponically developed necrotic spots on their leaves under excess Mn, while the Mn toxicity symptoms disappeared by the addition of Si, which was attributed to Si-mediated microdistribution of Mn in barley leaves (Williams and Vlamis 1957a, b). In a later study, it was shown that barley, rye (*Secale cereale*), rice and ryegrass (*Lolium perenne*) all developed Mn toxicity symptoms (i.e. brown necrotic spots) in the absence of Si, but Si addition prevented the development of necrotic spots (Vlamis and Williams 1967). Okuda and Takahashi (1962) demonstrated that the Si-mediated reduction of Fe and Mn toxicity in rice plants could be attributed to a decrease in uptake of Fe and Mn by the plants in the presence of Si. They also indicated that the reduced forms of Fe and Mn (i.e. Fe^{2+} and Mn^{2+}), the major forms of Fe and Mn, were much more readily oxidized by the roots in the Si-fed rice than in the Si-deficient rice, thus forming more insoluble coatings as Fe or Mn plaque on the surface of the Si-fed rice roots. Thus, Si is believed to promote the oxidative power of rice roots with the consequent apoplastic deposition of Fe and Mn oxides. This effect depended upon Si within the rice plants but not upon Si in the external solution.

Over the past two to three decades, more progress has been made in demonstrating *in planta* mechanisms of Si-mediated detoxification of metals in plants (e.g. Cocker et al. 1998a, b; Horiguchi and Morita 1987; Iwasaki and Matsumura 1999;

Iwasaki et al. 2002a, b; Rogalla and Römheld 2002; Wang et al. 2004; Liang et al. 2005; Nwugo and Huerta 2008a,b; Song et al. 2009; Song et al. 2011; Li et al. 2012a; Dragisic Maksimovic et al. 2012). For example, Cocker et al. (1998a) demonstrated that Si could reduce the inhibitory effects of Al on root elongation under conditions that were neither the result of Si-induced increase in solution pH nor the formation of HAS. In the experiments with the Al-sensitive maize (*Zea mays* L.) cultivar Lixis, Wang et al. (2004) showed that Si treatment but not Si pretreatment mitigated Al-induced root injury as indicated by less root growth inhibition and callose formation and Si treatment did not alter monomeric Al concentrations in the nutrient solution, suggesting an *in planta* mechanism. More recently, Gu et al. (2012) showed that, compared with the high Zn treatment alone (200 μM Zn), the addition of 0.5 or 1.8 mM Si to the high Zn treatment had no significant impacts on the activity of free Zn^{2+} in the nutrient solutions, although the Si treatments mitigated Zn toxicity in rice seedling.

5.2.2.1 Uptake and Translocation of Metals

The question whether Si can suppress the uptake of metals or not is still under debate. Okuda and Takahashi (1962) postulated that Si decreased Fe and Mn uptake in rice plants by increasing the oxidizing power of the roots grown under flooded conditions. However, Williams and Vlamis (1957a, b) and Vlamis and Williams (1967) clearly demonstrated that, at least in graminaceous species, the uptake of Mn was not reduced by Si application, but Si caused Mn to be more evenly distributed instead of being concentrated in discrete necrotic spots. This finding was supported by subsequent experiments in dicots (Horst and Marschner 1978a; Iwasaki and Matsumura 1999; Rogalla and Römheld 2002; Führs et al. 2009). Iwasaki and Matsumura (1999) demonstrated that Si alleviated Mn toxicity through a co-localized accumulation of Mn with Si in a metabolically inactive form around the base of the trichomes on the leaf surface, while Si addition did not affect the plant Mn content. Horiguchi (1988) reported that Si treatment altered the distribution of Mn by increasing Mn concentration in roots and decreasing that in shoots, thus reducing the leaf symptoms of Mn toxicity in rice. In cucumber grown under Mn excess, the leaf concentration and total content of Mn were found to be independent or even higher in the Si-fed plants than in Si-deprived plants (Rogalla and Römheld 2002; Shi et al. 2005a). On the contrary, Li et al. (2012a) showed that the addition of Si decreased shoot Mn concentration, but greatly increased root Mn concentration in a Mn-sensitive rice cultivar (XXY) (Table 5.5). Similar to the findings of Horiguchi (1988), this phenomenon indicates that Si-enhanced Mn tolerance is not due to restriction of Mn uptake from solutions, but to restriction of root-to-shoot Mn transport. However, Li et al. (2012a) also showed that the addition of Si greatly decreased both shoot and root Mn concentration in a Mn-tolerant rice cultivar (ZLY), suggesting that Si-enhanced Mn tolerance is due to restriction of Mn uptake from the nutrient solutions (Table 5.5). This would suggest that Si-mediated Mn tolerance is genotype dependent.

Table 5.5 Mn concentrations in shoots and roots of rice plants grown in nutrient solution containing normal (6.7 μM) or high (2.0 mM) Mn with or without Si supply

Variety	Mn treatment	Si supply	Shoot ($\mu g\ g^{-1}$ DW)			Root ($\mu g\ g^{-1}$ DW)		
XXY	Normal	–	415.80 ± 44.70			135.93 ± 5.26		
		+	490.51 ± 54.25			187.35 ± 12.86		
	High	–	4515.82 ± 175.58			2069.20 ± 39.76		
		+	4340.81 ± 62.06			3804.40 ± 137.82		
ZLY	Normal	–	335.11 ± 4.70			193.64 ± 12.96		
		+	445.79 ± 60.17			291.83 ± 61.98		
	High	–	5754.97 ± 82.31			3532.94 ± 345.31		
		+	2088.34 ± 185.40			1481.43 ± 81.31		
			Shoot			Root		
			df	*P*	$LSD_{0.05}$	*df*	*P*	$LSD_{0.05}$
Source of variation	Cultivar		1	<0.001	149.918	1		200.263
	Mn		1	<0.001	149.918	1	<0.001	200.263
	Si		1	<0.001	149.918	1		200.263
	Cultivar × Mn		1	<0.05	212.017	1		283.215
	Cultivar × Si		1	<0.001	212.017	1	<0.001	283.215
	Mn × Si		1	<0.001	212.017	1		283.215
	Cultivar × Mn × Si		1	<0.001	299.837	1	<0.001	400.526

From Li et al. (2012a). Data are means ± SD of 3 replicates

Apart from Mn, Si-mediated uptake and transport of Al and Cd are two typical case studies that have been extensively examined. Using X-ray microanalysis, Hodson and Sangster (1993) provided the first direct evidence showing that Al toxicity may be decreased through co-precipitation or co-deposition of Al with Si in the root epidermal walls of sorghum. Subsequent work in the same laboratory indicated that Al and Si were co-localized in the epidermal and hypodermal cells of the roots of wheat treated with Al and Si (Cocker et al. 1997). As is well-known, the root cell wall, which is mainly composed of cellulose and matrix polysaccharides including pectins and hemicelluloses, plays a crucially important role in Al tolerance and Al resistance (Horst 1995; Yang et al. 2008). It is in the cell wall that Al and Si interact actively, forming co-precipitation and/or co-deposition of Al with Si and leading to less Al transport into the symplast and less toxic effects of Al on cell morphology (Cocker et al. 1998a).

Chen et al. (2000) reported that the higher pH and increasing available Si from the furnace slag contributed to the reduced Cd uptake by rice and wheat, leading to a more effective mitigation of Cd toxicity by furnace slag than by calcium carbonate or steel sludge. The role of Si in minimizing uptake and root-to-shoot transport of metal ions has been confirmed in both monocotyledonous and dicotyledonous plants exposed to toxic Cd concentrations (Shi et al. 2005b; Liang et al. 2005; Zhang et al. 2008; Song et al. 2009; Shi et al. 2010; Farooq et al. 2013; Zhang et al. 2013a). In pot experiments with maize, Liang et al. (2005) showed that although soil pH and availability of Cd were unaffected by incorporation of Si at a rate of 50 mg Si kg^{-1},

Cd concentration in shoots of Si-treated plants was significantly decreased, but the total Cd uptake in shoots and roots was significantly increased due to its significantly higher shoot and root biomass compared with the Si-untreated plants. Shi et al. (2005b) reported that Si also significantly reduced the root-to-shoot transport of the apoplastic fluorescence tracer PTS (tri-sodium-8-hydroxy-1, 3, 6-pyrenesulphonate), suggesting that the heavy deposition of silica in the endodermis might partially physically block the apoplast bypass flow across the roots, thus restraining the apoplastic transport of Cd. After analysing Cd content in roots and comparing the total Cd content in plants, Zhang et al. (2008) found both Cd content in roots and the ratio increased significantly by Si treatment, indicating that the exogenous Si supply significantly enhanced Cd retention in the roots and inhibited the root-to-shoot Cd translocation.

Si-mediated Cd uptake and translocation were observed not only in Si-accumulating plant species such as rice and maize but also in non-Si accumulators such as pakchoi (Song et al. 2009) and strawberry (*Fragaria vesca*) (Treder and Cieslinski 2005). As can be seen in Fig. 5.7, the addition of 1.5 mM Si significantly decreased Cd concentration in shoots but increased Cd concentration in roots of pakchoi grown hydroponically with 0.5 or 5.0 mg Cd L^{-1} (Song et al. 2009).

It was also reported that Si used as soil amendment prior to planting was effective in preventing excessive Cd uptake by strawberry plants grown on sandy soil, and this effect was observed in stems, leaves and fruits but not in roots (Treder and Cieslinski 2005). However, foliar application of potassium silicate did not reduce the concentration of Cd in the individual parts of strawberry plants (Treder and Cieslinski 2005). More importantly, such inhibitory effects of Si on root-to-shoot Cd translocation were species dependent. For example, working with a

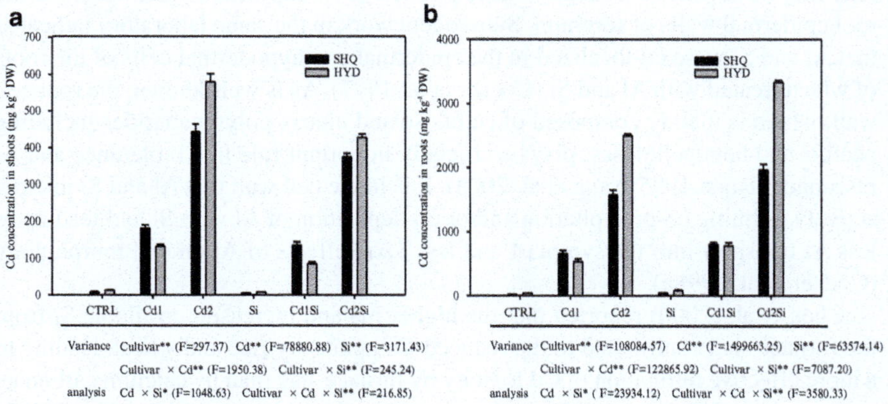

Fig. 5.7 Cd concentrations in shoots (**a**) and roots (**b**) of pakchoi plants grown hydroponically with various levels of Cd with (1.5 mM) or without Si for 7 days (From Song et al. 2009). Data are means ± SD ($n = 3$); *P*-value indicates significance level based on three-way ANOVA; $*P < 0.05$, $**P < 0.01$. *CTRL* treatment with neither Cd nor Si, *Cd1* treatment with 0.5 mg L^{-1} Cd, *Cd2* treatment with 5.0 mg Cd L^{-1}, *Si* treatment with 1.5 mM Si, *Cd1Si* treatment with 0.5 mg L^{-1} Cd plus 1.5 mM Si, *Cd2Si* treatment with 5.0 mg Cd L^{-1} plus 1.5 mM Si

heavy metal hyperaccumulator, *Solanum nigrum*, grown hydroponically with Cd, Liu et al. (2013a) have recently reported that the Cd concentration in both roots and shoots was significantly reduced by the addition of Si, especially in expanding and old leaves, while the relative proportion of ethanol-extractable Cd, water-extractable Cd and NaCl-extractable Cd in roots were increased, and the root-to-shoot Cd translocation was not decreased by the addition of Si. The differences in Cd translocation affected by Si between this study and those performed previously (e.g. Shi et al. 2005b; Liang et al. 2005; Zhang et al. 2008; Rizwan et al. 2012) could be attributed partly to the different plant species used. *Solanum nigrum* is a Cd hyperaccumulator, but none of rice, cucumber or durum wheat is not.

Inhibition of As uptake and root-to-shoot transport by Si has also been proposed as Si-mediated mitigation of As toxicity in rice (Guo et al. 2005; Tripathi et al. 2013). Guo et al. (2005) demonstrated that As concentrations in shoots and roots and total As uptake by rice seedlings grown with toxic level of As were significantly decreased by the inclusion of Si into the nutrient solution. The inhibitory effect of indigenous silicic acid in the soil solution on As uptake by rice was clearly shown, implying that soils with high plant-available Si contents resulted in low plant As contents and that Si application to soils may decrease the As content of rice (Bogdan and Schenk 2008).

Uptake and transport of Zn in maize and rice (Kaya et al. 2009; Song et al. 2011), Pb in banana (Li et al. 2012b), Cr in rice (Tripathi et al. 2012b) and Cu in *Erica andevalensis* (Oliva et al. 2011) have recently been reported to be significantly decreased by the incorporation of Si. Tripathi et al. (2012b) showed that Si addition alleviated Cr toxicity and promoted growth of rice by decreasing Cr accumulation and root-to-shoot Cr transport. Zn toxicity in rice (Song et al. 2011) and maize (Kaya et al. 2009) mitigated by Si added was also attributed partly to silicon-suppressed uptake and transport of Zn in plants. For example, Song et al. (2011) reported that the addition of Si significantly decreased Zn concentrations in shoots but increased Zn concentrations in roots of both Zn-sensitive (cv. FYY 326) and Zn-resistant (cv. TY 167) rice cultivars grown hydroponically with high Zn supply (2 mM Zn) (Fig. 5.8). On the contrary, addition of Si considerably increased root Zn concentrations in both cultivars to different degrees, especially in the resistant cultivar (Song et al. 2011). Oliva et al. (2011) showed that the addition of Si to high-Cu nutrient solutions significantly decreased leaf Cu concentration and increased Cu concentration in roots of *E. andevalensis*. EDX analysis of phytoliths isolated from leaves further showed that the phytoliths consisting of silica deposits were associated with Cu and other elements (K, Ca and P). The leaf phytoliths formed in Si-treated plants might have partly contributed to tolerance by Cu immobilization and inactivation. However, contrasting results have recently been reported by Collin et al. (2014) that Cu concentrations in leaves, stems or roots of bamboo grown with 100 μM Cu in a hydroponics system were not affected by Si, but the proportion of Cu adsorbed on roots was 61.5 % in the Cu plus Si treatment compared to 42 % in the Cu treatment only, suggesting that biologically active Cu was reduced by Si supplementation.

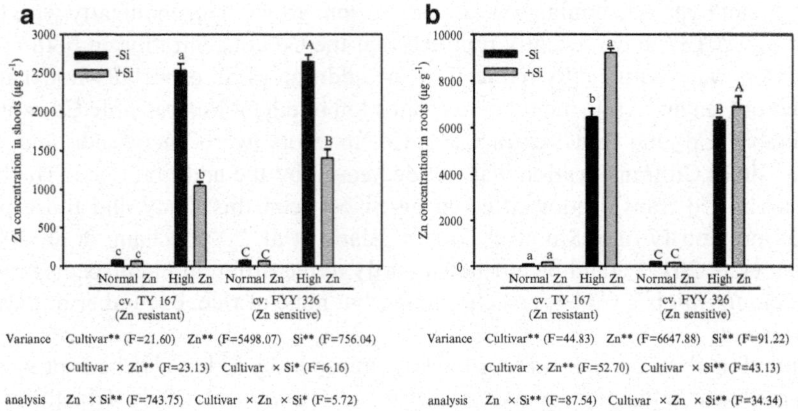

Fig. 5.8 Zn concentration in shoots (**a**) and roots (**b**) of two rice cultivars grown hydroponically in either normal (0.15 μM) or high (2 mM) Zn supplied nutrient solutions with (1.5 mM) or without Si for 7 days (From Song et al. 2011). Data are means ± SD of three replicates; *P*-values indicate significance level based on three-way ANOVA. **P*<0.05, ** *P*<0.01; data followed by different letters within the same cultivar are significantly different (*P*<0.05)

5.2.2.2 Binding Properties of Cell Walls

As is well known, cell walls consisting of cellulose and matrix polysaccharides (hemicellulose and pectins) play an important role in metal retention and immobilization in plants and thus in metal tolerance and/or resistance, as the cell wall is the first barrier or compartment for any ions including heavy metals to enter the cells (Horst 1995; Yang et al. 2008; Lux et al. 2011). Horst et al. (1999) observed that Si lowered the apoplastic Mn concentration in cowpea, suggesting that Si may modify the cation-binding capacity of the cell wall. Working on cowpea plants grown hydroponically with high Mn (50 μM), Iwasaki et al. (2002a,b) found that the Mn concentration in the leaf apoplastic washing fluids (AWFs) was significantly higher in plants receiving a continuous supply of Si and not showing typical Mn toxicity symptoms than in plants not receiving Si supply and showing distinct Mn toxicity symptoms. These results suggest that Si alleviates Mn toxicity not only by decreasing the concentration of soluble apoplastic Mn through enhanced binding of Mn to the cell walls, but soluble Si plays a role in the detoxification of apoplastic Mn. By working on Si and Mn tolerance in cucumber, Rogalla and Römheld (2002) reported that Mn concentrations in the AWF were significantly higher in Si-untreated plants than in the Si-treated plants, especially in the $BaCl_2$- and DTPA-extractable fraction of the leaf apoplast although the total Mn content in the leaves was unaffected by Si treatment. The Mn concentration of the AWF was correlated positively with the severity of Mn toxicity symptoms but negatively with the Si supply. Furthermore, less Mn was located in the symplast (<10 %) and more Mn was bound to the cell walls (>90 %) in the leaves of Si-supplied plants, while a uniform distribution of Mn (about 50 % in each compartment) was found in the Si-untreated plants. These authors concluded that the Si-mediated tolerance to excess Mn is a consequence of stronger binding of Mn to the cell walls which lowers Mn concentration within the symplast. However, the Si-mediated strong binding of Mn to the cell wall was detectable only when the plants received a simultaneous supply of Si and high Mn concentrations in the nutrient solution, and no alteration in Mn binding capacity was observed in Si-treated plants when isolated cell walls were loaded with Mn during a short period of time (Rogalla and Römheld 2002). Moreover, supply of Si to roots did not change the CEC of the cell wall materials isolated from cucumber leaves in the high Mn-treated plants (Dragisic Maksimovic et al. 2012). On the other hand, studies in cowpea suggest that the alleviation of Mn toxicity cannot be explained only by a decrease in free leaf apoplastic Mn through its enhanced binding by the cell wall macromolecules in Si-treated plants (Iwasaki et al. 2002b; Führs et al. 2009). It is not possible, thus, to generalize that Si-enhanced cell wall binding of Mn demonstrated in cucumber (Rogalla and Römheld 2002; Wiese et al. 2007; Dragisic Maksimovic et al. 2012) constitutes the universal mechanism of Si alleviation of Mn toxicity in all plant species. Sequestration of Mn into the vacuoles might play an important role in Si-mediated Mn tolerance in some plant species such as bean (Horst and Marschner 1978b), but again this mechanism has yet to be confirmed in others such as cowpea (Horst et al. 1999).

There is convincing evidence showing Al and Si co-deposition in plant cell walls (Hodson and Evans 1995; Wang et al. 2004). Al and Si are co-localized in the epidermis of the needles of white spruce (Hodson and Sangster 1998). By using electron energy loss spectroscopy, Turnau et al. (1993) demonstrated that Al and Si co-deposition in polyphosphate granules in the fungal symbiont *Paxillus involutus* associated with the roots of *Pinus sylvestris* is the possible mechanism of mycorrhizal amelioration of Al toxicity. Wang et al. (2004) also drew a conclusion from their data that Si treatment led to the formation of hydroxyaluminosilicates (HASs) in the apoplast of the root apex, thus detoxifying Al. Neumann and zur Nieden (2001)) also reported that Zn was co-precipitated with Si as zinc silicate in the leaf epidermal cell walls of a Zn-tolerant plant, *Minuartia verna*, naturally grown in a mining area. In a solution culture experiment with maize grown with Cd supply, more Cd was bound to the root cell walls in Si-treated plants than in non-Si-treated plants (Liang et al. unpublished), suggesting an important role of the root apoplast in Si-mediated detoxification of excess Cd. More recently, Zhang et al. (2013a) have reported the effect of Si on Cd subcellular distribution in the leaves and root tips of mangrove seedlings. They found that the mechanisms of Si amelioration of Cd stress were tissue dependent. In the leaves and root tips, Si reduced Cd concentration in subcellular fractions, Cd mobility and the concentration of biologically active Cd in the cell wall. Si did not change the distribution of Cd between compartments in the leaves, but it increased the proportion of Cd in the cell walls and reduced the proportion of Cd in the symplast of the root tips. This result is consistent with the earlier report by Wang et al. (2004) and Prabagar et al. (2011), who suggested that Si treatment reduced the concentration of biologically active Al within the cell walls of maize and Norway spruce, which decreased Al penetration and cell damage. In a vermiculite pot experiment, Ye et al. (2012) also found that Cd concentrations in different fractions extracted from the cell walls of root tips of

Kandelia obovata were greatly affected by Si supply. Silicon treatment restricted the apoplastic transport of Cd with more Cd adsorbed on the root cell walls of *K. obovata*, thus leading to a reduction of the ratio of Cd in the symplast and alleviation of Cd toxicity to the cytoplasm. This result suggests that Si enhances the Cd-binding capacity of the cell walls and restricts the apoplastic transport of Cd, resulting in Si-mediated amelioration of Cd toxicity in plants. More recently, using suspended rice cells and protoplasts as materials, Liu et al. (2013a) have demonstrated that most Si accumulated in the cell walls was present as a wall-bound organosilicon compound. At moderate concentrations of Cd in the culture medium, the protoplast Cd concentrations in Si-accumulating cells were significantly decreased compared to those from Si-limited cells (Liu et al. 2013a). In situ analysis of cellular fluxes of the Cd^{2+} in suspension cells and root cells of rice exposed to Cd^{2+} and/or Si treatments showed that + Si cells significantly inhibited the net Cd^{2+} influx, compared to –Si cells. All these findings suggest that co-deposition of Si and Cd in the cell walls via a [Si-hemicellulose] Cd co-complexation may explain why Cd ion uptake in rice can be suppressed by Si (Liu et al. 2013a; Ma et al. 2015). Working on alleviative effect of Si on Zn toxicity in rice grown hydroponically and its underpinning mechanism, Gu et al. (2012) showed that, compared with Zn treatment only, the addition of Si to the high-Zn treatment resulted in a lower Zn loading in the free symplastic and free apoplastic space and cytoplasm/vacuole and higher share of Zn in the cell-wall-bound fraction of the whole rice plant. This is further supported by the Zinpyr-1 fluorescence test and energy-dispersive X-ray spectroscopy analysis showing that the concentrations of biologically active Zn^{2+} were decreased, with Zn and Si co-localized in the cell wall of metabolically less active tissues, especially in the root sclerenchyma (Gu et al. 2012).

However, contrasting results have been reported (Shi et al. 2005b; Collin et al. 2014). For instance, Collin et al. (2014) have recently observed that Si supplementation decreased the visible damage in bamboo grown hydroponically with high Cu exposure (100 μM Cu) for 4 months, but Si was not directly involved in Cu speciation in roots, which is contrary to the findings of Wang et al. (2004) who associated Si-ameliorated Al toxicity in rice to the formation of HAS in the roots (Wang et al. 2004). Collin et al. (2014) also showed that Si increased the proportion of Cu(I) S-ligands (organic and inorganic) in leaves of bamboo exposed to high Cu stress and may have enhanced the capacity of the plant to produce S ligand available to bind Cu, thereby increasing the Cu sequestration in a less toxic form. In an earlier report, Shi et al. (2005b) showed that most of the total root Cd (87 %) was localized in the symplast of rice grown hydroponically with Cd, whereas the apoplast Cd fraction accounted for only 13 % of the total Cd irrespective of Si treatment. Furthermore, it was shown that the distribution ratios of Cd in both symplast and apoplast of the leaves of Cd-stressed rice were not changed significantly by Si treatment. X-ray microanalysis (EDX) showed that Cd was deposited mainly in the vicinity of the endodermis and epidermis, and Si deposition was heavier in the endodermis than in the epidermis (Shi et al. 2005b). The results from EDX, however, cannot explain the mechanisms underlying the Si-enhanced Cd tolerance in rice seedlings because such low Cd concentrations in the roots were far below the detection limit of EDX (Shi et al. 2005b).

Recently, Khandekar and Leisner (2011) have investigated the expression of three MT (metallothionein) genes, *MT1a*, *MT2a* and *MT2b* encoding for cysteine-rich metal-binding proteins, in *Arabidopsis* treated with toxic level of Cu and Si and found that while all three MT genes were induced by Cu, the levels of expression either remained elevated or were boosted to even higher levels in plants subjected to treatment with Si and phytotoxic doses of Cu. The elevated production of MT in the Si-treated plants under excessive Cu stress may suggest that Si may be promoting gene expression of MTs to keep the levels of soluble Cu low within the cell to alleviate Cu toxicity.

5.2.2.3 Si-Mediated Synthesis and Release of Organic Acids for Chelation of Metals

Al-induced release of organic acids (e.g. malic acid and oxalic acid) has been considered a potential mechanism of Al-tolerant plants as the acids have the potential to chelate Al ions (Ma et al. 1997; Cocker et al. 1998b; Larsen et al. 1998; Klug and Horst 2010). Barceló et al. (1993) report that the concentration of malic acid was substantially higher in roots of teosinte (a wild maize) grown hydroponically with 60 μM Al and 4 μM Si than in roots of plants with supply of 60 μM Al only. These roots containing high malic acid neither showed Al-induced growth reduction nor exhibited Al-toxicity symptoms, suggesting that chelation of Al by malic acid within plant roots contributed partly to Al tolerance of teosinte grown in Si-containing nutrient solutions. Cocker et al. (1998b) suggested that exudation of malate (or other organic compounds) into the cell walls may promote the formation of HA and/or HAS. However, Kidd et al. (2001) reported that Si-enhanced exudation of phenolic compounds (catechol and flavonoid-type phenolics: catechin and quercetin), leading to complexation and detoxification of Al, was responsible for the Si-mediated enhancement of Al resistance in an Al-resistant maize cultivar. By contrast, Wang et al. (2004) concluded that the Al-induced exudation of organic acid anions and phenols from the root apices was unaffected by Si treatment, whereas Si accumulation in the cell wall fraction was greatly enhanced by Al treatment, thereby reducing the mobility of apoplastic Al. Cocker et al. (1998b) established a schematic model showing five processes involved in the proposed mechanism of reduction of Al toxicity by Si in root tips. One of these processes inferred that malate and other organic compounds (such as phenolic acids) were secreted into the bulk media and the cell walls of roots under Al stress and some of these organic compounds promoted the formation of AS and/or HAS. This way, far less Al was available for uptake into the symplast thereby alleviating Al toxicity.

Apart from the extensive research on Si-mediated synthesis and release of organic acids for chelation of Al, little information is available regarding the Si effects on other heavy metals. It is unknown whether Si can mediate the metabolism of organic acids, amino acids, polyamine, phenolic acids and phytochelatins (PCs) and metallothioneins (MTs) to chelate heavy metal ions via *ex* or *in planta* mechanisms. More recently, Khandekar and Leisner (2011) reported that the relative transcript levels of *MT2a*, *MT2b* (the type 2 metallothioneins genes) and *PCS1* (the major phytochelatin synthase gene) in the leaves of *Arabidopsis thaliana* Sf-1 plants under Cu stress were higher in the presence of Si than in the absence of Si, but a reverse result was obtained for relative transcript levels of *MT1a* (the type 1 metallothioneins gene).

5.2.2.4 Si-Mediated Antioxidative Defence Responses

It was reported that Si could increase antioxidant defence activity and decrease oxidative damage in cucumber under Mn stress (Shi et al. 2005a). Several subsequent studies confirmed this finding in Mn-stressed rice (Li et al. 2012a) and cucumber (Dragisic Maksimovic et al. 2012). For instance, Li et al. (2012a) showed that in Mn-sensitive rice cultivar, high Mn significantly increased superoxide dismutase (SOD), catalase (CAT) and ascorbate peroxidase (APX) activities but decreased nonprotein thiols (NPT) and glutathione (GSH) concentrations, leading to accumulation of hydrogen peroxide (H_2O_2) and MDA. The addition of Si significantly counteracted high Mn-elevated MDA and H_2O_2 concentrations and enhanced plant growth in the Mn-sensitive cultivar. In contrast, in Mn-tolerant rice cultivar, high Mn considerably raised SOD activities and GSH concentrations, thus leading to relatively low oxidative damage, while the addition of Si mainly affected nonenzymatic antioxidants. Thus, it seems that the role of Si in mediating antioxidative defence activities in high-Mn-stressed plants is genotype dependent. Dragisic Maksimovic et al. (2012) investigated the role of Si in reducing Mn toxicity caused by elevated production of hydroxyl radicals (OH) in the leaf apoplast of cucumber grown under excessive Mn concentrations. The apoplastic concentration of free Mn^{2+} and H_2O_2 in high-Mn-treated plants was significantly decreased by Si supply, whereas the addition of Si suppressed the Mn-induced increased abundance of POD isoforms in the leaf apoplastic fluid and rapidly suppressed guaiacol-POD activity under excess Mn (Fig. 5.9). Although supplying Si markedly decreased the accumulation of $\cdot OH$ in the leaf apoplast with excess Mn, adding silicic acid to the Mn^{2+}/H_2O_2 reaction mixture did not directly affect the Fenton reaction ($Mn^{2+} + H_2O_2 \rightarrow Mn^{3+} + OH^- + \cdot OH$) in vitro. These results indicate that Si contributes indirectly to a decrease in $\cdot OH$ in the leaf apoplast by decreasing the free apoplastic Mn^{2+}, thus regulating the Fenton reaction. A direct inhibitory effect of Si on guaiacol-POD activity (also demonstrated in vitro) may also contribute to decreasing the POD-mediated generation of $\cdot OH$.

Dragisic Maksimovic et al. (2007) also showed that the concentrations of phenolic compounds, such as coniferyl alcohol and coumaric and ferulic acids, in the leaf extracts tended to be lower in the Si-treated plants at high Mn supply. On the other hand, application of Si induced a significant increase in the concentrations of chlorogenic and caffeic acids in the leaf extracts of high Mn-treated plants. The POD and PPO activities were enhanced by the high Mn supply in both root and leaf extracts, while the root application of Si decreased POD and PPO activities in both roots and leaves. These results suggested that Si nutrition modulated the metabolism

Fig. 5.9 Separation of the POD isoforms from the LAF by IEF (From Dragisic Maksimovic et al. 2012). PODs were detected by staining with 9.2 mM guaiacol and 5 mM H_2O_2. The cucumber plants were grown in nutrient solutions with normal (0.5 μM) and high (100 μM) supply of Mn, with (1.5 mM) or without $Si(OH)_4$

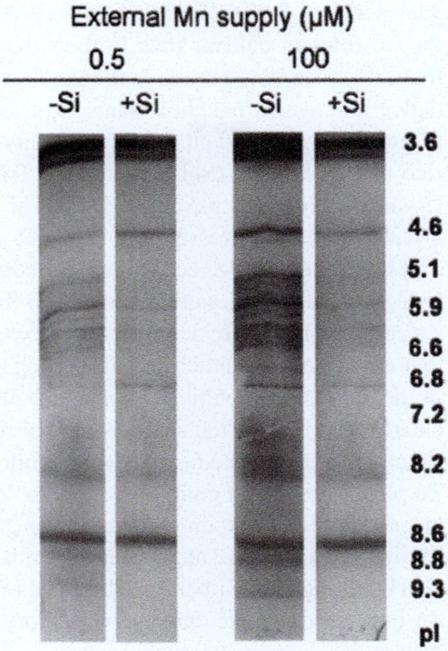

and utilization of phenols mainly at the leaf level by stimulating the formation of Si polyphenol complexes. Concomitantly, lower concentrations of phenolic compounds available to act as substrates for PPO and POD in Si-treated Mn-stressed plants may thus be responsible for depressing the generation of potential toxic intermediates that induce leaf browning (Dragisic Maksimovic et al. 2007). It was reported by Führs et al. (2009) that in Si-treated cowpea, a high Mn supply (not showing Mn toxicity symptoms) led to increased concentrations of benzoic acid in the apoplastic washing fluid (AWF) fraction and to decreased abundance of ferulic acid and ρ-hydroxybenzoic acid compared with non-Si-treated plants, which showed symptoms of Mn toxicity.

Numerous experiments have recently been performed to examine whether Si-mediated antioxidant defence capacity observed in Mn-stressed plants is a universal mechanism for the alleviation of metal toxicity in plants (Nwugo and Huerta 2008a; Song et al. 2009; Feng et al. 2010; Shi et al. 2010; Song et al. 2011; Zeng et al. 2011; Khandekar and Leisner 2011; Tripathi et al. 2012a, b, 2013; Liu et al. 2013a, b). Song et al. (2009) investigated the roles of Si in enhancing Cd tolerance in two contrasting non-Si-accumulating pakchoi cultivars grown hydroponically with 0.5 and 5 mg L^{-1} Cd without or with Si supply. Their results indicated that SOD, CAT and APX activities decreased, but MDA and H_2O_2 concentrations increased at the higher Cd level, which were counteracted by the addition of Si. Furthermore, ascorbic acid (AsA), GSH and NPT concentrations increased at the higher Cd level and even more so with the addition of Si. The effects of Si and Cd on the antioxidant enzyme activity were further verified by SOD and CAT isoenzyme analysis (Fig. 5.10). The Si effect was more effective in enhancing Cd tolerance in the Cd-tolerant cultivar than in the Cd-sensitive cultivar. All these results suggest that Si can mediate antioxidant defence capacity to reduce membrane lipid peroxidative damage in pakchoi plants exposed to Cd stress. Similar results regarding Si-enhanced antioxidant defence capacity were also obtained in Cd-stressed peanut, rice and *S. nigrum* (Shi et al. 2010; Tripathi et al. 2012a; Liu et al. 2013b), in Cr-stressed rice and pakchoi (Zeng et al. 2011; Tripathi et al. 2012b; Zhang et al. 2013b) and in Zn-stressed rice (Song et al. 2011). Song et al. (2011) reported that SOD, CAT and APX activities were significantly increased, whereas MDA and H_2O_2 concentrations were decreased in Si-supplied plants of both Zn-sensitive and Zn-resistant rice cultivars exposed to excess Zn. These alleviative effects of Si, further confirmed by histochemical staining methods (Fig. 5.11), were more prominent in the Zn-resistant cultivar than in the Zn-sensitive one. From Fig. 5.11, it can be clearly seen that for the Zn-resistant cultivar (TY-167), the roots treated with 2 mM Zn alone were strained extensively, while the roots became lighter straining in the Zn plus Si treatment compared with the Zn treatment alone, suggesting that Si had prophylactic effects on cell membranes against Zn-induced oxidative damage. Similar changes were also observed in the roots of the Zn-sensitive FYY-326 (Fig. 5.11) although the Si beneficial effects on the protection of cell membrane against Zn-induced oxidative damage were more significant in the Zn-resistant plant roots than in the Zn-sensitive plant roots.

Taken together, Si appears to enhance enzymatic and nonenzymatic antioxidant defence activities in plants to protect the cell organs such as chloroplasts, mitochondria and nuclei from oxidative damage, thereby leading to Si-mediated detoxification of heavy metals (see Nwugo and Huerta 2008; Song et al. 2009; Shi et al. 2010).

5.2.2.5 Molecular Aspects of Si-Mediated Metal Resistance

So far, interactive effects of Si on metal toxicity in plants have been extensively investigated at the physiological and/or biochemical level and rapid progress has been made in this field. However, very limited information is available on molecular aspects of Si and metal interaction.

Li et al. (2008) investigated two genes related to Cu transport in *Arabidopsis*, Cu transporter 1 (*COPT1*) and heavy metal ATPase subunit 5 (*HMA5*), which were induced by high levels of Cu, and found that the RNA levels of *COPT1* and *HMA5* were significantly decreased by elevated Si supply (Li et al. 2008). Khandekar and Leisner (2011) found that Cu/Zn-SOD activity was induced by Cu toxicity but SOD activity was increased even more if plants were provided with extra Si and toxic levels of Cu. This is in line with the findings showing that expression of both *CSD1* and *CSD2* is induced by high levels of Cu, while expression of both *CSD1* and *CSD2* was brought to even higher levels if plants were provided with elevated doses of Cu and Si. This is interesting, since the *CSD1* gene product is cytoplasmic, whereas that of *CSD2* is localized in the chloroplast. The increases in SOD RNA levels matched well with increases in enzyme activity, suggesting that at high doses of Cu, some of the heavy metal must be entering *Arabidopsis* leaf cells, and

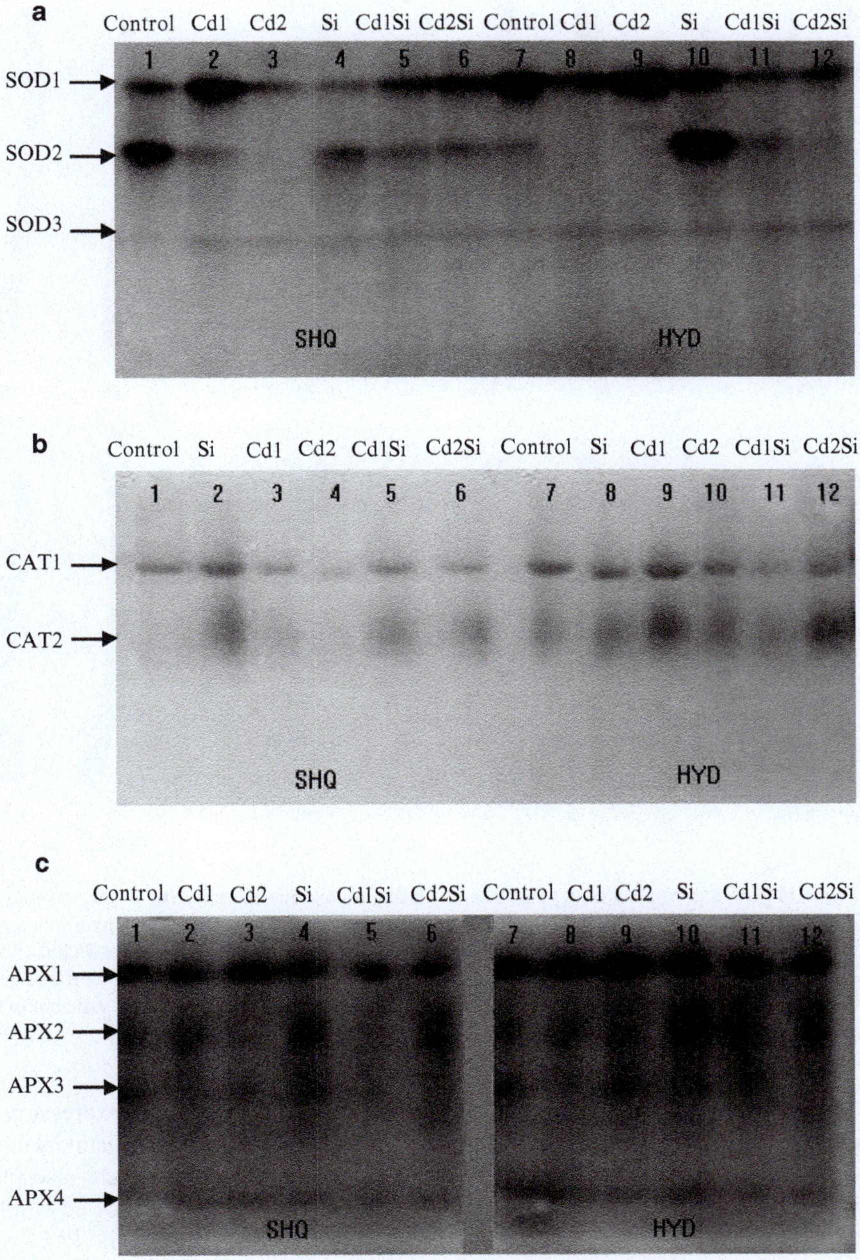

Fig. 5.10 Effect of Si on SOD (**a**), CAT (**b**) and APX (**c**) isoform activities of the leaves of two pakchoi cultivars grown under Cd stress (From Song et al. 2009). Proteins from leaves were extracted and loaded into the native PAGE. Following the electrophoresis, the gels were strained and photographed. Lanes 1–6 represent the Cd-sensitive cultivar (SHQ), while lanes 7–12 represent the Cd-tolerant cultivar (HYD). *Control*: treatment with neither Cd nor Si. *Cd1*: treatment with Cd at 0.5 mg L^{-1}. *Cd2*: treatment with Cd at 5.0 mg L^{-1}. Si: treatment with 1.5 mM Si. *Cd1Si*: treatment with 0.5 mg Cd L^{-1} plus 1.5 mM Si. *Cd2Si*: treatment with 5.0 mg Cd L^{-1} plus 1.5 mM Si

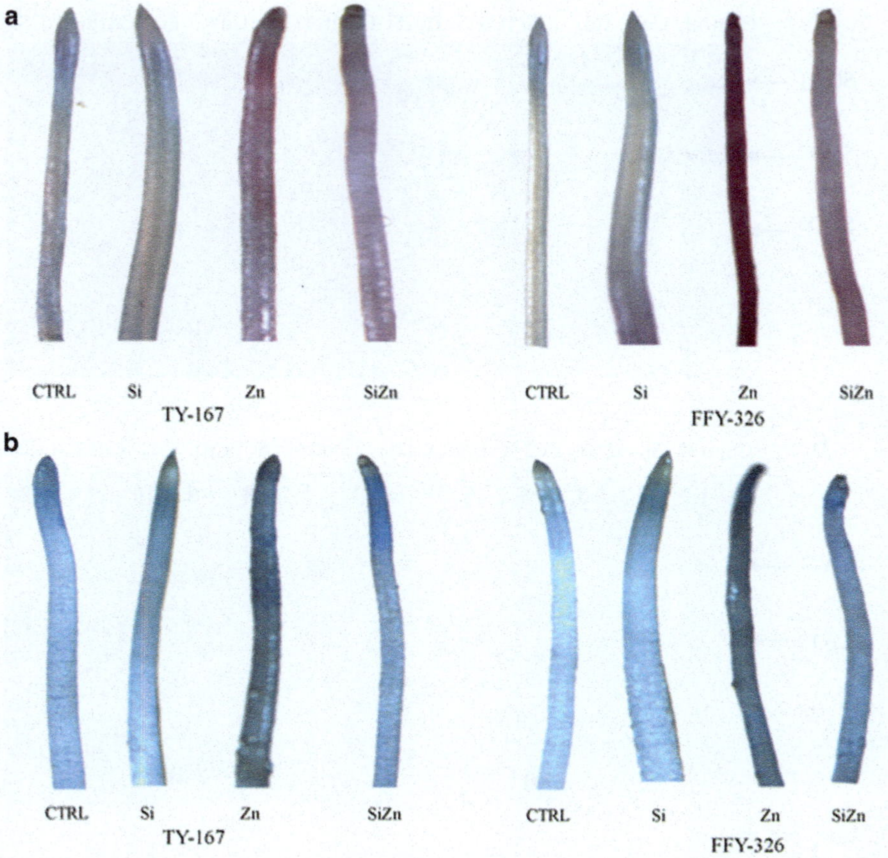

Fig. 5.11 Lipid peroxidation (**a**) and loss of plasma membrane integrity (**b**) in the root tips of Zn-sensitive rice cultivar (FFY-326) and Zn-tolerant rice cultivar (TY-167) grown hydroponically with either normal (0.15 μM) and high (2 mM) Zn, with (1.5 mM) or without Si for 7 days (Song et al. 2011). The roots were rinsed with 0.5 M HCl solution and were stained with Schiff's reagent (**a**) or Evans blue (**b**) and immediately photographed under a light microscope. The scale bar in the graph indicates 1.0 cm

Si is permitting plants to more effectively deal with Cu by enhancing expression of detoxifying genes. Plants treated with elevated Cu showed increased phenylalanine ammonia-lyase (PAL) activity that was reduced when the plants were also provided with extra Si. The three genes, *PAL1*, *PAL2* and *PAL3*, all showed similar patterns of gene expression that matched previous enzymatic data, while *PAL4* was elevated by the presence of high Cu irrespective of whether Si was present or not (Khandekar and Leisner 2011). Kim et al. (2014a) investigated the expressions of *HMA* genes (*OsHMA2* and *OsHMA3*) in rice plants grown hydroponically for 1, 5 and 10 days with six treatments (i.e. control, 1.0 mM Si, 100 μM Cu, 100 mM Cd,

100 μM Cu plus 1.0 mM Si and 100 μM Cd plus 1.0 mM Si). Their results showed that compared with control and Si treatment only, the mRNA expression of *OsHMA2* and *OsHMA3* encoding heavy metal transporters (*OsHMA2* and *OsHMA3*) was significantly upregulated by treatment with Cu or Cd, but was significantly and greatly downregulated by the combined treatment with Cu/Cd with Si. These expression patterns of *OsHMA2* and *OsHMA3* are in line with the significantly lower concentration of Cu and Cd in the combined treatment with Cu/Cd with Si as compared with the Cu/Cd treatment alone. To elucidate the mechanisms of Si involved in Cd detoxification at the molecular level, Nwugo and Huerta (2011) investigated the effect of Si on the leaf proteome of rice plants exposed to Cd stress. In total 60 protein spots were identified that were differentially regulated by Cd and/or Si treatments. Among these, 18 matching to 11 proteins were related to CO_2 assimilation/photosynthesis processes, representing the largest functional category group of proteins that were differentially expressed due to Cd and/or Si treatments. More recently, based on the results regarding Si-improved photosynthesis parameters and chloroplast ultrastructure under high-Zn (2 mM) stress, Song et al. (2014) have analysed the results of high-throughput sequencing and selected the main photosynthesis-related genes that were differentially expressed (see Fig. 5.12). According to the results of high-throughput sequencing (Fig. 5.12), Song et al. (2014) investigated the expression levels of the photosynthesis-related genes by using fluorescent real-time qPCR and showed that the expression levels of photosynthesis-related genes including *Os08g02630* (*PsbY*), *Os05g48630* (*PsaH*), *Os07g37030* (*PetC*), *Os03g57120* (*PetH*), *Os09g26810* and *Os04g38410* decreased in Si-deprived rice plants under high-Zn (0.2 mM Zn) stress, while the addition of 1.5 mM Si increased the expression levels of these genes in plants under high-Zn stress at 72 h, and the expression levels were higher in Si-treated plants than in Si-deprived plants. Similar work was also done by Li et al. (2011). They showed that a group of genes responsible for photosynthesis showed differential expression in rice grown under excess Mn (2 mM). The relative expression levels of *PsbP* and *ATPase* protein genes were significantly increased in plants treated with excess Mn, but were maintained at even higher levels in plants treated with both excess Mn and 1.5 mM Si. By contrast, expression levels of *HemD*, *Lhcb*, *pyrophosphatase* and *phosphoribulokinase* were significantly decreased under excess Mn, while the addition of Si significantly increased the expression level of these genes. Expression levels of *PsaH* were significantly increased by excess Mn, but were significantly reduced by Si.

It seems that more direct and convincing evidence needs to be accumulated to uncover the mechanisms of Si effects on plant resistance to heavy metals at a transcriptional or proteomic level. It may be interesting to investigate whether Si is involved in stimulating the efflux and inhibiting the influx of heavy metals in plasma membranes under heavy metal stress (Wu et al. 2013).

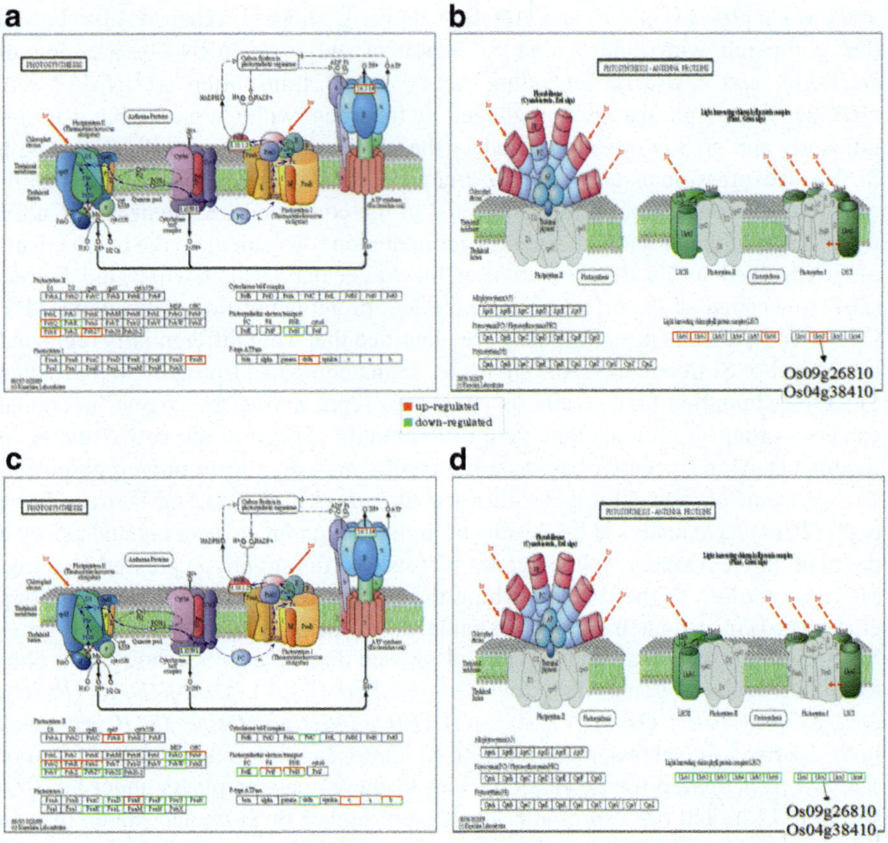

Fig. 5.12 The pathway analysis of photosynthesis (**a** and **c**) and photosynthesis-antenna proteins (**b** and **d**) of Zn-sensitive rice cultivar grown hydroponically with either normal (0.15 μM) or high (2 mM) Zn with (1.5 mM) or without Si for 72 h through high-throughput sequencing (From Song et al. 2014)

5.3 Conclusion and Perspectives

In a review paper, Ma (2001) summarized the possible beneficial effects of Si on crop growth in relation to biotic and abiotic stresses. The beneficial roles of Si were attributed mainly to silica gel deposited in the leaves and stems of plants and partially to the interactions between silicic acid and other elements such as Al, but no evidence was then available showing the involvement of Si in plant metabolism (Ma 2001; Ma et al. 2001). As discussed above, however, rapid progress has recently been made in demonstrating that Si not only plays some mechanical roles in providing resistance to stressful environments but, more importantly, is actively involved in physiological and biochemical processes. Growing evidence suggests that Si is

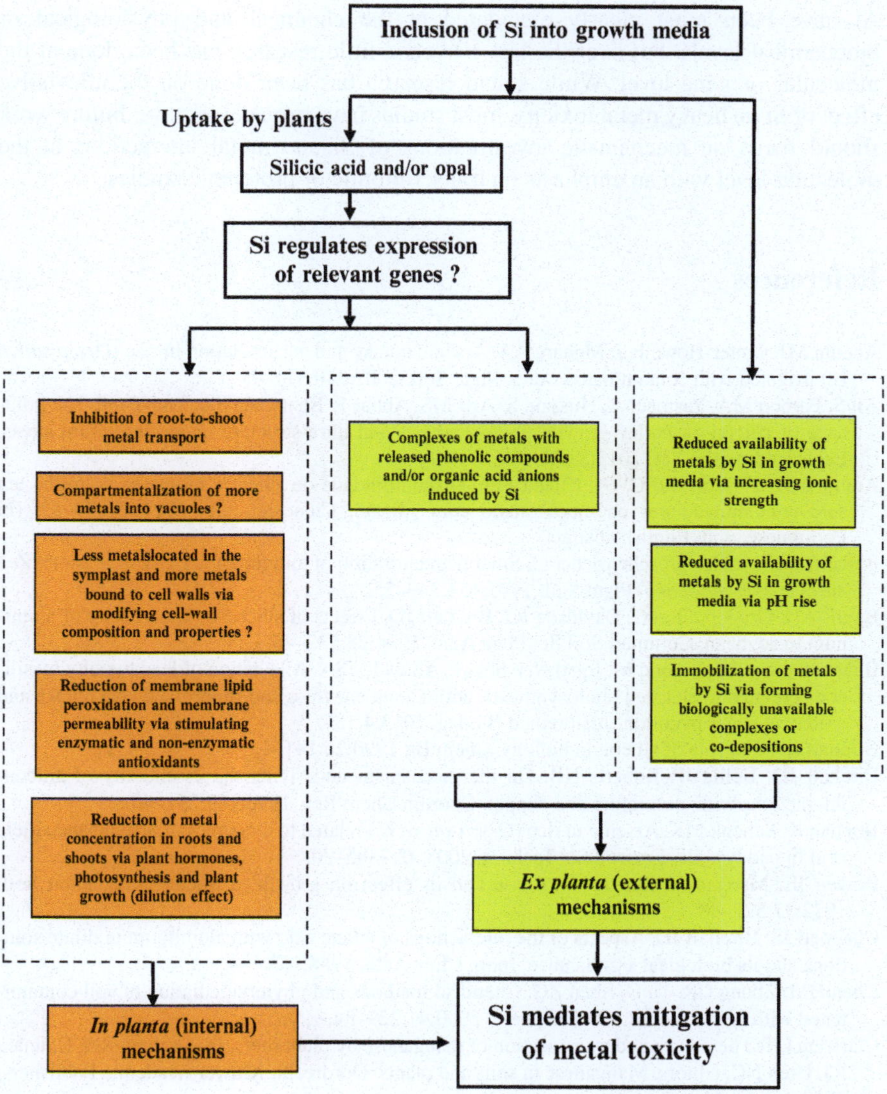

Fig. 5.13 Schematic diagram of proposed mode of action of Si-mediated alleviation of metal stress in plants

involved in regulating the expression of genes responsible for many plant metabolic processes, especially under heavy metal stress conditions. Based on the current information in the literature (Epstein 1994, 1999; Ma 2001; Liang et al. 2007; Wu et al. 2013), a schematic diagram of action model of Si-mediated alleviation of metal stress in plants is summarized in Fig. 5.13. The key question is to ascertain whether Si can regulate the expression of genes related to the metabolic processes in plants exposed to excess metals as shown in Fig. 5.6. The interactions of Si with

Al have been most clearly elucidated at the chemical and physiological or biochemical levels, but over the last 10 years, little research has been done at the molecular or gene level. While sound research has been done on the alleviative effect of Si on heavy metal toxicity, most studies have been descriptive. Future work should focus on mechanistic investigations of Si and metal interactions at the molecular level with an emphasis on transcriptomic or proteomic studies.

References

Abedin MJ, Cotter-Howells J, Meharg AA. Arsenic uptake and accumulation in rice (*Oryza sativa* L.) irrigated with contaminated water. Plant Soil. 2002;240:311–9.

Ali S, Farooq MA, Yasmeen T, Hussain S, Arif MS, Abbas F, Bharwana SA, Zhang GP. The influence of silicon on barley growth, photosynthesis and ultra-structure under chromium stress. Ecotoxol Environ Saf. 2013;89:66–72.

Araujo JCT, Nascimento CWA, Cunha FFF. Disponibilidade de silício e biomassa de milho em solo contaminado por chumbo tratado com silicato. Ciênc agrotec; 2011;35:878–83. (in Portuguese, with English abstract)

Barceló J, Guevara P, Poschenrieder C. Silicon amelioration of aluminum toxicity in teosinte, *Zea mays* L. *ssp. mexicana*. Plant Soil. 1993;154:249–55.

Baylis AD, Gragopoulou C, Davidson KJ, Birchall JD. Effects of silicon on the toxicity of aluminum to soybean. Commun Soil Sci Plant Anal. 1994;25:537–46.

Bharwana S, Ali S, Farooq M, Iqbal N, Abbas F, Ahmad MSA. Alleviation of lead toxicity by silicon is related to elevated photosynthesis, antioxidant enzymes and suppressed lead uptake and oxidative stress in cotton. J Bioremed Biodeg. 2013;4:187.

Birchall JD. The role of silicon in biology. Chem Br. 1990;26:141–4.

Birchall JD, Bellia JP, Roberts NB. On the mechanism underlying the essentiality of silicon-interactions with aluminium and copper. Coordin Chem Rev. 1996;149:231–40.

Bogdan K, Schenk MK. Arsenic in rice (*Oryza sativa* L.) related to dynamics of arsenic and silicic acid in paddy soils. Environ Sci Technol. 2008;42:7885–90.

Bowen JE. Manganese-silicon interaction and its effect on growth of Sudan grass. Plant Soil. 1972;37:577–88.

Chappell JS, Birchall JD. Aspects of the interactions of silicic acid with aluminium in dilute solutions and its biological significance. Inorg Chim Acta. 1988;153:1–4.

Chen HM, Zheng CR, Tu C, Shen ZG. Chemical methods and phytoremediation of soil contaminated with heavy metals. Chemosphere. 2000;41:229–34.

Clarkson DT. The uptake and translocation of manganese by plant roots. In: Graham RD, Hannam RJ, Uren NC, editors. Manganese in soils and plants. Dordrecht: Kluwer Academic Publishers; 1988. p. 101–11.

Clements HF. The roles of calcium silicate slags in sugarcane growth. Repts Hawaiian Sugarc Technol. 1965;25:103–26.

Cocker KM, Hodson MJ, Evans DE, Sangster AG. The interaction between silicon and aluminum in *Triticum aestivum* L. (cv. Celtic). Isr J Plant Sci. 1997;45:285–92.

Cocker KM, Evans DE, Hodson MJ. The amelioration of aluminium toxicity by silicon in higher plants: solution chemistry or an in planta mechanism? Physiol Plant. 1998a;104:608–14.

Cocker KM, Evans DE, Hodson MJ. The amelioration of aluminium toxicity by silicon in wheat (*Triticum aestivum* L.): malate exudation as evidence for an in planta mechanism. Planta. 1998b;204:318–23.

Collin B, Doelsch E, Keller C, Cazevieille P, Mc T, Chaurand P, Panfili F, Hazemann JL, Meunier JD. Evidence of sulfur-bound reduced copper in bamboo exposed to high silicon and copper concentrations. Environ Pollut. 2014;187:22–30.

Corrales I, Poschenrieder C, Barceló J. Influence of silicon pretreatment on aluminium toxicity in maize roots. Plant Soil. 1997;190:203–9.

Cullen WR, Reimer KJ. Arsenic speciation in the environment. Chem Rev. 1989;89:713–64.

Cunha KPV, Nascimento CWA. Silicon effects on metal tolerance and structural changes in maize (Zea mays L.) grown on a cadmium and zinc enriched soil. Water Air Soil Pollut. 2009;197:323–30.

Cunha KPV, Nascimento CWA, Silva AJ. Silicon alleviates the toxicity of cadmium and zinc for maize (*Zea mays* L.) grown on a contaminated soil. J Plant Nutr Soil Sci. 2008;171:849–53.

Dhar RK, Biswas BK, Samanta G, Mandal BK, Chakraborti D, Roy S, Jafar A, Islam A, Ara G, Kabir S, Khan AW, Ahmed SA, Hadi SA. Groundwater arsenic calamity in Bangladesh. Curr Sci India. 1997;73:48–59.

Dixit V, Pandey V, Shyam R. Chromium ions inactivate electron transport and enhance superoxide generation in vivo in pea (*Pisum sativum* L. cv. Azad) root mitochondria. Plant Cell Environ. 2002;25:687–90.

Doncheva S, Poschenrieder C, Stoyanova Z, Georgieva K, Velichkova M, Barceló J. Silicon amelioration of manganese toxicity in Mn-sensitive and Mn-tolerant maize varieties. Environ Exp Bot. 2009;65:189–97.

Dragisic Maksimovic J, Bogdanovic J, Maksimovic V, Nikolic M. Silicon modulates the metabolism and utilization of phenolic compounds in cucumber (*Cucumis sativus* L.) grown at excess manganese. J Plant Nutr Soil Sci. 2007;170:739–44.

Dragisic Maksimovic J, Mojovic M, Maksimovic V, Römheld V, Nikolic M. Silicon ameliorates manganese toxicity in cucumber by decreasing hydroxyl radical accumulation in the leaf apoplast. J Exp Bot. 2012;63:2411–20.

Dufey I, Gheysens S, Ingabire A, Lutts S, Bertin P. Silicon application in cultivated rices (*Oryza sativa* L and *Oryza glaberrima* Steud) alleviates iron toxicity symptoms through the reduction in iron concentration in the leaf tissue. J Agron Crop Sci. 2014;200:132–42.

Edwards DG, Asher CJ. Tolerance of crop and pasture species to manganese toxicity. In: Scaife A, editor. Proceedings of the ninth plant nutrition colloquium. Warwick: Commonwealth Agricultural Bureaux; 1982. p. 145–50.

El-Jaoual T, Cox DA. Manganese toxicity in plants. J Plant Nutr. 1998;21:353–86.

Epstein E. The anomaly of silicon in plant biology. Proc Natl Acad Sci U S A. 1994;91:11–7.

Epstein E. Silicon. Annu Rev Plant Physiol Plant Mol Biol. 1999;50:641–64.

Exley C, Birchall JD. Hydroxyaluminosilicate formation in solutions of low total aluminium concentration. Polydedron. 1992;11:1901–7.

Exley C, Birchall JD. A mechanism of hydroxyaluminosilicate formation. Polydedron. 1993; 12:1007–17.

Farooq MA, Ali S, Hameed A, Ishaque W, Mahmood K, Iqbal Z. Alleviation of cadmium toxicity by silicon is related to elevated photosynthesis, antioxidant enzymes; suppressed cadmium uptake and oxidative stress in cotton. Ecotoxol Environ Saf. 2013;96:242–9.

Feng JP, Shi QH, Wang XF. Effects of exogenous silicon on photosynthetic capacity and antioxidant enzyme activities in chloroplast of cucumber seedlings under excess manganese. Agric Sci China. 2009;8:40–50.

Feng JP, Shi QH, Wang XF, Wei M, Yang FJ, Xu HN. Silicon supplementation ameliorated the inhibition of photosynthesis and nitrate metabolism by cadmium (Cd) toxicity in *Cucumis sativus* L. Sci Hortic. 2010;123:521–30.

Finnegan PM, Chen WH. Arsenic toxicity: the effects on plant metabolism. Front Physiol. 2012; 3:182.

Fox RL, Silva JA, Younge OR, Plucknett DL, Sherman GD. Soil and plant silicon and silicate response by sugarcane. Soil Sci Soc Am Proc. 1967;31:775–9.

Foy CD, Chaney RL, White MC. The physiology of metal toxicity in plants. Annu Rev Plant Physiol. 1978;29:511–67.

Foy CD, Weil RR, Coradetti CA. Differential manganese tolerances of cotton genotypes in nutrient solution. J Plant Nutr. 1995;18:685–706.

Frantz JM, Khandekar S, Leisner S. Silicon differentially influences copper toxicity response in silicon-accumulator and non-accumulator species. J Am Soc Hortic Sci. 2011;136:329–38.

Fu YQ, Shen H, Wu DM, Cai KZ. Silicon-mediated amelioration of Fe^{2+} toxicity in rice (*Oryza sativa* L.) roots. Pedosphere. 2012;22:795–802.

Führs H, Götze S, Specht A, Erban A, Gallien S, Heintz D, Van Dorsselaer A, Kopka J, Braun H-P, Horst WJ. Characterization of leaf apoplastic peroxidases and metabolites in *Vigna unguiculata* in response to toxic manganese supply and silicon. J Exp Bot. 2009;60:1663–78.

Galvez L, Clark RB. Effects of silicon on growth and mineral composition of sorghum (*Sorghum bicolor*) grown with toxic levels of aluminium. In: Wright RJ, Baligar VC, Murrmann RP, editors. Plant-soil interactions at low pH. Dordrecht: Kluwer Academic Publishers; 1991. p. 815–23.

Galvez L, Clark RB, Gourley LM, Maranville JW. Silicon interactions with manganese and aluminum toxicity in sorghum. J Plant Nutr. 1987;10:1139–47.

Garg N, Singla P. Arsenic toxicity in crop plants: physiological effects and tolerance mechanisms. Environ Chem Lett. 2011;9:303–21.

Gu HH, Qiu H, Tian T, Zhan SS, Deng THB, Chaney RL, Wang SZ, Tang YT, Morel JL, Qiu RL. Mitigation effects of silicon rich amendments on heavy metal accumulation in rice (*Oryza sativa* L.) planted on multi-metal contaminated acidic soil. Chemosphere. 2011;83:1234–40.

Gu HH, Zhan SS, Wang SZ, Tang YT, Chaney RL, Fang XH, Cai XD, Qiu RL. Silicon-mediated amelioration of zinc toxicity in rice (*Oryza sativa* L.) seedlings. Plant Soil. 2012; 350:193–204.

Guo W, Hou YL, Wang SG, Zhu YG. Effect of silicate on the growth and arsenate uptake by rice (*Oryza sativa* L.) seedlings in solution culture. Plant Soil. 2005;272:173–81.

Hammond KE, Evans DE, Hodson MJ. Aluminium/silicon interactions in barley (Hordeum vulgare L.) seedlings. Plant Soil. 1995;173:89–95.

Hara T, Gu MH, Koyama H. Ameliorative effect of silicon an aluminum injury in the rice plant. Soil Sci Plant Nutr. 1999;45:929–36.

Hodson MJ, Evans DE. Aluminium/ silicon interactions in higher plants. J Exp Bot. 1995; 46:161–71.

Hodson MJ, Sangster AG. The interaction between silicon and aluminium in *Sorghum bicolor* (L.) Moench: growth analysis and X-ray microanalysis. Ann Bot. 1993;72:389–400.

Hodson MJ, Sangster AG. Mineral deposition in the needles of white spruce [*Picea glauca* (Moench) Voss]. Ann Bot. 1998;82:375–85.

Horiguchi T. Mechanism of manganese toxicity and tolerance of plants. 4. Effect of silicon on alleviation of manganese toxicity of rice plants. Soil Sci Plant Nutr. 1988;34:65–73.

Horiguchi T, Morita S. Mechanism of manganese toxicity and tolerance of plants. 6. Effect of silicon on alleviation of manganese toxicity of barley. J Plant Nutr. 1987;10:2299–310.

Horst WJ. The role of the apoplast in aluminum toxicity and resistance of higher-plants – a review. Z Pflanzenernahr Bodenkd. 1995;158:419–28.

Horst WJ, Marschner H. Effect of silicon on manganese tolerance of bean plants (*Phaseolus vulgaris* L.). Plant Soil. 1978a;50:287–303.

Horst WJ, Marschner H. Einfluβ von Silizium auf den Bindungszustand von Mangan im Blattgewebe von Bohnen (*Phaseolus vulgaris*). Z Pflanzenernaehr Bodenkd. 1978b; 141:487–97.

Horst WJ, Fecht M, Naumann A, Wissemeier AH, Maier P. Physiology of manganese toxicity and tolerance in *Vigna unguiculata* (L.) Walp. J Plant Nutr Soil Sci. 1999;162:263–74.

IRRI, International Rice Research Institute. Annual report 1964. Los Banos: Laguna; 1965.

IRRI, International Rice Research Institute. Annual report 1965. Los Banos: Laguna; 1966.

Iwasaki K, Matsumura A. Effect of silicon on alleviation of manganese toxicity in pumpkin (*Cucurbita moschata* Duch cv. Shintosa). Soil Sci Plant Nutr. 1999;45:909–20.

Iwasaki K, Maier P, Fecht M, Horst WJ. Effects of silicon supply on apoplastic manganese concentrations in leaves and their relation to manganese tolerance in cowpea (*Vigna unguiculata* (L.) Walp.). Plant Soil. 2002a;238:281–8.

Iwasaki K, Maier P, Fecht M, Horst WJ. Leaf apoplastic silicon enhances manganese tolerance of cowpea (*Vigna unguiculata*). J Plant Physiol. 2002b;159:167–73.

Júnior LAZ, Fontes RLF, Neves JCL, Korndörfer GH, de Ávila VT. Rice grown in nutrient solution with doses of manganese and silicon. R Bras Ci Solo. 2010;34:1629–39.

Kaya C, Tuna AL, Sonmez O, Ince F, Higgs D. Mitigation effects of silicon on maize plants grown at high zinc. J Plant Nutr. 2009;32:1788–98.

Khandekar S, Leisner S. Soluble silicon modulates expression of *Arabidopsis thaliana* genes involved in copper stress. J Plant Physiol. 2011;168:699–705.

Kidd P, Proctor J. The growth response of ecotypes of *Holcus lanatus* L. from different soil types in northwestern Europe to aluminium and silicon. Fresenius Environ Bull. 2001;10:590–4.

Kidd PS, Llugany M, Poschenrieder C, Gunse B, Barcelo J. The role of root exudates in aluminium resistance and silicon-induced amelioration of aluminium toxicity in three varieties of maize (*Zea mays* L.). J Exp Bot. 2001;359:1339–52.

Kim YH, Khan AL, Waqas M, Shim JK, Kim DH, Lee KY, Lee IJ. Silicon application to rice root zone influenced the phytohormonal and antioxidant responses under salinity stress. J Plant Growth Regul. 2014a;33:137–49.

Kim YH, Khan AL, Kim DH, Lee SY, Kim KM, Waqas M, Jung HY, Shin JH, Kim JG, Lee IJ. Silicon mitigates heavy metal stress by regulating P-type heavy metal ATPases, *Oryza sativa* low silicon genes, and endogenous phytohormones. BMC Pant Biol. 2014b;14:13.

Klug B, Horst WJ. Oxalate exudation into the root-tip water free space confers protection from aluminum toxicity and allows aluminum accumulation in the symplast in buckwheat (*Fagopyrum esculentum*). New Phytol. 2010;187:380–91.

Larsen PB, Degenhardt J, Tai C-Y, Stenzler LM, Howell SH, Kochian LV. Aluminum-resistant Arabidopsis mutants that exhibits altered patterns of aluminum accumulation and organic release from roots. Plant Physiol. 1998;117:9–18.

Li YC, Alva AK, Sumner ME. Response of cotton cultivars to aluminum in solutions with varying silicon concentrations. J Plant Nutr. 1989;12:881–92.

Li YC, Sumner ME, Miller WP, Alva AK. Mechanism of silicon induced alleviation of aluminum phytotoxicity. J Plant Nutr. 1996;19:1075–87.

Li J, Lesiner SM, Frantz J. Alleviation of copper toxicity in Arabidopsis thaliana by silicon addition to hydroponic solutions. J Am Soc Hortic Sci. 2008;133:670–7.

Li P, Song AL, Li ZJ, Fan FL, Liang YC. Silicon ameliorates manganese toxicity by regulating physiological processes and expression of genes related to photosynthesis in rice (Oryza sativa L.). Proceedings of 5th International Conference on Silicon in Agriculture; September 13–18, Beijing; 2011. p 103.

Li P, Song AL, Li ZJ, Fan FL, Liang YC. Silicon ameliorates manganese toxicity by regulating manganese transport and antioxidant reactions in rice (*Oryza sativa* L.). Plant Soil. 2012a;354:407–19.

Li LB, Zheng C, Fu YQ, Wu DM, Yang XJ, Shen H. Silicate-mediated alleviation of Pb toxicity in banana grown in Pb-contaminated soil. Biol Trace Elem Res. 2012b;145:101–8.

Liang YC, Yang CG, Shi HH. Effects of silicon on growth and mineral composition of barley grown under toxic levels of aluminum. J Plant Nutr. 2001;24:229–43.

Liang YC, Wong JWC, Wei L. Silicon-mediated enhancement of cadmium tolerance in maize (*Zea mays* L.) grown in cadmium contaminated soil. Chemosphere. 2005;58:475–83.

Liang YC, Sun WC, Zhu YG, Christie P. Mechanisms of silicon- mediated alleviation of abiotic stresses in higher plants: a review. Environ Pollut. 2007;147:422–8.

Liu J, Ma J, He CW, Li XL, Zhang WJ, Xu FX, Lin YJ, Wang LJ. Inhibition of cadmium ion uptake in rice (*Oryza sativa*) cells by a wall-bound form of silicon. New Phytol. 2013a;200:691–9.

Liu JG, Zhang HM, Zhang YX, Chai TY. Silicon attenuates cadmium toxicity in *Solanum nigrum* L. by reducing cadmium uptake and oxidative stress. Plant Physiol Biochem. 2013b;68:1–7.

Lux A, Martinka M, Vaculik M, White PJ. Root responses to cadmium in the rhizosphere: a review. J Exp Bot. 2011;62:21–37.

Ma JF. Role of silicon in enhancing the resistance of plants to biotic and abiotic stresses. Soil Sci Plant Nutr. 2001;50:11–8.

Ma JF, Sasaki M, Matsumoto H. Al-induced inhibition of root elongation in corn, *Zea mays* L. is overcome by Si addition. Plant Soil. 1997;188:171–6.

Ma JF, Miyake Y, Takahashi E. Silicon as a beneficial element for crop plants. In: Datnoff LE, Snyder GH, Korndörfer GH, editors. Silicon in agriculture. Amsterdam: Elsevier; 2001. p. 17–39.

Ma J, Cai HM, He CW, Zhang WJ, Wang LJ. A hemicellulose-bound form of silicon inhibits cadmium ion uptake in rice (*Oryza sativa*) cells. New Phytol. 2015;206:1063–74.

Mandal BK, Suzuki KT. Arsenic round the world: a review. Talanta. 2002;58:201–35.

Marschner H. Mineral nutrition of higher plants. 2nd ed. London: Academic Press; 1995.

Masarovic D, Slovakova L, Bokor B, Bujdos M, Lux A. Effect of silicon application on *Sorghum bicolor* exposed to toxic concentration of zinc. Biologia. 2012;67:706–12.

Meharg AA, Williams PN, Adomako A, Lawgali YY, Deacon C, Villada A, Cambell RCJ, Sun G, Zhu YG, Feldmann J, Rabb A, Zhao FJ, Islam R, Hossain S, Yanai J. Geographical variation in total and inorganic arsenic content of polished (white) rice. Environ Sci Technol. 2009;43:1612–7.

Morikawa CK, Saigusa M. Si amelioration of Al toxicity in barley (*Hordeum vulgare* L.) growing in two Andosols. Plant Soil. 2002;240:161–8.

Morris HD, Pierre WH. Minimum concentrations of manganese necessary for injury to various legumes in culture solutions. Agron J. 1949;41:107–12.

Neumann D, zur Nieden U. Silicon and heavy metal tolerance of higher plants. Phytochemistry. 2001;56:685–92.

Nordstrom DK. Public health - worldwide occurrences of arsenic in ground water. Science. 2002;296:2143–5.

Nwugo CC, Huerta AJ. Effects of silicon nutrition on cadmium uptake, growth and photosynthesis of rice plants exposed to low-level cadmium. Plant Soil. 2008;311:73–86.

Nwugo CC, Huerta AJ. The effect of silicon on the leaf proteome of rice (*Oryza sativa* L.) plants under cadmium-stress. J Proteome Res. 2011;10:518–28.

Okuda A, Takahashi E. Effect of silicon supply on the injuries due to excessive amounts of Fe, Mn, Cu, As, Al, Co of barley and rice plant. Jpn J Soil Sci Plant Nutr. 1962;33:1–8 (in Japanese).

Okuda A, Takahashi E. The role of silicon. In: The mineral nutrition of the rice plant. Madison: Johns Hopkins Press; 1965. p. 146–56.

Oliva SR, Mingorance MD, Leidi EO. Effects of silicon on copper toxicity in Erica andevalensis Cabezudo and Rivera: a potential species to remediate contaminated soils. J Environ Monitor. 2011;13:591–6.

Panda SK. Chromium-mediated oxidative stress and ultrastructural changes in root cells of developing rice seedlings. J Plant Physiol. 2007;164:1419–28.

Peaslee DE, Frink CR. Influence of silicic acid on uptake of Mn, Al, Zn, and Cu by tomatoes (*Lycopersicum esculentum*) grown on an acid soil. Soil Sci Soc Am Proc. 1969;33:569–71.

Prabagar S, Hodson MJ, Evans DE. Silicon amelioration of aluminium toxicity and cell death in suspension cultures of Norway spruce (*Picea abies* (L.) Karst.). Envand Exp Bot. 2011; 70:266–76.

Rahman MT, Kawamura K, Koyama H, Hara T. Varietal differences in the growth of rice plants in response to aluminum and silicon. Soil Sci Plant Nutr. 1998;44:423–31.

Rizwan M, Meunier JD, Miche H, Keller C. Effect of silicon on reducing cadmium toxicity in durum wheat (*Triticum turgidum* L. cv. Claudio W.) grown in a soil with aged contamination. J Hazard Mater. 2012;209–210:326–34.

Rogalla H, Römheld V. Role of leaf apoplast in silicon-mediated manganese tolerance of *Cucumis sativus* L. Plant Cell Environ. 2002;25:549–55.

Rorison IH. The calciole-calcifuge problem. II The effects of mineral nutrition on seedling growth in solution culture. J Ecol. 1960;48:679–88.

Ryder M, Gerard F, Evans DE, Hodson MJ. The use of root growth and modelling data to investigate amelioration of aluminium toxicity by silicon in *Picea abies* seedlings. J Inorg Biochem. 2003;97:52–8.

Shanker AK, Cervantes C, Loza-Tavera H, Avudainayagam S. Chromium toxicity in plants. Environ Int. 2005;31:739–53.

Shi MC. Effects of silicon on arsenic uptake and toxicity to rice plants (*Oryza sativa*). MSc thesis, Guangxi University; 2008.

Shi QH, Bao ZY, Zhu ZJ, He Y, Qian QQ, Yu JQ. Silicon-mediated alleviation of Mn toxicity in *Cucumis sativus* in relation to activities of superoxide dismutase and ascorbate peroxidase. Phytochemistry. 2005a;66:1551–9.

Shi XH, Zhang CC, Wang H, Zhang FS. Effect of Si on the distribution of Cd in rice seedlings. Plant Soil. 2005b;272:53–60.

Shi GR, Cai QS, Liu CF, Wu L. Silicon alleviates cadmium toxicity in peanut plants in relation to cadmium distribution and stimulation of antioxidative enzymes. Plant Growth Regul. 2010;61:45–52.

Singh VP, Tripathi DK, Kumar D, Chauhan DK. Influence of exogenous silicon addition on aluminium tolerance in rice seedlings. Biol Trace Elem Res. 2011;144:1260–74.

Song AL, Li ZJ, Zhang J, Xue GF, Fan FL, Liang YC. Silicon-enhanced resistance to cadmium toxicity in *Brassica chinensis* L. is attributed to Si-suppressed cadmium uptake and transport and Si-enhanced antioxidant defense capacity. J Hazard Mater. 2009;172:74–83.

Song AL, Li P, Li ZJ, Fan FL, Nikolic M, Liang YC. The alleviation of zinc toxicity by silicon is related to zinc transport and antioxidative reactions in rice. Plant Soil. 2011;344:319–33.

Song AL, Li P, Fan FL, Li ZJ, Liang YC. The effect of silicon on photosynthesis and expression of its relevant genes in rice (*Oryza sativa* L.) under high-zinc stress. Plos One. 2014;9(11), e113782. doi:10.1371/journal.pone.0113782.

Taylor PD, Jugdaohsingh R, Powell JJ. Soluble silica with high affinity for aluminum under physiological and natural conditions. J Am Chem Soc. 1997;119:8852–6.

Tanaka A, Park YD. Significance of the absorption and distribution of silica in the growth of the rice plant. Soil Sci Plant Nutr. 1966;12:23–8.

Tavakkoli E, English P, Guppy CN. Interaction of silicon and phosphorus mitigate manganese toxicity in rice in a highly weathered soil. Comm Soil Sci Plant Anal. 2011;42:503–13.

Treder W, Cieslinski G. Effect of silicon application on cadmium uptake and distribution in strawberry plants grown on contaminated soils. J Plant Nutr. 2005;28:917–29.

Tripathi DK, Singh VP, Kumar D, Chauhan DK. Rice seedlings under cadmium stress: effect of silicon on growth, cadmium uptake, oxidative stress, antioxidant capacity and root and leaf structures. Chem Ecol. 2012a;28:281–91.

Tripathi DK, Singh VP, Kumar D, Chauhan DK. Impact of exogenous silicon addition on chromium uptake, growth, mineral elements, oxidative stress, antioxidant capacity, and leaf and root structures in rice seedlings exposed to hexavalent chromium. Acta Physiol Plant. 2012b;34:279–89.

Tripathi P, Tripathi RD, Singh RP, Dwivedi S, Goutam D, Shri M, Trivedi PK, Chakrabarty D. Silicon mediates arsenic tolerance in rice (*Oryza sativa* L.) through lowering of arsenic uptake and improved antioxidant defence system. Ecol Eng. 2013;52:96–103.

Turnau K, Kortke I, Oberwinkler F. *Paxillus involutus: Pinus sylvestris* mycorrhizae from a heavily polluted forest. I. Element localization using electron energy loss spectroscopy and imaging. Botan Acta. 1993;106:213–9.

Vaculík M, Landberg T, Greger M, Luxová M, Stoláriková LA. Silicon modifies root anatomy, and uptake and subcellular distribution of cadmium in young maize plants. Ann Bot. 2012; 110:433–43.

Vatehova Z, Kollarova K, Zelko I, Richterova-Kucerova D, Bujdos M, Liskova D. Interaction of silicon and cadmium in *Brassica juncea* and *Brassica napus*. Biologia. 2012;67:498–504.

Vlamis J, Williams DE. Manganese and silicon interaction in the gramineae. Plant Soil. 1967;27:131–40.

Wagner F. The importance of silicic acid for the growth of some cultivated plants, their metabolism, and their susceptibility to true mildews. Phytopathol Z. 1940;12:427–79.

Wang YX, Stass A, Horst WJ. Apoplastic binding of aluminum is involved in silicon-induced amelioration of aluminum toxicity in maize. Plant Physiol. 2004;136:3762–70.

Watanabe T, Osaki M, Tadano T. Aluminum-induced growth stimulation in relation to calcium, magnesium, and silicate nutrition in *Melastoma malabathricum* L. Soil Sci Plant Nutr. 1997;43:827–37.

Welch RM. Micronutrient nutrition of plants. Crit Rev Plant Sci. 1995;14:49–82.

Wiese H, Nikolic M, Römheld V. Silicon in plant nutrition. In: Sattelmacher B, Horst WJ, editors. The apoplast of higher plants: compartment of storage. Transport and reactions. Heidelberg: Springer; 2007. p. 33–47.

Williams DE, Vlamis J. Manganese toxicity in standard culture solutions. Plant Soil. 1957a; 8:183–93.

Williams DE, Vlamis J. The effect of silicon on yield and manganese-54 uptake and distribution in the leaves of barley plants grown in culture solutions. Plant Physiol. 1957b;32:404–9.

Wu JW, Shi Y, Zhu YX, Wang YC, Gong HJ. Mechanisms of enhanced heavy metal tolerance in plants by silicon: a review. Pedosphere. 2013;23:815–25.

Yang YH, Chen SM, Chen Z, Zhang HY, Shen HG, Hua ZC, Li N. Silicon effects on aluminum toxicity to mungbean seedling growth. J Plant Nutr. 1999;22:693–700.

Yang CG, Dou H, Liang YC, Lou YS. Influence of silicon on cadmium availability and cadmium uptake by maize in cadmium-contaminated soil. Sci Agric Sin. 2005;38:116–21 (In Chinese with English abstract).

Yang JL, Li YY, Zhang YJ, Zhang SS, Wu YR, Wu P, Zheng SJ. Cell wall polysaccharides are specifically involved in the exclusion of aluminum from the rice root apex. Plant Physiol. 2008;146:602–11.

Ye J, Yan CL, Liu JC, Lu HL, Liu T, Song ZF. Effects of silicon on the distribution of cadmium compartmentation in root tips of Kandelia obovata (S., L.) Yong. Environ Pollut. 2012; 162:369–73.

Yoshida S, Ohnishi Y, Kitagishi K. The chemical nature of silicon in rice plant. Soil Plant Food (Tokyo). 1959;5:23–7.

Zayed AM, Terry N. Chromium in the environment: factors affecting biological remediation. Plant Soil. 2003;249:139–56.

Zeng FR, Zhao FS, Qiu BY, Ouyang YN, Wu FB, Zhang GP. Alleviation of chromium toxicity by silicon addition in rice plants. Agric Sci China. 2011;10:1188–96.

Zhang CC, Wang LJ, Nie Q, Zhang WX, Zhang FS. Long-term effects of exogenous silicon on cadmium translocation and toxicity in rice (*Oryza sativa* L.). Environ Exp Bot. 2008; 62:300–7.

Zhang Q, Yan CL, JC, Lu HL, Duan HH, Du JN, Wang WY. Silicon alleviation of cadmium toxicity in mangrove (*Avicennia marina*) in relation to cadmium compartmentation. J Plant Growth Regul. 2013a;33:233–42.

Zhang SR, Li SY, Ding XD, Li FB, Liu CP, Liao XR, Wang RP. Silicon mediated the detoxification of Cr on pakchoi (*Brassica chinensis* L.) in Cr-contaminated soil. J Food Agric Environ. 2013b;11:814–9.

Zhang YX, Yang PQ, Xu T, Yu YX, Li XF. Al-detoxification by Si in *Stylosanthes*. Genomics Appl Biol. 2009;1:57–61 (In Chinese with English abstract).

Zsoldos F, Vashegyi A, Pecsvaradi A, Bona L. Influence of silicon on aluminium toxicity in common and durum wheats. Agronomie. 2003;23:349–54.

Chapter 6
Silicon-Mediated Tolerance to Salt Stress

Abstract Silicon (Si) has widely been reported to increase plant tolerance and/or resistance to salt (salinity) stress, but the underlying mechanisms remain poorly understood. This chapter reviews the updated knowledge concerning Si and salt (salinity) interactions in higher plants. Based on the current literature, it seems that (1) silica deposited in the apoplast as SiO_2 opal or phytolith can enhance water retention by inhibiting transpirational water loss, thus reducing salt-induced osmotic stress, and (2) soluble Si in the symplast may be actively involved in physiological and biochemical metabolisms by regulating the expression of genes related to the biosynthesis of hormones (ABA and JA etc.), antioxidant defence enzymes, H^+-pumps and osmolytes to rebalance ion stoichiometry, reduce membrane permeability, and improve membrane structure and stability, hence improving salt tolerance in plants. However, further work is needed to better understand Si-mediated tolerance and/or resistance to salt stress at the molecular level.

Keywords Osmotic stress • Oxidative damage • Photosynthesis • Salt stress • Silicon • Salt ion toxicity

6.1 Photosynthetic Parameters and Plant Growth

Soil salinization is one of the major edaphic factors that severely restrict crop growth and compromise the development of sustainable agriculture in the coastal and inland areas worldwide (Liang et al. 1996; Zhu and Gong 2014). It is estimated that approximately 7 % of the land on the Earth and 20 % of the total arable land are adversely affected by salinity (Liang et al. 1996; Rasool et al. 2013). The major approach to improving saline soils is to leach salts down the soil profile by over-irrigating. However, this is frequently restricted both in arid and semiarid areas and in coastal areas because of limited freshwater supply or increased salinity in streams (Liang et al. 1996).

Accumulation of excessive salt or salinity in saline conditions may significantly reduce plant growth and development and crop yield; it inhibits net assimilation by restricting the CO_2 supply due to partial closure of stomata, by impairing the biochemical mechanisms of CO_2 fixation or by both. Salt (salinity) stress may also

Y. Liang et al., *Silicon in Agriculture*, DOI 10.1007/978-94-017-9978-2_6

affect electron transport, photophosphorylation and enzymatic activities (Fadzilla et al. 1997; Liang 1999; Steduto et al. 2000). Inclusion of Si has widely been reported to significantly increase the growth of many plant species through increasing photosynthetic activity, leaf area and chlorophyll content and improving chloroplast structure in salt-stressed plants (Liang et al. 1996; Liang 1998; Yeo et al. 1999; Al-Aghabary et al. 2004; Zhu et al. 2004; Romero-Aranda et al. 2006; Murillo-Amador et al. 2007; Tuna et al. 2008; Savvas et al. 2009; Hashemi et al. 2010; Lee et al. 2010; Tahir et al. 2012; Kafi and Rahimi 2011; Bae et al. 2012; Yin et al. 2013; Soundararajan et al. 2013; Haghighi and Pessarakli 2013). Indeed, Liang et al. (1996) showed that the addition of Si could significantly enhance dry matter yields of both salt-sensitive (cv. Kepin No. 7) and salt-tolerant (cv. Jian No. 4) barley (*Hordeum vulgare*) cultivars grown under salt-stressed conditions. They demonstrated that Si improved the specific leaf area and CO_2 assimilation rate, which were decreased due to salt stress (also see Fig. 6.1). In a later study, Liang (1998) showed that though leaf chlorophyll content was consistently higher in the salt-tolerant barley cultivar than in the salt-sensitive one, the net leaf CO_2 assimilation rate increased significantly in both cultivars tested under salt stress when Si was provided. The addition of Si to the salt treatment was found to improve the cell ultrastructure of leaves. Under salt stress conditions, the double membranes of chloroplasts were degraded, but the membrane integrity was markedly improved with supplementation of Si. Yeo et al. (1999) also reported that when rice (*Oryza sativa*) plants were grown in the nutrient solution containing 50 mM NaCl, the addition of Si significantly counteracted the growth reduction caused by salt stress through increasing both the CO_2 assimilation rate and the stomatal conductance. Tuna et al. (2008) showed that chlorophyll contents were considerably decreased in both the salt-tolerant and salt-sensitive wheat (*Triticum aestivum*) cultivars grown under high salt stress (100 mM NaCl), and the addition of Si completely restored chlorophyll content to the level of non-stressed plants. It was also reported that addition of Si significantly enhanced leaf concentrations of chlorophyll a and b in wheat plants grown hydroponically with supply of 100 mM NaCl (Tahir et al. 2012). While Si application per se (in the absence of salt stress) did not influence the net photosynthetic rate, stomatal conductance and transpiration rate in hydroponically grown sorghum (*Sorghum bicolor*) seedlings, Si supplementation caused a significant increase in these parameters when plants were subjected to salt stress at 100 mM NaCl (Yin et al. 2013). Ashraf et al. (2010) showed that the yield and yield attributes of sugarcane (*Saccharum officinarum*) under salt stress were significantly higher in the presence of Si than in the absence of Si. Juice quality characteristics were significantly improved by the application of Si in salt-sensitive and salt-tolerant sugarcane genotypes tested. Bae et al. (2012) reported that the addition of Si significantly increased the chlorophyll and carotenoid contents of the salt-stressed Kentucky bluegrass (*Poa pratensis*). Silicon (Si) can improve not only the photosynthetic parameters and the growth of monocotyledonous plant species subjected to salt or salinity stress but also those of some dicotyledonous plants. For instance, Lee et al. (2010) demonstrated that both leaf chlorophyll content and total fresh/dry biomass of soybean (*Glycine max*) plants grown under salt stress (80 mM NaCl) significantly increased

Fig. 6.1 Effect of 1.0 mM Si addition on the growth of salt-sensitive (cv. Kepin No. 7, *upper*) and salt-tolerant (cv. Jian No. 4, *lower*) barley cultivars grown hydroponically with 120 mM NaCl (Photography by Yongchao Liang)

with the addition of Si. Zuccarini (2008) also reported that the presence of 1.5 mM Si in the nutrient solutions containing 30 or 60 mM NaCl significantly increased stomatal conductance, net photosynthetic rate, shoot dry weight and leaf area in bean (*Phaseolus vulgaris*) plants grown hydroponically for 30 days. Hashemi et al. (2010) showed that under high salt stress (150 mM NaCl), canola (*Brassica napus*) plants grown hydroponically without Si had approximately 50 % lower total dry weight compared with plants supplied with 2 mM Si. The decline in shoot dry weight was stronger than the decline in root dry weight under salt stress, and consequently the ratio of shoot to root decreased by more than 60 %, which was partly recovered by Si supply. Although tomato (*Lycopersicon esculentum*) and faba bean (*Vicia faba*) are considered Si-excluding species (Liang et al. 2005a; Nikolic et al. 2007), benefits of Si supplementation on salt toxicity have been reported. For example, Romero-Aranda et al. (2006) demonstrated that the addition of 80 mM NaCl significantly reduced the stomatal conductance to water, the net CO_2 assimilation rate and the transpiration rate in tomato plants, while supplemental Si recovered the values of gas exchange parameters in salt-stressed plants to the levels in non-stressed plants.

Si not only alleviates growth reduction and improves photosynthetic parameters in non-halophytic plants exposed to salt stress but also plays a significant role in raising tolerance of halophytes to high salinity. The treatment with salt (680 mM NaCl) substantially reduced total dry matter weight and mean relative growth rate of a halophytic grass (*Spartina densiflora*) grown hydroponically, but these parameters were not affected when 0.5 mM Si was supplied to the salt-stressed plants (Mateos-Naranjo et al. 2013). The addition of Si significantly improved net photosynthetic rate, stomatal conductance, intercellular CO_2 concentration, and the contents of photosynthetic pigments (chlorophyll a and b and carotenoid) in the salt-stressed *Spartina* plants. The same authors showed that the values of the maximum quantum efficiency of PSII photochemistry (Fv/Fm), maximum fluorescence (Fm), and quantum efficiency of PSII (FPSII) were all significantly higher, but the value of non-photochemical quenching (NPQ) was significantly lower in the salt plus Si treatment than in the salt treatment without Si supply.

6.2 Silicon-Mediated Osmotic Stress and Water Use Efficiency

High salt or salinity stress can negatively affect plant growth, development and even survival. The mechanisms involved include water or osmotic stress, ion toxicity and nutritional disorders. Osmotic stress is the first phase of salt stress, which starts immediately after the salt concentration around the roots increases beyond a threshold level, and is followed by a significant reduction of shoot growth rate. Therefore, overcoming osmotic stress or physiological water deficit is one of the most important strategies of plant adaptation to salt and/or salinity stressful environments. Si has been shown to effectively prevent plants from water loss by inhibiting the transpiration rate (Savant et al. 1999) or by reducing water loss through the cuticles due to silica deposition beneath the epidermal cells of leaves and stems (Matoh et al. 1986; Bélanger et al. 1995; Liang et al. 1999; Raven 2003; Trenholm et al. 2004). It is important to consider the role Si plays in adjusting plant water status because the initial reduction of the growth of plants subjected to salt stress is due to the osmotic effect of salts (Zhu 2002). Liang et al. (1999) reported that addition of Si markedly decreased leaf cell-sap concentration in the leaves of water-cultured or potted barley plants under salt stress, which strongly suggests that Si may play a positive role in enhancing the water-retention capacity in salt-stressed plants, thereby alleviating water stress induced by salinity. They agree that the Si-enhanced water-retention capacity under salt stress may be associated with not only the reduction of transpirational water loss due to silica deposition beneath the epidermal cells of leaves and stems but also the role of potassium in regulation of stomatal closure since more potassium was taken up by salt-stressed plants in the presence of Si than in the absence of Si (Liang et al. 1999). Romero-Aranda et al. (2006) reported that the plant water content in tomato, calculated as the difference between fresh and dry weight, was reduced with salinity by 54 %; when plants subjected to 80 mM NaCl were treated with 2.5 mM Si, their plant water content increased by 40 % compared to Si-deprived stressed plants. Moreover, leaf water potential was reduced by NaCl treatment and was even lower in NaCl-stressed plants with added Si. Supply of Si lowered osmotic potential (more negative) and increased turgor of the leaves in NaCl-stressed tomato plants (Romero-Aranda et al. 2006). Yin et al. (2013) also demonstrated that, under salt stress conditions, the osmotic potential of sorghum plants was significantly lower and the turgor pressure was significantly higher in the presence of Si than in the absence of Si. Meanwhile, the sucrose and fructose levels, but not the proline levels, were significantly higher in the Si-fed plants than in the Si-deprived plants. Tuna et al. (2008) and Tahir et al. (2012) reported that the relative water content (RWC) was significantly lower in both salt-tolerant and salt-sensitive wheat plants grown under high salt stress (100 mM NaCl), while addition of Si completely restored RWC to the levels recorded in the non-stressed plants. A similar effect of Si on leaf RWC was also demonstrated in hydroponically grown bean plants supplied with high NaCl (Zuccarini 2008). Water use efficiency (WUE) is raised by Si also in halophytic species (adapted to saline environments) under salt stress. For instance, WUE was significantly increased by the addition of Si in *Spartina densiflora* (a halophytic grass) grown under 680 mM NaCl (Mateos-Naranjo et al. 2013).

6.3 Silicon-Mediated Transport and Distribution of Salt Ions

Ion toxicity is the second phase of plant response to salinity, which starts when salt accumulates to toxic concentrations in the older leaves (Munns and Tester 2008). For most plant species, Na^+ appears to be more toxic than Cl^-, so that most studies have focused on Na^+ exclusion and the control of Na^+ transport within the plant tissues. Reducing Na^+ uptake or increasing Na^+ exclusion remains a primary determinant of variability in salinity tolerance within a species (Liang 1999; Liang et al. 1999; Munns and Tester 2008). On the other hand, increasing K^+ uptake to maintain a high K^+/Na^+ ratio is also an important factor in determining salt tolerance of plants (Liang et al. 1996, 1999; Liang 1999). The addition of Si to the nutrient solution cultures was reported to significantly diminish the Na^+ content in shoots of rice, wheat and barley (Matoh et al. 1986; Ahmad et al. 1992; Liang et al. 1996, 1999; Liang 1999). This was attributed to Si-induced reduction in the transpiration rate

(Matoh et al. 1986) and to the partial blockage of the transpirational bypass flow in rice in saline conditions (Yeo et al. 1999; Gong et al. 2006). Liang (1999) first demonstrated that added Si significantly decreased Na^+ concentration and increased K^+ uptake in shoots and roots of both salt-sensitive and salt-tolerant barley plants grown hydroponically with 120 mM NaCl, resulting in a higher K^+/Na^+ ratio in Si-fed plants subjected to salt/salinity stress. The mechanism of increased uptake and transport of K^+ and decreased uptake and transport of Na^+ from roots to shoots in barley was thought to be attributed to Si-induced stimulation of the root plasma membrane H^+-ATPase under salt stress (Liang 1999; Liang et al. 2006). This was based on evidence that selective uptake and transport of salt ions is largely dependent on the activity of plasma membrane H^+-ATPase (proton pump), which is the driving force for ions to be translocated across the membranes (Marschner 1995), and the inhibition of Na^+ influx and active Na^+ efflux have been proposed as mechanisms of salt tolerance in plants (Greenway and Munns 1980; Marschner 1995; Yue et al. 2012). Liang (1999) further investigated into the effect of exogenous Si on H^+-ATPase activity in roots of barley exposed to NaCl stress and found that addition of 1.0 mM Si to the salt treatment increased plasma membrane H^+-ATPase activity in both salt-tolerant and salt-sensitive cultivars as compared with the plants treated with 120 mM NaCl only, which contributed partially to the reduced Na^+ concentration in the shoots. On the other hand, it has been also well documented that transportation of Na^+ from the cytoplasm into the vacuole via the tonoplast Na^+/H^+ antiporter is an important mechanism of salt tolerance in plants (Chen et al. 1999; Blumwald 2000). The H^+-ATPase on the tonoplast, coupled with H^+-pyrophosphatase (H^+-PPase), is the major source of H^+-electrochemical potential gradient, which generates the driving force for removing Na^+ from the cytosols or compartmentalizing Na^+ into the vacuoles (Staal et al. 1991; Marschner 1995; Blumwald 2000; Yue et al. 2012). Working again on Si-mediated salt tolerance in barley, Liang et al. (2005b) made an in-depth study into the effects of Si on H^+-ATPase and H^+-PPase activity of tonoplast vesicles isolated from roots of salt-stressed barley and found that 1.0 mM Si to salt treatments significantly reversed both H^+-ATPase and H^+-PPase activities of two contrasting barley plants differing in salt tolerance. However, such effects of Si on plasma membrane or tonoplast H^+-ATPase and H^+-PPase activities under salt stress may be indirect or secondary, as confirmed by an in vitro experiment (Liang et al. 2006) showing that Si added or not to the reaction medium containing plasma membrane vesicles isolated from leaves of plants exposed to NaCl or not did not affect the activity of in vitro H^+-ATPase. On the other hand, it remains unknown whether such stimulating effects of Si on plasma membrane or tonoplast H^+-ATPase and H^+-PPase activities under salt stress can facilitate compartmentalization of salt ions such as Na^+ and/or Cl^- into vacuoles, thus needing clarifying in the future research. Another important issue to address in the future research is whether the genes encoding for H^+-ATPase and H^+-PPase located on plasma membranes and/or tonoplasts are differentially expressed during such responses of H^+-ATPase and H^+-PPase to Si and salt interactions.

Such findings of Si-mediated increased uptake and transport of K^+ and decreased uptake and transport of Na^+ (Liang et al. 1996, 1999; Liang 1999) and/or Cl^- (Shi et al. 2013) from roots to shoots under salt stress can be further explained and supported by the results obtained from an energy dispersive X-ray analysis (Liang and Ding 2002; also see Figs. 6.2, 6.3, 6.4 and 6.5). It can be clearly seen that higher Cl^- and Na^+ X-ray peaks were recorded in the root epidermal, cortical and stelar cells of roots for the NaCl treatment with the majorities of Na^+ and Cl^- ions accumulated in epidermal and cortical cells, while relatively low K^+ peaks were observed. By contrast, considerably higher K^+ peaks were detected in the epidermal, cortical and stelar cells of the roots for the NaCl plus Si treatment, but lower Cl^- and Na^+ peaks were also observed for this treatment with both Cl^- and Na^+ ions evenly distributed in the epidermal, cortical and stelar cells. These findings provided evidence

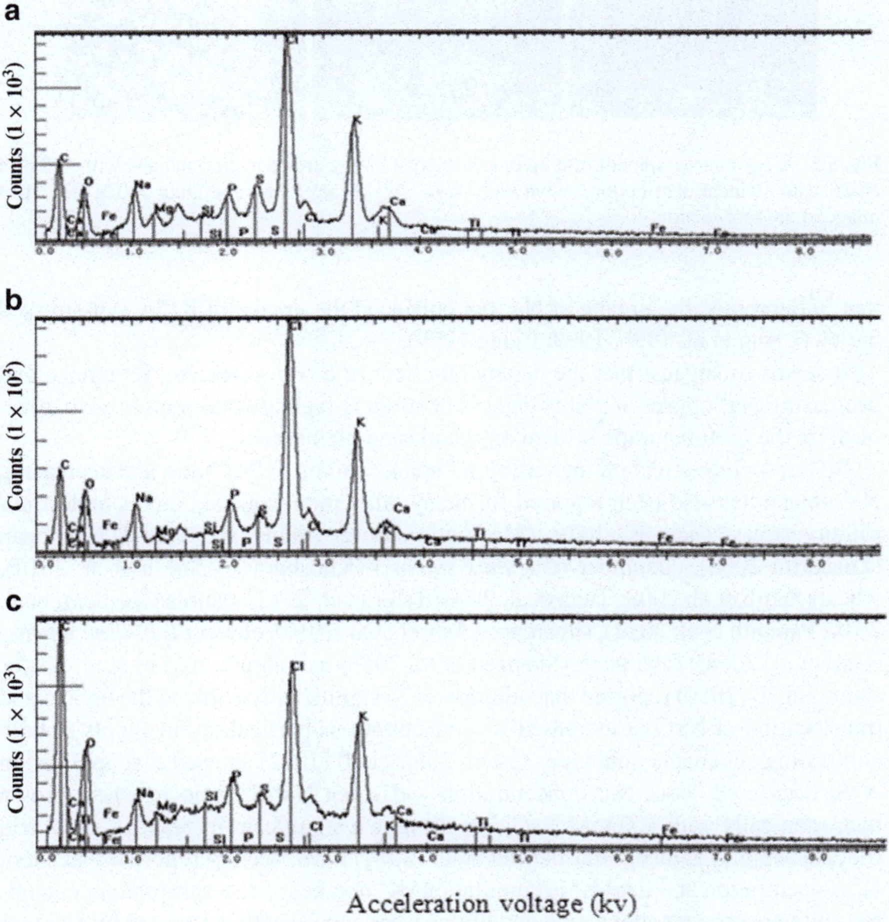

Fig. 6.2 EDX microanalysis of root tips of barley (cv. Jian No. 4) grown hydroponically with 120 mM NaCl (From Liang and Ding 2002). (**a**) epidermis; (**b**) cortex; (**c**) stele

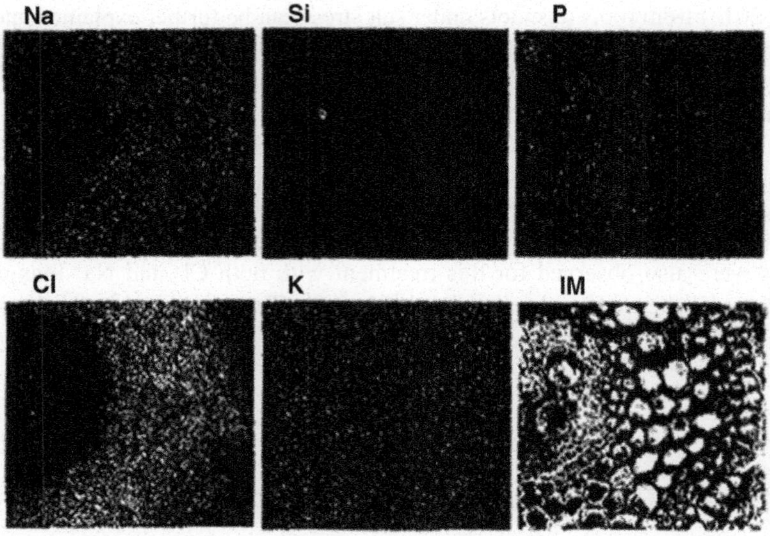

Fig. 6.3 X-ray micrography of root tip of salt tolerant barley grown hydroponically with 120 mM NaCl, with element distribution shown with *white dots* (From Liang and Ding 2002). IM: SEM image of the root section

that Si depressed the uptake of Na^+ but enhanced the uptake of K^+ by salt-stressed barley (Liang et al. 1996, 1999; Liang 1999).

It seems to suggest that the beneficial effect of Si on increasing K^+ uptake and decreasing Na^+ uptake in plants under salt stress is regulated through in vivo effect of Si on the proton pumps located on the plasma membranes.

So far, such Si effect on increasing K^+ uptake and/or K^+/Na^+ ratio and decreasing Na^+ uptake has also been reported for many other monocots and dicots under salt/salinity stress, such as alfalfa (*Medicago sativa*) (Wang and Han 2007), bean (Zuccarini 2008), cucumber (*Cucumis sativus*) (Khoshgoftarmanesh et al. 2014), wheat (Saqib et al. 2008; Tuna et al. 2008; Tahir et al. 2012), canola (Hashemi et al. 2010; Farshidi et al. 2012), sugarcane (Ashraf et al. 2010), clover (*Trifolium repens*) (Guo et al. 2013), faba bean (Shahzad et al. 2013) and sorghum (Yin et al. 2013). Ashraf et al. (2010) reported that addition of Si significantly reduced the uptake and translocation of Na^+ but increased K^+ concentrations particularly in shoots of both contrasting sugarcane genotypes tested. Tahir et al. (2012) showed that application of Si decreased shoot Na^+ concentration and shoot Na^+/K^+ ratio in wheat grown hydroponically with 100 mM NaCl, resulting in a significant increase in shoot and root dry weight. Khoshgoftarmanesh et al. (2014) have recently reported that shoot K^+ concentration and total K^+ accumulation (K^+ uptake) of two salt-tolerant cucumber cultivars grown under saline conditions were significantly increased by addition of a tiny amount of Si (1.0 μM) Si as sodium silicate), while shoot Na^+ concentration and total Na^+ accumulation (Na^+ uptake) were significantly and greatly decreased, resulting in a significantly greater shoot K^+/Na^+ ratio in the presence of

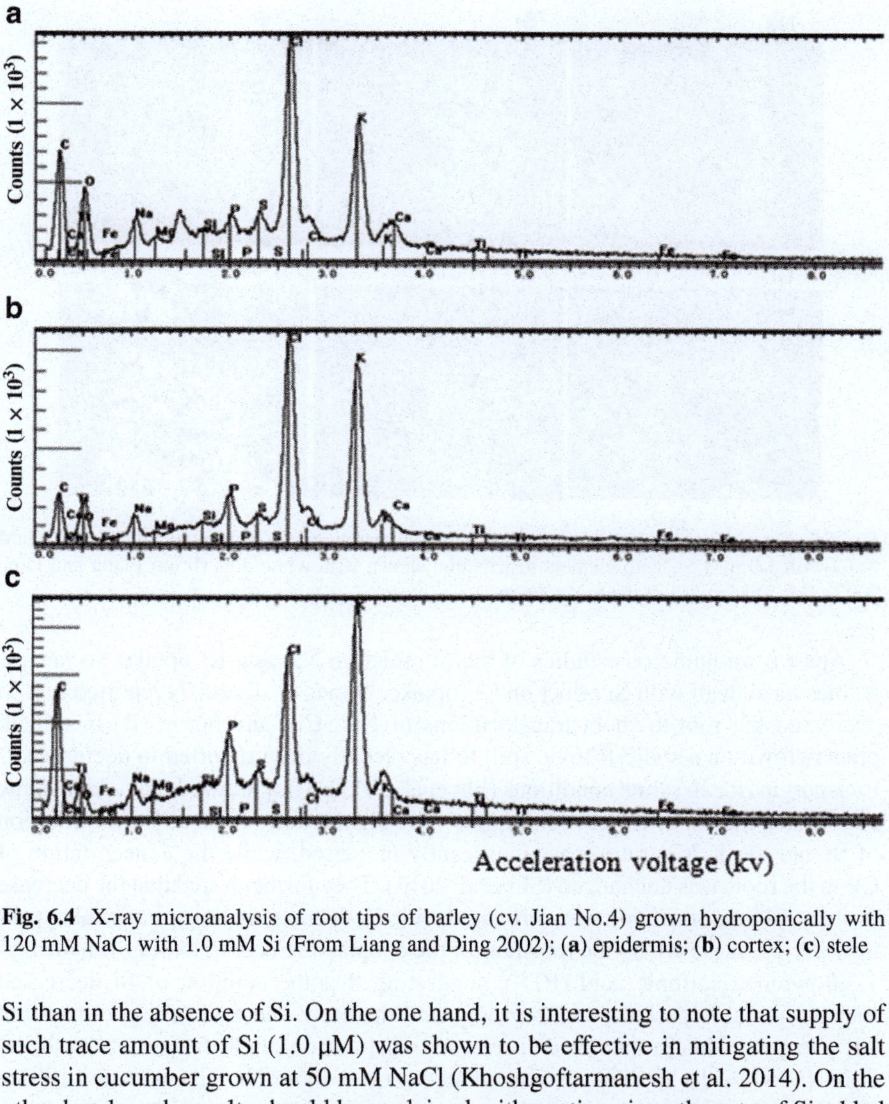

Fig. 6.4 X-ray microanalysis of root tips of barley (cv. Jian No.4) grown hydroponically with 120 mM NaCl with 1.0 mM Si (From Liang and Ding 2002); (**a**) epidermis; (**b**) cortex; (**c**) stele

Si than in the absence of Si. On the one hand, it is interesting to note that supply of such trace amount of Si (1.0 μM) was shown to be effective in mitigating the salt stress in cucumber grown at 50 mM NaCl (Khoshgoftarmanesh et al. 2014). On the other hand, such results should be explained with caution since the rate of Si added that was normally reported to be effective in the majorities of experiments is at least 100-fold greater reported than used in this experiment (Liang et al. 1996; Liang 1999; Liang et al. 1999; Zhu et al. 2004; Ashraf et al. 2010; Farshidi et al. 2012; Tahir et al. 2012); whether Si is the sole contributing factor of mitigation against salt stress in cucumber still needs confirming because no data were provided on Si uptake and accumulation in plants (Khoshgoftarmanesh et al. 2014). Therefore, from the discussion above, it can be concluded that the suppression of Na^+ uptake and stimulation of K^+ uptake in salt-stressed plants contribute significantly to the Si enhancement of salt tolerance.

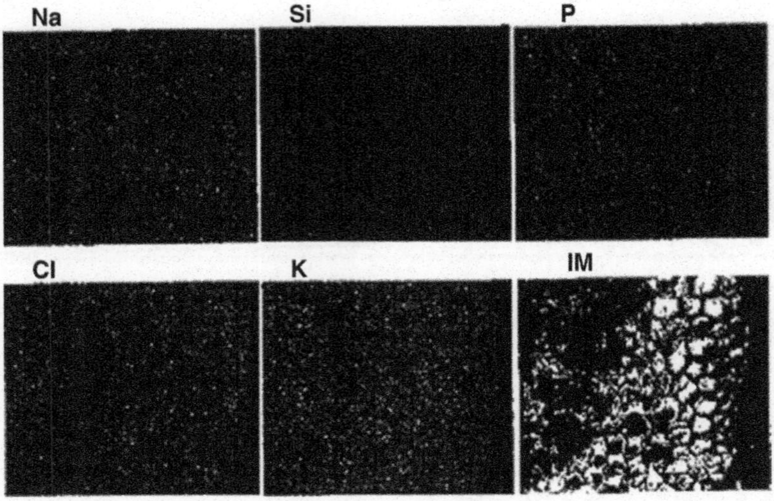

Fig. 6.5 X-ray micrography of root tip of salt tolerant barley grown hydroponically with 120 mM NaCl with 1.0 mM Si, with element distribution shown with white dots (From Liang and Ding 2002). IM: SEM image of the root section

Apart from numerous studies of the Si effect on Na^+ and K^+ uptake, so far few studies have dealt with Si effect on Cl^- uptake. Gunes et al. (2007) reported that Si decreased the root-to-shoot translocations of Na^+, Cl^-, and boron (B) in tomato plants grown on a sodic-B toxic soil. Si has recently been reported to decrease Cl^- transport in rice in saline conditions (Shi et al. 2013). They showed that under saline conditions, the shoot Cl^- concentrations were significantly reduced by the addition of Si and the K^+/Cl^- ratio was significantly increased, while the concentration of Cl^- in the roots was unchanged (Shi et al. 2013). They further found that the decrease in shoot Cl^- concentration was correlated with the decrease in transpirational bypass flow in rice, as shown by the transport of the apoplastic tracer trisodium-8-hydroxy-1,3,6-pyrenetrisulfonic acid (PTS), suggesting that the addition of Si decreased transpirational bypass flow in the roots, and therefore decreased the root-to-shoot Cl^- transport. This Si-mediated decreased translocation of Cl^- from roots to shoots is in line with the finding on Si-mediated root-to-shoot Na^+ translocation in rice (Yeo et al. 1999; Gong et al. 2006).

However, it should be noted that further research is needed to verify whether Si-mediated uptake of Na^+ and K^+ (and/or Cl^-) through Si-mediated promotion of the activities of proton pumps as observed in barley under salt stress can be applied to explain Si effect on Na^+ and K^+ uptake by other crops.

6.4 Silicon-Mediated Uptake of Nutrients

Apart from Si-mediated increased uptake and transport of K^+ and decreased uptake and transport of Na^+ (Liang et al. 1996, 1999; Liang 1999; Wang and Han 2007; Zuccarini 2008; Saqib et al. 2008; Tuna et al. 2008; Ashraf et al. 2010; Hashemi et al. 2010; Farshidi et al. 2012; Tahir et al. 2012; Guo et al. 2013; Shahzad et al. 2013; Yin et al. 2013; Khoshgoftarmanesh et al. 2014), and/or Cl^- (Shi et al. 2013) from roots to shoots under salt stress as discussed above, Si seems to affect acquisition of other essential nutrients such as phosphorus (P), nitrogen (N) and calcium (Ca) and other micronutrients as well (Liang et al. 1999; Wang and Han 2007; Farshidi et al. 2012), thereby improving the growth of plants and the overall tolerance against salt stress. Liang et al. (1999) reported that P concentration and total P content in solution-cultured and potted barley plants were increased by addition of Si under saline conditions. The possible causes for this may be associated with both Si-stimulated root activity indicated by root dehydrogenase activity (Liang et al. 1999) and Si-improved P bioavailability in soils due to the chemical competition between $H_2PO_4^-$ and silicate ($H_3SiO_4^-$) anions for the sorption sites (Obihara and Russell 1972; Smyth and Sanchez 1980; Liang et al. 1999). An increased nitrogen uptake in salt-stressed barley plants was also recorded possibly due to the Si-stimulated root activity and plant vigor (Liang et al. 1999). The concentrations of Fe^{3+}, Mg^{2+} and Zn^{2+} in the roots, shoots and leaves of salt-sensitive alfalfa (cv. Defor) and in the roots and shoots of (salt-tolerant) alfalfa (cv. Zhongmu No. 1) were not significantly affected by application of Si under NaCl stress, but the Ca^{2+} concentration in the roots of Zhongmu No.1 and the Mn^{2+} concentration in the shoots and leaves of both cultivars were increased, while the Ca^{2+} and Cu^{2+} concentrations of the shoots and leaves of both cultivars were decreased by Si under salt stress (Wang and Han 2007). It was also reported that the iron concentration in roots and shoots of Si-fed salt-grown canola plants was about 64 % and 74 %, respectively, greater than that in non-Si-fed salt-grown plants (Farshidi et al. 2012). Si nutrition significantly decreased the concentration of B in shoots under normal conditions as compared to the Si-deprived control plants, but salinity decreased the concentration of B irrespective of Si application (Farshidi et al. 2012). Khoshgoftarmanesh et al. (2014) showed that in the two salt-tolerant cucumber cultivars, salt treatment at 50 mM NaCl reduced both shoot Ca concentration and total Ca uptake, while addition of small amount of Si (1.0 μM) significantly increased Ca concentration and total Ca uptake in shoots of cucumber plants exposed to 50 mM NaCl.

Thus it seems to suggest that Si can mitigate salt stress toxicity through affecting nutrient acquisition by plants, thereby rebalancing the mineral ions and improving plant growth and salt tolerance.

6.5 Silicon-Mediated Antioxidative Responses

Oxidative damage through membrane lipid peroxidation is thought to be one of the most important mechanisms of salt toxicity in higher plants (Hernandez et al. 1993; Fadzilla et al. 1997). Increasing evidence shows that salt stress-induced plasma membrane injury is associated closely with an increased production of highly toxic oxygen free radicals such as superoxide anion, hydroxyl radical, hydrogen peroxide and singlet oxygen (Singha and Choudhuri 1990; Hernandez et al. 1993; Liang et al.

2003). The activities of some key antioxidant defence enzymes, which can scavenge toxic oxygen free radicals, such as superoxide dismutase (SOD), catalase (CAT), and peroxidase (POD), are decreased in plants exposed to salt stress (Hernandez et al. 1993; Singha and Choudhuri 1990), and as a result, the end product of membrane lipid peroxidation, malondialdehyde (MDA), accumulates rapidly (Fadzilla et al. 1997; Hernandez et al. 1993; Lutts et al. 1996), which indicates an increase in membrane permeability and instability of plasma membranes. Liang (1999) first reported that Si could increase SOD activity and decrease MDA concentration in salt-stressed barley, leading to less oxidative damage caused by membrane lipid peroxidation. In a later study, Liang et al. (2003) demonstrated the beneficial effects of Si on both the major antioxidant enzyme activities and nonenzymatic antioxidants and lipid peroxidation in roots of salt-stressed barley using two contrasting cultivars differing in salt tolerance. They showed that the activities of the antioxidant defence enzymes in roots of salt-stressed barley were significantly higher but the levels of MDA and electrolytic leakage percentage were significantly lower in the presence of Si than in the absence of Si. The reduced glutathione (GSH) concentration in plants treated with 120 mM NaCl was significantly depressed but was significantly increased by added Si (Liang et al. 2003). These results suggest that Si-enhanced salt tolerance in plants can be attributed partly to Si-mediated enhancement of antioxidant defence capacity and suppression of membrane lipid peroxidation, which is closely related to cell membrane permeability, integrity, stability and function in plants (see 6.5). Similar results were obtained in barley exposed to a sodic-B toxic soil (Gunes et al. 2007), and in salt-stressed maize (Moussa 2006), and were confirmed in some other dicotyledonous plants under salt stress including cucumber (Zhu et al. 2004; Khoshgoftarmanesh et al. 2014), tomato (Al-Aghabary et al. 2004), grapevine rootstocks (Soylemezoglu et al. 2009) and canola (Hashemi et al. 2010; Farshidi et al. 2012). These results clearly demonstrate that Si is involved in mediating enzymatic and nonenzymatic antioxidant defence systems not only in monocotyledonous plant species but also in dicotyledonous plant species under salt stress. However, how Si mediates this secondary metabolism in plants under salt stress remains to be further investigated at a molecular level.

6.6 Silicon-Mediated Plasma and Tonoplast Membrane Integrity, Stability and Functions

It is generally recognized that an integrated and stable membrane is crucially important for the cellular membrane to function normally, especially under stressful environments (Marschner 1995). The membrane fluidity, stability and functions can be affected by fatty acid composition of membrane lipids, which, as the structural constituents of membranes, play an important role in the tolerance to physiological stress such as water-deficit stress and salt stress (Parida and Das 2005; Liang et al. 2005b). Under salt stress, membrane permeability and membrane lipid peroxidation were increased, as indicated by higher electrolytic leakage percentage and higher MDA concentration, but were significantly counteracted by the addition of Si (Liang et al. 1996, 2003; Liang 1999; Zhu et al. 2004; Al-Aghabary et al. 2004; Khoshgoftarmanesh et al. 2014). Khoshgoftarmanesh et al. (2014) showed that the concentration of MDA was positively correlated with total Na uptake by cucumber but was negatively correlated with total Ca and K uptake and that Si increased K and Ca uptake but decreased Na uptake into shoots under salt stress. These results suggest that Si plays an important role in protecting cell membranes from salt-induced lipid peroxidation and oxidative damage although how Si affects this secondary metabolism is still ambiguous and needs further investigation.

Unsaturated fatty acids of lipids are susceptible to oxidative attack by lipid peroxidation under salt stress, thus leading to enhanced membrane permeability and instability (Parida and Das 2005). It was reported that the ratio of unsaturated to saturated fatty acids (U/S) of the tonoplast vesicles isolated from the roots of barley plants increased under salt stress (120 mM NaCl) for both salt-sensitive (cv. Kepin No. 7) and salt-tolerant (cv. Jian No. 4) cultivars used; however, addition of 1.0 mM Si to salt treatment increased the ratio of U/S in the salt-tolerant cultivar but did not in the salt-sensitive cultivar (Liang et al. 2005b). The salt treatment decreased tonoplast fluidity of roots of barley significantly compared with the control treatment. However, root tonoplast fluidity was significantly lowered by the presence of Si in the salt treatment. The research by Liang et al. (2006) also showed that the ratio of phospholipids to protein in plasma membranes was significantly increased in leaves of the two barley cultivars under salt stress, a phenomenon that was reversed by Si. Furthermore, the membrane fluidity of plasma membrane vesicles isolated from the leaves of plants was significantly lower in the salt treatment than in the control, but was brought back to normal by Si. Thus, a close relationship between plasma membrane H^+-ATPase activity and membrane fluidity was found; plasma membrane fluidity together with H^+-ATPase activity decreased when plants were treated with NaCl but was reversed by addition of Si (Liang et al. 2006).

These results clearly suggest that Si may help maintain the membrane integrity, stability and functions of salt-stressed barley and decrease membrane permeability via a decrease in membrane lipid oxidative damage, thus mitigating salt toxicity in barley plants. However, more in-depth work is needed to verify whether this finding can be applied to Si effects on other plants exposed to salt stress.

Si added diminished the NaCl-induced increase in ratio of phospholipids to protein and maintained this ratio at a normal level similar to control, thus maintaining the optimal membrane fluidity which was reduced by salt treatment. The evidence therefore suggests that Si maintain the integrity, stability and functioning of plasma membranes in salt-stressed barley.

6.7 Silicon Regulates Biosynthesis of Proline, Lignin and Plant Hormones

In response to stressful environments, plants usually synthesize and accumulate many low-molecular-weight organic osmolytes or compatible solutes such as proline (Gzik 1997; Balibrea et al. 1997; Mansour 1998; Ashraf and Foolad 2007), glycine betaine (Rhodes and Hanson 1993; Mansour 1998; Ashraf and Foolad 2007), carbohydrates (Balibrea et al. 1997) and polyols (Kumar and Bandhu 2005). The roles these low-molecular-weight organic compounds may play under stressful conditions are reportedly contradictory. These solutes do not interfere with normal biochemical or metabolic processes in plant tissues, but rather, they are indicators of plant adaptation to stressful conditions or they may alleviate the inhibitory effects of high ion concentrations on enzyme activity by stabilizing proteins, protein complexes or membranes under environmental stresses (Bohnert and Shen 1999; Ashraf and Foolad 2007). They may also function as oxygen radical scavengers to reduce oxidative damage in plants under salt stress (Seckin et al. 2009) and exogenous application of these compounds can increase plant tolerance to different stresses (Ashraf and Foolad 2007). Si has widely been reported to decrease proline concentrations under saline stress in barley, wheat, grapevine, soybean and sorghum (Gunes et al. 2007; Tuna et al. 2008; Soylemezoglu et al. 2009; Lee et al. 2010; Yin et al. 2013; see also the review by Zhu and Gong, 2014). For instance, Soylemezoglu et al. (2009) showed that proline concentration for rootstock 41B and 1103P was, respectively, 63 % and 45 % lower ($P<0.05$) in the saline plus Si treatment than in the saline treatment only, which is in line with a lower concentration of MDA and H_2O_2 but higher shoot weight in the saline plus Si treatment. According to Lee et al. (2010), the leaf free proline content of soybean was 10-fold higher in plants grown with 80 mM NaCl without Si supply than in control plants, while the free proline concentration in salt-stressed soybean was approximately 30 % lower in the presence of Si than in the absence of Si and the plant dry weight of Si-fed plants was about 40 % higher than that of Si-deprived plants. These results reflect that a lower concentration of free proline accumulated in Si-fed plants is an indication of less severe salt toxicity due to presence of Si. Yin et al. (2013) also found that short-term application of 0.83 mM Si could significantly increase the levels of sucrose and fructose in sorghum under salt stress, suggesting that Si can alleviate salt-induced osmotic stress. However, how Si affects the metabolism of these compatible solutes in relation to water-deficit-stress adaptation and/or tolerance remains to be explored as Si has been found to mitigate salt, drought and freezing stress in plants.

As early as 1978, Si was reported to be bound with the cell wall as an ester-like derivative of silicic acid (R^1-O-Si-O-R^2) acting as a bridge in the structural organization of polyuronides (Jones 1978). Si seems to influence phenolic compounds (Parry and Kelso 1975; Maksimovic et al. 2007). Si interacts with the primary cell wall constituents (Inanaga et al. 1995; He et al. 2013, 2015) such as pectins, hemicellulose and polyphenols, and these cross-links obviously increase cell wall elasticity during extension growth (Emadian and Newton 1989). Si would increase the extensibility of cell walls in the apical elongation (AE) and apical maturation (AM) zone, but significantly decrease the cell wall extensibility of stele tissues in the basal maturation (BM) zone. Si treatment significantly decreased viscoelastic parameters of the AE and AM zone, but increased those of BM zone. These Si-induced changes in cell wall extensibility and viscoelastic parameters provide resistance to biotic and abiotic stress (Hattori et al. 2003). Recently, Fleck et al. (2011) have reported that silicic acid supply enhanced the lignification of sclerenchyma in rice. Si may therefore affect the stability of higher plants not only as an inert deposition in lignified cell walls but also by modulating lignin biosynthesis (Marschner 1995). More recently, He et al. (2013, 2015) have reported that Si was present in a hemicellulose-bound form in the cell walls of suspension-cultured rice cells, providing the potential to improve the mechanical properties of the cell walls against stresses (see Chaps. 1 and 5). Several researchers have reported the effects of Si on the composition of cell walls (e.g. lignin) in plants under salt stress (Hashemi et al. 2010). Hashemi et al. (2010) found that lignin concentration in roots of canola exposed to 150 mM NaCl was significantly increased compared to the control treatment but was significantly decreased by the addition of Si. They believe that cell wall loosening and plant growth would be impaired as a result of lignification as shown in salt-stressed canola. The decreased lignification in the canola plants, particularly in the shoots would result from the partial substitution of lignin by Si or the formation of Si–polyphenol complexes (Maksimovic et al. 2007) in the cell walls, which may facilitate cell wall loosening and cell extension and promote the growth of plants under salt stress conditions (see Zhu and Gong 2014). However, the lower concentration of lignin observed in the roots of Si-treated canola under salt stress was due to the "dilution effect" as it can be calculated that no significant difference in total lignin accumulated in roots existed between the Si-treated and Si-untreated canola (Hashemi et al. 2010). Thus, it remains to be further clarified whether and how Si can affect the modifications of cell walls (e.g. lignification) under salt stress.

Silicon has also been reported to affect plant growth hormones under salt stress (Lee et al. 2010; Kim et al. 2014). Lee et al. (2010) report that abscisic acid content increased in soybean plants grown hydroponically with salt stress but decreased by the addition of Si. By contrast, the reverse is true for another plant hormone, gibberellin (Lee et al. 2010). The gibberellin level decreased in salt-stressed soybean but increased significantly in the presence of Si. Kim et al. (2014) showed that stress- and defence-related phytohormones such as jasmonic acid (JA) were significantly down-regulated by application of Si under salinity stress. Conversely, ABA level was significantly higher after 6 and 12 h in Si-treated plants under salinity stress. After 6 and 12 h, the ABA biosynthesis-related genes (*OsZEP* encoding for zeaxanthin epoxidase, and *OsNCED1* and *4* encoding for 9-cis-epoxycarotenoid dioxygenase 1 and 4) were up-regulated by Si treatment under salinity stress, whereas at 24 h after Si treatment the expression of these genes was significantly downregulated. The expression of *OsNCED3* significantly increased at 6 and 24 h after Si treatment. These results indicate that increasing the Si concentrations for longer periods of time can regulate the salinity-induced stress by modulating phytohormonal responses. This infers that Si mediates salt tolerance in plants through regulating the expression of genes related to the biosynthesis of plant hormones in plants under salt stress.

6.8 Conclusions

A schematic diagram of the proposed mode of action of Si-mediated alleviation of salt stress in plants, based on the currently available literature, is summarized in Fig. 6.6. From this, it is clear that more direct and convincing evidence at the transcriptomic and proteomic level is needed to demonstrate whether soluble Si present in symplast, which accounts only for a small proportion of Si taken up by plants, can be involved in so many metabolic functions. Another important issue to address in the near future is whether the Si-stimulated H^+-ATPase and H^+-PPase activities really affect the compartmentalization of toxic salt ions into vacuoles to enhance plant tolerance to salt stress. From the discussion in this chapter, it can be concluded that Si can effectively mitigate salt (salinity) stress in plants, but the mechanisms behind this phenomenon are still ambiguous and need further investigation.

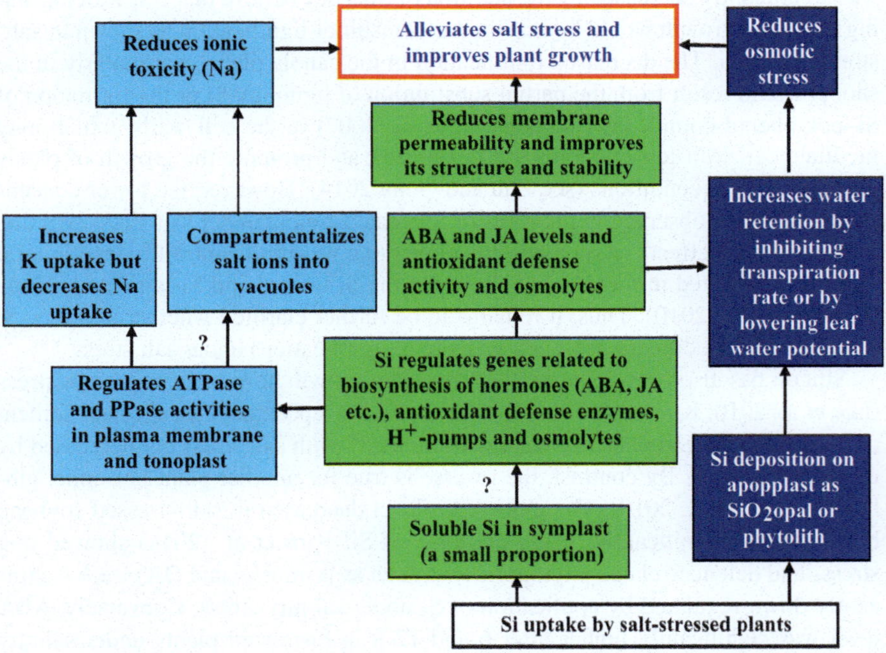

Fig. 6.6 Schematic diagram of the proposed mode of action of Si-mediated alleviation of salt stress in plants

References

Ahmad R, Zaheer SH, Ismail S. Role of silicon in salt tolerance of wheat (*Triticum aestivum* L.). Plant Sci. 1992;85:43–50.

Al-Aghabary K, Zhu Z, Shi Q. Influence of silicon supply on chlorophyll content, chlorophyll fluorescence and anti-oxidative enzyme activities in tomato plants under salt stress. J Plant Nutr. 2004;27:2101–15.

Ashraf M, Foolad MR. Roles of glycine betaine and proline in improving plant abiotic stress resistance. Environ Exp Bot. 2007;59:206–16.

Ashraf M, Rahmatullah AM, Ahmed R, Mujeeb F, Sarwar A, Ali L. Alleviation of detrimental effects of NaCl by silicon nutrition in salt-sensitive and salt-tolerant genotypes of sugarcane (*Saccharum officinarum* L.). Plant Soil. 2010;326:381–91.

Bae EJ, Lee KS, Huh MR, Lim CS. Silicon significantly alleviates the growth inhibitory effects of NaCl in salt-sensitive 'Perfection' and 'Midnight' Kentucky bluegrass (*Poa pratensis* L). Hortic Environ Biotech. 2012;53:477–83.

Balibrea ME, Rus-alvarez AM, Bolarfn MC, Pérez-alfocea F. Fast changes in soluble carbohydrates and proline contents in tomato seedlings in response to ionic and non ionic iso-osmotic stresses. J Plant Physiol. 1997;151:221–6.

Bélanger RR, Bowen PA, Ehret DL, Menzies JG. Soluble silicon: its role in crop and disease management of greenhouse crops. Plant Dis. 1995;79:329–36.

Blumwald E. Sodium transport and salt tolerance in plants. Curr Opin Cell Biol. 2000;12:431–4.

Bohnert HJ, Shen B. Transformation and compatible solutes. Sci Hortic. 1999;78:237–60.

Chen Q, Zhang W, Liu Y. Effect of NaCl, glutathione and ascorbic acid on function of tonoplast vesicles isolated from barley leaves. J Plant Physiol. 1999;155:685–90.

Emadian SF, Newton RJ. Growth enhancement of loblolly pine (*Pinus taeda* L.) seedlings by silicon. J Plant Physiol. 1989;134:98–103.

Fadzilla NM, Finch RP, Burdon RH. Salinity, oxidative stress and antioxidant responses in shoot cultures of rice. J Exp Bot. 1997;48:325–31.

Farshidi M, Abdolzadeh A, Sadeghipour HR. Silicon nutrition alleviates physiological disorders imposed by salinity in hydroponically grown canola (*Brassica napus* L.) plants. Acta Physiol Plant. 2012;34:1779–88.

Fleck AT, Nye T, Repenning C, Stahl F, Zahn M, Schenk MK. Silicon enhances suberization and lignification in roots of rice (*Oryza sativa*). J Exp Bot. 2011;62:2001–11.

Gong HJ, Randall DP, Flowers TJ. Silicon deposition in the root reduces sodium uptake in rice (*Oryza sativa* L.) seedlings by reducing bypass flow. Plant Cell Environ. 2006;29:1970–9.

Greenway H, Munns R. Mechanisms of salt tolerance in nonhalophytes. Annu Rev Plant Physiol. 1980;31:149–90.

Gunes A, Inal A, Bagci EG, Coban S. Silicon-mediated changes on some physiological and enzymatic parameters symptomatic of oxidative stress in barley grown in sodic-B toxic soil. J Plant Physiol. 2007;164:807–11.

Guo Q, Meng L, Mao PC, Tian XX. Role of silicon in alleviating salt-induced toxicity in white clover. Bull Environ Contam Toxicol. 2013;91:213–6.

Gzik A. Accumulation of proline and pattern of α-amino acids in sugar beet plants in response to osmotic, water and salt stress. Environ Exp Bot. 1997;36:29–38.

Haghighi M, Pessarakli M. Influence of silicon and nano-silicon on salinity tolerance of cherrytomatoes (*Solanum lycopersicum* L.) at early growth stage. Sci Hortic. 2013;161:111–7.

Hashemi A, Abdolzadeh A, Sadeghipour HR. Beneficial effects of silicon nutrition in alleviating salinity stress in hydroponically grown canola, *Brassica napus* L., plants. Soil Sci Plant Nutr. 2010;56:244–53.

Hattori T, Inanaga S, Tanimoto E, Lux A, Luxova M, Sugimoto Y. Silicon-induced changes in viscoelastic properties of sorghum root cell walls. Plant Cell Physiol. 2003;44:743–9.

He CW, Wang LJ, Liu J, Liu X, Li XL, Ma J, Lin YJ, Xu FS. Evidence for 'silicon' within the cell walls of suspension-cultured rice cells. New Phytol. 2013;200:700–9.

He CW, Ma J, Wang LJ. A hemicellulose-bound form of silicon with potential to improve the mechanical properties and regeneration of the cell wall of rice. New Phytol. 2015;206:1051–62.

Hernandez JA, Corpass FJ, Gomez M, del Rio LA, Sevilla F. Salt-induced oxidative stress mediated by active oxygen species in pen leaf mitochondria. Physiol Plant. 1993;89:103–10.

Inanaga S, Okasaka A, Tanaka S. Does silicon exist in association with organic compounds in rice plant? Soil Sci Plant Nutr. 1995;41:111–7.

Jones LHP. Mineral components of plant cell walls. Am J Clin Nutr. 1978;31:94–8.

Kafi M, Rahimi Z. Effect of salinity and silicon on root characteristics, growth, water status, proline content and ion accumulation of purslane (*Portulaca oleracea* L.). Soil Sci Plant Nutr. 2011;57:341–7.

Khoshgoftarmanesh AH, Khodarahmi S, Haghighi M. Effect of silicon nutrition on lipid peroxidation and antioxidant response of cucumber plants exposed to salinity stress. Arch Agron Soil Sci. 2014;60:639–53.

Kim YH, Khan AL, Kim DH, Lee SY, Kim KM, Waqas M, Jung HY, Shin JH, Kim JG, Lee IJ. Silicon mitigates heavy metal stress by regulating P-type heavy metal ATPases, *Oryza sativa* low silicon genes, and endogenous phytohormones. BMC Pant Biol. 2014;14:13.

Kumar AP, Bandhu AD. Salt tolerance and salinity effects on plants: a review. Ecotoxicol Environ Saf. 2005;60:324–49.

Lee SK, Sohn EY, Hamayun M, Yoon JY, Lee IJ. Effect of silicon on growth and salinity stress of soybean plant grown under hydroponic system. Agrofor Syst. 2010;80:333–40.

Liang YC. Effects of Si on leaf ultrastructure, chlorophyll content and photosynthetic activity in barley under salt stress. Pedosphere. 1998;8:289–96.

Liang YC. Effects of silicon on enzyme activity, and sodium, potassium and calcium concentration in barley under salt stress. Plant Soil. 1999;209:217–24.

Liang YC, Ding RX. Influence of silicon on microdistribution of mineral ions in roots of salt-stressed barley as associated with salt tolerance in plants. Sci China (Series C). 2002;45:298–308.

Liang YC, Shen QR, Shen ZG, Ma TS. Effects of silicon on salinity tolerance of two barley cultivars. J Plant Nutr. 1996;19:173–83.

Liang YC, Ding RX, Liu Q. Effects of silicon on salt tolerance of barley and its mechanism. Sci Agric Sin. 1999;32(6):75–83 (In Chinese with English abstract).

Liang YC, Chen Q, Liu Q, Zhang WH, Ding RX. Exogenous silicon (Si) increases antioxidant enzyme activity and reduces lipid peroxidation in roots of salt-stressed barley (*Hordeum vulgare* L.). J Plant Physiol. 2003;160:1157–64.

Liang YC, Si J, Römheld V. Silicon uptake and transport is an active process in *Cucumis sativus*. New Phytol. 2005a;167:797–804.

Liang YC, Zhang WH, Chen Q, Ding RX. Effects of silicon on H^+-ATPase and H^+-PPase activity, fatty acid composition and fluidity of tonoplast vesicles from roots of salt-stressed barley (*Hordeum vulgare* L.). Environ Exp Bot. 2005b;53:29–37.

Liang YC, Zhang WH, Chen Q, Liu YL, Ding RX. Effect of exogenous silicon (Si) on H^+-ATPase activity, phospholipids and fluidity of plasma membrane in leaves of salt-stressed barley (*Hordeum vulgare* L.). Environ Exp Bot. 2006;57:212–9.

Lutts S, Kinet JM, Bouharmont J. NaCl-induced senescence in leaves of rice (*Oryza sativa* L.) cultivars differing in salinity resistance. Ann Bot. 1996;78:389–98.

Mansour MMF. Protection of plasma membrane of onion epidermal cells by glycinebetaine and proline against NaCl stress. Plant Physiol Biochem. 1998;36:767–72.

Marschner H. Mineral nutrition of higher plants. 2nd ed. London: Academic Press; 1995.

Mateos-Naranjo E, Andrades-Moreno L, Davy AJ. Silicon alleviates deleterious effects of high salinity on the halophytic grass *Spartina densiflora*. Plant Physiol Biochem. 2013;63:115–21.

Maksimovic JD, Bogdanovic J, Maksimovic V, Nikolic M. Silicon modulates the metabolism and utilization of phenolic compounds in cucumber (*Cucumis sativus* L.) grown at excess manganese. J Plant Nutr Soil Sci. 2007;170:739–44.

Matoh T, Kairusmee P, Takahashi E. Salt-induced damage to rice plants and alleviation effect of silicate. Soil Sci Plant Nutr. 1986;32:295–304.

Moussa HR. Influence of exogenous application of silicon on physiological response of salt-stressed maize (Zea mays L.). Int J Agric Biol. 2006;8:293–7.

Munns R, Tester M. Mechanisms of salinity tolerance. Annu Rev Plant Biol. 2008;59:651–81.

Murillo-Amador B, Yamada S, Yamaguchi T, Rueda-Puente E, Avila-Serrano N, Garcia-Hernandez JL, Lopez-Aguilar R, Troyo-Dieguez E, Nieto-Garibay A. Influence of calcium silicate on growth, physiological parameters and mineral nutrition in two legume species under salt stress. J Agron Crop Sci. 2007;193:413–21.

Nikolic M, Nikolic N, Liang Y, Kirkby EA, Römheld V. Germanium-68 as an adequate tracer for silicon transport in plants. Characterization of silicon uptake in different crop species. Plant Physiol. 2007;143:495–503.

Obihara CH, Russell EW. Specific adsorption of silicate and phosphorus by soils. J Soil Sci. 1972;23:105–17.

Parida AK, Das AB. Salt tolerance and salinity effects on plants: a review. Ecotox Environ Saf. 2005;60:324–49.

Parry DW, Kelso M. The distribution of silicon deposits in the root of *Molinia caerulea* (L.) Moench and *Sorghum bicolour* (L.) Moench. Ann Bot. 1975;39:995–1001.

Rasool S, Hameed A, Azooz MM, Muneeb-u-Rehman STO, Parvaiz Ahmad P. Salt stress: causes, types and responses of plants. In: Ahmad P, Azooz MM, Prasad MNV, editors. Ecophysiology and responses of plants under salt stress. New York: Springer; 2013. p. 1–24.

Raven JA. Cycling silicon–the role of accumulation in plants. New Phytol. 2003;158:419–21.

Rhodes D, Hanson AD. Quaternary ammonium and tertiary sulfonium compounds in higher plants. Annu Rev Plant Physiol Plant Mol Biol. 1993;44:357–84.

Romero-Aranda MR, Jurado O, Cuartero J. Silicon alleviates the deleterious salt effect on tomato plant growth by improving plant water status. J Plant Physiol. 2006;163:847–55.

Saqib M, Zörb C, Schubert S. Silicon-mediated improvement in the salt resistance of wheat (*Triticum aestivum*) results from increased sodium exclusion and resistance to oxidative stress. Funct Plant Biol. 2008;35:633–9.

Savant NK, Korndorfer GH, Datnoff LE, Snyder GH. Silicon nutrition and sugarcane production. A review. J Plant Nutr. 1999;22:1853–903.

Savvas D, Giotis D, Chatzieustratiou E, Bakea M, Patakioutas G. Silicon supply in soilless cultivations of zucchini alleviates stress induced by salinity and powdery mildew infections. Environ Exp Bot. 2009;65:11–7.

Seckin B, Sekmen AH, Türkan İ. An enhancing effect of exogenous mannitol on the antioxidant enzyme activities in roots of wheat under salt stress. J Plant Growth Regul. 2009;28:12–20.

Shahzad M, Zorb C, Geilfus CM, Muhling KH. Apoplastic Na^+ in *Vicia faba* leaves rises after short-term salt stress and is remedied by silicon. J Agron Crop Sci. 2013;199:161–70.

Shi Y, Wang YC, Flowers TJ, Gong HJ. Silicon decreases chloride transport in rice (*Oryza sativa* L.) in saline conditions. J Plant Physiol. 2013;170:847–53.

Singha S, Choudhuri MA. Effect of salinity (NaCl) stress on H_2O_2 metabolism in *Vigna* and *Oryza* seedlings. Biochem Physiol Pfl. 1990;186:69–74.

Smyth TJ, Sanchez PA. Effects of lime, silicate, and phosphorus applications to an Oxisol on phosphorus sorption and ion retention. Soil Sci Soc Am J. 1980;44:500–5.

Soundararajan P, Sivanesan I, Jo EH, Jeong BR. Silicon promotes shoot proliferation and shoot growth of *Salvia splendens* under salt stress in vitro. Hortic Environ Biotechnol. 2013;54:311–8.

Soylemezoglu G, Demir K, Inal A, Gunes A. Effect of silicon on antioxidant and stomatal response of two grapevine (*Vitis vinifera* L.) rootstocks grown in boron toxic, saline and boron toxic-saline soil. Sci Hortic. 2009;123:240–6.

Staal M, Maathuis FJM, Elzenga TM, Overberk M, Prins HBA. Na^+/H^+ antiport activity in tonoplast vesicles from roots of the salt-tolerant *Plantago maritime* and the salt-sensitive *Plantago media*. Physiol Plant. 1991;82:179–84.

Steduto P, Albrizio R, Giorio P, Sorrentino G. Gas-exchange response and stomatal and non-stomatal limitations to carbon assimilation of sunflower under salinity. Environ Exp Bot. 2000;44:243–55.

Tahir MA, Aziz T, Farooq M, Sarwar G. Silicon-induced changes in growth, ionic composition, water relations, chlorophyll contents and membrane permeability in two salt-stressed wheat genotypes. Arch Agron Soil Sci. 2012;58:247–56.

Trenholm LE, Datnoff LE, Nagara RT. Influence of silicon on drought and shade tolerance of *St.* Augustinegrass Hortic Technol. 2004;14:487–90.

Tuna AL, Kaya C, Higgs D, Murillo-Amador B, Aydemir S, Girgin AR. Silicon improves salinity tolerance in wheat plants. Environ Exp Bot. 2008;62:10–6.

Wang XS, Han JG. Effects of NaCl and silicon on ion distribution in the roots, shoots and leaves of two alfalfa cultivars with different salt tolerance. Soil Sci Plant Nutr. 2007;53:278–85.

Yeo AR, Flowers SA, Rao G, Welfare K, Senanayake N, Flowers TJ. Silicon reduces sodium uptake in rice (*Oryza sativa* L.) in saline conditions and this is accounted for by a reduction in the transpirational bypass flow. Plant Cell Environ. 1999;22:559–65.

Yin LN, Wang SW, Li JY, Tanaka K, Oka M. Application of silicon improves salt tolerance through ameliorating osmotic and ionic stresses in the seedling of *Sorghum bicolour*. Acta Physiol Plant. 2013;35:3099–107.

Yue Y, Zhang M, Zhang JC, Duan LS, Li ZH. *SOS1* gene overexpression increased salt tolerance in transgenic tobacco by maintaining a higher K^+/Na^+ ratio. J Plant Physiol. 2012;169:255–61.

Zhu JK. Salt and drought stress signal transduction in plants. Annu Rev Plant Biol. 2002;53:247–73.

Zhu YX, Gong HJ. Beneficial effects of silicon on salt and drought tolerance in plants. Agron Sustain Dev. 2014;34:455–72.

Zhu ZJ, Wei GQ, Li J, Qian QQ, Yu JP. Silicon alleviates salt stress and increases antioxidant enzymes activity in leaves of salt-stressed cucumber (*Cucumis sativus* L.). Plant Sci. 2004;167:527–33.

Zuccarini P. Effects of silicon on photosynthesis, water relations and nutrient uptake of *Phaseolus vulgaris* under NaCl stress. Biol Plant. 2008;52:157–60.

Chapter 7
Silicon-Mediated Tolerance to Drought and Low-Temperature Stress

Abstract Drought and low temperature are two of the major adverse climatic factors that restrain plant growth and sustainable agricultural development. Silicon (Si) application can alleviate various abiotic stresses including drought and low temperature, suggesting its potential application in agricultural practices in these adverse soil and climate conditions. The mechanisms for Si-mediated increases of tolerance to drought and low-temperature stresses include physiological, biochemical and physical aspects. These include promoting photosynthetic enzymatic activities, photochemical efficiency and photosynthetic rate; maintaining nutrient balance; improving water retention by decreasing water loss from leaves and increasing water uptake by roots; and scavenging reactive oxygen species by improving the capabilities of antioxidant defence. Although some progress has been made in the understanding of mechanisms for Si-mediated drought and low-temperature stresses at the physiological, biochemical and physical levels, information on the molecular aspect is still lacking. The use of modern technologies, such as transcriptomics and proteomics approaches, may help clarify the transcriptional and post-transcriptional regulatory mechanisms of Si-mediated tolerance in plants, which will provide a theoretical basis for Si application in agricultural production.

Keywords Antioxidant defence activity • Drought stress • Low temperature • Photosynthesis • Silicon • Water retention

7.1 General

Drought is one of the major abiotic factors that limit plant growth and productivity worldwide, especially in arid and semiarid areas (Eneji et al. 2008). Approximately 30 % of the world land surface is arid or semiarid (Huang et al. 2013). According to the prediction of the current climate change models, the frequency and severity of drought will increase in several regions around the world (Shen et al. 2010). Low temperature is another major adverse climatic factor that restrains plant growth, distribution and sustainable agricultural development, causing considerable economic losses worldwide. With the progressing growth of the world population, it is

Y. Liang et al., *Silicon in Agriculture*, DOI 10.1007/978-94-017-9978-2_7

urgent to increase the drought and cold tolerance of crops to maintain food supplies.

Drought and freezing stress can cause deleterious effects on plant growth and physiological processes, such as photosynthetic assimilation, water relations and antioxidant defence capacity (Cattivelli et al. 2008; Yadav 2010; Ashraf and Harris 2013; Hasanuzzaman et al. 2013). Many reports have claimed that silicon (Si) applications can increase the tolerance to drought and freezing stress in plants, and the mechanisms for this protection include physiological, biochemical and physical aspects, as summarized in Fig. 7.1. Although great progress has been made in the understanding of mechanisms for Si-mediated drought and freezing stress at the physiological and biochemical levels, yet information on the molecular aspect is still unclear. Transcriptome and proteome platforms are important technologies to reveal tolerance mechanisms in plants (Ahsan et al. 2009; Wang et al. 2009). The use of these technologies may help shed light on the transcriptional and post-transcriptional regulatory mechanisms of Si-mediated tolerance in plants, which will provide a theoretical basis for dissecting the whole picture of Si-mediated drought and freezing tolerance.

In this chapter, the recent advances on the beneficial effects of Si in plant tolerance to drought and freezing stress and the underlying mechanisms are reviewed.

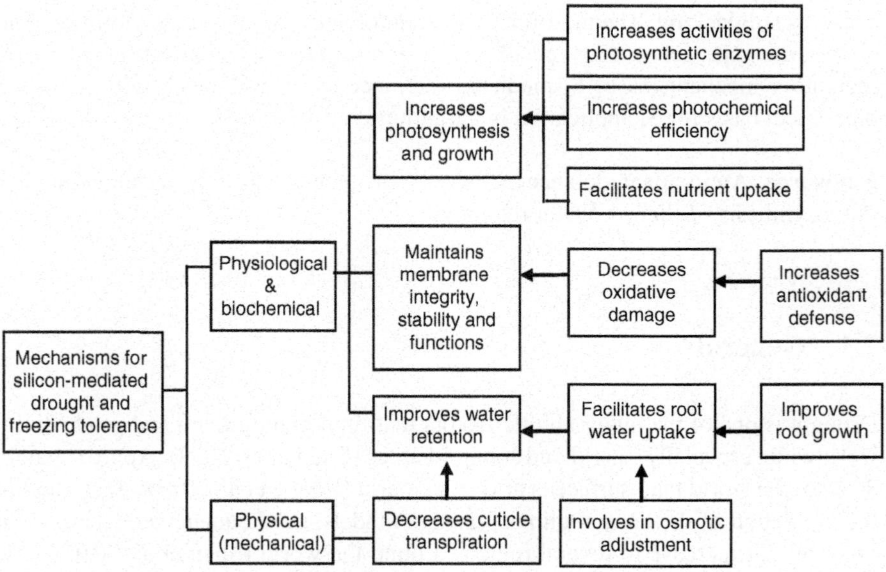

Fig. 7.1 Possible mechanisms for Si-mediated drought and freezing tolerance in plants

7.2 Photosynthesis and Plant Growth

Under environmental stress, photosynthesis is inhibited (Ashraf and Harris 2013), and exogenous Si has been found to revert this inhibition and therefore restore normal plant growth under drought and low-temperature conditions. Gong et al. (2005) observed that, under drought conditions, Si-supplied wheat seedlings had higher net CO_2 assimilation rate as compared to untreated plants. A similar phenomenon was observed in other drought-stressed plants such as sorghum (*Sorghum bicolor*) (Hattori et al. 2005; Liu et al. 2014) and rice (*Oryza sativa*) (Chen et al. 2011). For instance, Hattori et al. (2005) showed that the decrease in net assimilation rate (NAR) in drought-stressed sorghum was ameliorated by application of Si, while Si addition had no effect on NAR in wet conditions. The photosynthetic rate of sorghum in drought conditions was maintained at a significantly higher level in the presence of Si than in the absence of Si. As could be seen clearly, supplemental Si led to higher dry matter production under drought conditions (Hattori et al. 2005; see Fig. 7.2). Under freezing conditions (<−5 °C), Zhu et al. (2006) found that

Fig. 7.2 Effect of Si application on the growth of sorghum under dry conditions (From Hattori et al. 2005). Sorghum plants (cv. Gadambalia) were grown in Si-applied soil (*left*) and non-applied soil (*right*). Plants were 47-day old (22 days after the initiation of dry treatment)

application of Si could increase the net photosynthetic rate of wheat (*Triticum aestivum*), with the improvement being more obvious in the low-temperature-sensitive cultivar.

7.2.1 Stomatal and Non-stomatal Limitations of Photosynthesis

The inhibition of photosynthesis under drought stress has been attributed to both stomatal and non-stomatal limitations (Yordanov et al. 2000). Stomatal closure is the first response of plants subjected to serious water deficit, and it is generally considered to be the main limiting factor for photosynthesis (Reddy et al. 2004; Farooq et al. 2009). The stomatal closure appears to be the first line of defence against desiccation since it is much faster than, e.g. changes in root growth, leaf area, chloroplast ultrastructure and pigment proteins (Yordanov et al. 2000). Although stomatal closure usually occurs under adverse environmental conditions, in some stress conditions, non-stomatal limitation, i.e. decreased capacity of C fixation by chloroplast, may inhibit photosynthesis. Gong et al. (2005) observed that drought stress depressed the wheat photosynthetic rate but that Si application reinstated a normal rate. They found no difference in the intercellular CO_2 concentration of leaves between the well-watered wheat and drought-stressed plants, regardless of Si application, suggesting that the stomatal factor was not the main factor for inhibition of photosynthesis and that the non-stomatal factor contributed to Si-mediated improvement of photosynthesis under drought conditions. Similar results were also obtained in drought-stressed sorghum (Hattori et al. 2005). Zhu et al. (2006) investigated the photosynthetic gas exchange of wheat leaves under freezing conditions and observed that the introduction of Si in the culture solution significantly increased the net photosynthesis of the stressed wheat seedlings. They found that although the leaf stomatal conductance was decreased under freezing stress and was increased by Si introduction, the intercellular CO_2 concentration was not changed (in the tolerant cultivar), or it was even increased (in the sensitive cultivar) under freezing stress without applied Si, while Si addition slightly decreased the concentration of intercellular CO_2 of the stressed plants. These results suggest that the photosynthesis inhibition and Si-mediated improvement under freezing stress could both chiefly be attributed to non-stomatal factor. In rice, Chen et al. (2011) suggested that both stomatal and non-stomatal factors were involved in Si-mediated improvement of photosynthesis of drought-stressed plants. Gong and Chen (2012) investigated the diurnal changes of photosynthetic gas exchange in wheat under field drought conditions. They found that the leaf net photosynthetic rate and stomatal conductance were obviously decreased under drought stress, and Si application increased the photosynthetic rate during most of the daytime period. Si addition also increased the stomatal conductance of the leaves of plants exposed to drought stress in the afternoon. Calculation of leaf stomatal limitation showed that it was not altered in the morning while it was increased from midday to the afternoon under drought, with Si-treated plants being intermediate. These results suggest that both stomatal and non-stomatal factors were involved in the improvement of photosynthesis by Si in wheat under drought stress: in the early morning, the non-stomatal factor was the main contributor, 9:30 a.m. being a turning point, after which the stomatal factor was the main contributor.

Chlorophyll plays an important role in the photosynthesis and is responsible for light harvesting. Lobato et al. (2009) found that the addition of Si could maintain the level of chlorophyll in capsicum (*Capsicum annuum*) under water-deficit stress, suggesting that Si could alleviate water stress-induced damage in the photosynthetic systems and therefore the improvement of photosynthesis. Similar results were also observed in drought-stressed rice (Chen et al. 2011). In sorghum plants, Yin et al. (2014) observed Si-mediated enhancement in polyamine synthesis and increase in concentrations of chlorophyll, suggesting the involvement of polyamines in maintaining higher levels of chlorophyll and therefore retarding senescence. In freezing conditions, Zhu et al. (2006) observed that Si addition not only increased the chlorophyll content but also increased the ratio of chlorophyll a/b which indicates the extent of thylakoid stacking (Aro et al. 1993). It has been shown that there is an inverse linear relationship between the ratio of chlorophyll a/b and extent of photoinhibition (Aro et al. 1993). Therefore, Si-mediated increase in chlorophyll content and chlorophyll a/b ratio indicates a higher stacking extent of thylakoid and less photoinhibition under stress conditions. Moreover, since chlorophyll a is more sensitive to reactive oxygen species (ROS) than chlorophyll b (Powles 1984), the increase in chlorophyll content and ratio of chlorophyll a/b by Si may reflect a decreased oxidative damage under freezing stress. The decrease in oxidative damage may be attributed to Si-mediated increase in activities of antioxidant enzymes such as superoxide dismutase (SOD) and catalase (CAT) (Gong et al. 2005; Liang et al. 2008), which will be discussed later.

Chlorophyll fluorescence parameters can provide useful information about photosystem II activity of plants, and both basal quantum yield [variable fluorescence (F_v)/minimal fluorescence (F_0)] and maximum quantum efficiency of PSII photochemistry [F_v/maximal fluorescence (F_m)] are related to photosynthetic efficiency (Shangguan et al. 2000). Chen et al. (2011) found that added Si not only increased the content of photosynthetic pigments but also increased F_v/F_0 and F_v/F_m of rice plants subjected to drought stress, suggesting that Si application could increase the photosynthetic efficiency and alleviate photosynthetic apparutus damage under stress.

Si has been reported to be able to regulate not only the light reaction of photosynthesis but also C assimilation, although relevant information is very limited. Ribulose bisphosphate carboxylase oxygenase (RuBisCO) is a key enzyme in C assimilation, and it catalyses the formation of glycerate-3-phosphate from ribulose-1,5-bisphosphate (RuBP) and CO_2 (Tao et al. 2012). In control plants, Adatia and Besford (1986) reported that the addition of Si increased the RuBisCO activity in cucumber grown hydroponically. However, under drought stress, Gong and Chen (2012) did not find any such change in wheat leaves, and the activity was even slightly decreased in Si-supplied plants. They also observed that the addition of Si increased the phosphoenolpyruvate carboxylase (PEPC) activity, which was decreased under drought stress. Moreover, the ratio of PEPC/RuBisCO was higher in Si-supplied plants than in plants under drought. This indicates that under drought conditions, the improvement in photosynthetic activity by the addition of Si in wheat was due to the increase in C4 photosynthetic enzyme activities.

The content of inorganic phosphorus affects ATP synthesis in chloroplasts (dos Santos et al. 2006). Gong and Chen (2012) observed that the concentration of inorganic phosphorus in wheat leaves was decreased under drought stress, while it was increased by supply of Si. The increase in inorganic phosphorus level may facilitate the synthesis of ATP, which is required in the cycle of CO_2 assimilation (Doubnerová and Ryšlavá 2011). However, as Tao et al. (2012) suggested, the distribution of inorganic phosphorus between stroma of chloroplast and cytoplasm influences the synthesis of ATP: high levels of inorganic phosphorus in chloroplast stroma can enhance ATP synthesis, and low levels of inorganic phosphorus can enhance the synthesis of sucrose in the cytoplasm. It remains unclear whether Si regulates the distribution of inorganic phosphorus in the cell in stress conditions. The increase in inorganic phosphorus level in the leaves might be attributed to Si-mediated increase in phosphorus uptake (Sistani et al. 1997). A study on the possible regulative role of Si on C assimilation under freezing stress is still lacking.

Altogether, the previous studies suggest that Si is involved in both stomatal movement and photochemical reactions and therefore regulates photosynthesis. Further work is required to reveal the details of direct or indirect regulative roles of Si in photosynthetic reactions.

7.2.2 Nutrient Uptake and Plant Growth

The improvement of plant growth by the application of Si under environmental stress is not only related to increased photosynthesis but may be related to nutrient uptake and use. Water-deficit and other environmental stresses restrict the uptake of nutrients by roots and their transport to shoots and thereby reduce the availability of nutrients (Chen et al. 1983; Farooq et al. 2009). Si may play an important role in maintaining the balance of the uptake, transport and distribution of mineral nutrients in stressed plants and therefore improve plant growth in adverse environmental conditions.

The addition of Si was reported to increase the levels of Ca and K in the leaves of water-stressed maize (*Zea mays*) (Kaya et al. 2006). In water-stressed wheat, Pei et al. (2010) found that although Si decreased the Ca, K and Mg concentrations in the shoots caused by a dilution effect resulting from a better growth, their total contents were actually increased. Chen et al. (2011) also observed similar results in rice under drought conditions. The Ca level has been suggested to be closely related to the expression of osmotic stress responsive genes (Zhu 2002; Mahajan and Tuteja 2005), and K plays an important role in osmotic adjustments in plants under stress conditions (Ashraf et al. 2001). Therefore, increased uptake of Ca and K under environmental stress could contribute to stress tolerance. This increase may be attributed to the decrease in plasma membrane permeability and increase in plasma membrane H^+-ATP activity by added Si (Liang 1999; Kaya et al. 2006). There are contradictory reports about the relationship between Si application and both P and N uptake (See Chap. 8 for more details). For instance, Gao et al. (2004) observed a significant decrease of P concentration in the xylem sap of maize with Si addition compared with plants grown without added Si. However, Eneji et al. (2008) found a strong positive correlation between Si and P uptake in both wet and dry conditions in grasses. Gong and Chen (2012) also observed that the concentration of inorganic P in wheat leaves was decreased under drought, while it was increased by Si supply. In rice, Deren (1997) observed decreases in N concentrations in all plant parts with Si application. However, Detmann et al. (2012) reported that Si could increase both rice grain yield and N use efficiency. Therefore, more work is needed to investigate how Si regulates P and N uptake and what factors affect Si efficacy.

Nutrient uptake is related to root traits, such as root surface area and root length (Barber 1984). The beneficial effects of Si on the growth of roots under stress conditions have been observed in some studies. In drought stress conditions, Hattori et al. (2005) observed a lower shoot/root ratio and higher root dry mass accumulation in Si-applied sorghum, which suggests that Si facilitated root growth under drought stress. Ahmed et al. (2011) found that the addition of Si was mainly beneficial to the growth of sorghum roots, with more biomass being allocated to the root system. The improvement of root growth may provide larger areas that are available for uptake of diffusible ions (Barber 1984 and references therein) and therefore increase drought tolerance of plants. The growth stimulation of roots by Si may be related to root elongation, thereby enhancing the extensibility of cell walls in the growing zone, as observed in sorghum by Hattori et al. (2003). In some plants such as wheat and sunflower (*Helianthus annuus*) (Gong et al. 2003; Gunes et al. 2008; Pei et al. 2010), however, the beneficial effects of Si on root growth under drought stress were not observed. The differences between these studies may be explained by the differences in culture conditions and the plant species/cultivars used. Further investigation is needed to clarify how Si may affect root structure and both water and nutrient uptake.

7.3 Silicon-Mediated Water Retention

7.3.1 Transpiration

In adverse conditions such as drought and freezing stress, the water potential and water content of plants are substantially decreased (Siddique et al. 2001; Verslues et al. 2006; Farooq et al. 2009; Liang et al. 2008). The addition of Si can improve the water status of plants under drought or freezing conditions. Gong et al. (2008) found that the application of Si could increase the leaf water potential in wheat under field drought stress. Similar results were also obtained in wheat and sorghum seedlings under water-deficit stress induced by polyethylene glycol (Pei et al. 2010; Liu et al. 2014). In freezing conditions, Liang et al. (2008) found that the effect of Si on the water content of wheat plants was dependent on the Si dose and cultivar used. For the freezing-susceptible cultivar, addition of Si significantly increased the water content of stressed plants, with 1.0 mM Si being more effective than 0.1 mM Si; in the tolerant cultivar, no obvious effect on the water content was observed at both levels of Si application.

Transpiration rate can influence plant water relations (Farooq et al. 2009). Plants transpire mainly through leaves, via the cuticle and stomata. The positive role of Si in plant growth and water retention has long been associated with Si-mediated alleviation of environmental stresses. In the early days, it was suggested that the formation of a silica–cuticle double layer on leaf epidermal tissue might contribute to the reduction in leaf transpiration (Yoshida 1965; Wong et al. 1972; Matoh et al. 1991). Under drought stress, Gong et al. (2003) observed that the growth of wheat seedlings was improved, and that leaves were thicker as a result of Si addition. They speculated that Si might improve drought tolerance through reduction of water loss by transpiration, but the transpiration rate was not measured in that study. In maize, Gao et al. (2006) observed that the transpirational rate and conductance from the leaf cuticle were not changed by Si, while the transpirational rate and conductance from the stomata were significantly decreased. The results suggest that Si was involved in the regulation of stomatal movement. Hence, the Si-triggered decrease in transpiration is one of the mechanisms for Si-mediated increase of tolerance. Whether the decrease of transpiration is from the cuticle or stomata may be related to plant species or culture conditions; more work is thus needed to confirm this speculation.

However, supply of Si does not always decrease plant transpiration. For example, Hattori et al. (2005) observed an increase in stomatal conductance and transpiration rate of leaves in potted sorghum by supply of Si under drought stress. Similar results were also observed in drought-stressed wheat (Gong et al. 2005, 2008) and rice (Chen et al. 2011). Hattori et al. (2008b) observed that Si treatment changed neither the transpiration rate nor the stomatal conductance in cucumber (*Cucumis sativus*), whether under osmotic stress or not. These conflicting results in the effects of Si on transpiration rate may be due to differences among plant species in the contribution of cuticular transpiration to the total transpiration. For example, the proportion of cuticular transpiration in the total transpiration was approximately 20–40 % in barley (Millar et al. 1968), 25–39 % in rice (Matoh et al. 1991) and up to 50 % in *Acer syriacum* and *Rhododendron poticum* (Whiteman 1965). In addition, different plants have different capabilities of Si accumulation. Therefore, the observed decrease of cuticular transpiration as a result of formation of silica–cuticle double layer on plant surface relies on both Si-accumulating ability and proportion of cuticular transpiration in the total transpiration. As mentioned above, Si could also regulate the stomatal transpiration, as observed by Gao et al. (2006) in maize; they found that both transpirational rate and conductance from the stomata were decreased by the added Si. Further work is needed to clarify how and if Si regulates the stomatal movement and what possible factors are involved. Another possible explanation for the differences in transpiration rate may be associated with growth conditions, i.e. soil culture or solution culture, as suggested by Hattori et al. (2008b). Under drought soil conditions, plants generally increase root length and/or root area to achieve a better access to water. In solution culture, since the roots are always in contact with water, plants need to adapt to water stress by increasing their internal hydraulic conductivity (Hattori et al. 2008b). Therefore, the adaptive responses of plants under these two stress conditions (soil vs. solution) are different, which may affect Si-regulated root growth and transpiration.

7.3.2 Root Water Uptake

The complex relationship between transpiration and Si suggests that there might be some mechanism besides transpiration that contributes to Si-mediated tolerance to environmental stresses. Root water uptake is an important process that maintains the water balance in plants. Si may affect root growth and therefore regulate water relations. Chen et al. (2011) found that application of Si enhanced the root growth of rice under drought conditions. They observed that the total root length, root surface area, root volume and root activity were all increased in Si-applied stressed plants. Si-mediated enhancement in root growth has also been observed in sorghum under drought (Yin et al. 2014). Yin et al. (2014) suggested that the growth enhancement of roots was related to the modulation of root plasticity, which were regulated by Si-mediated increase in polyamine levels and decrease in ethylene levels. The improved root growth can enhance water absorption, which helps to increase plant tolerance to drought stress. However, Si-mediated improvement in root growth under stress conditions is not observed in every case. For example, Gong et al. (2003) observed a positive effect of Si on the shoot growth but not on root growth in wheat. Similar results were also observed in cucumber (Hattori et al. 2008b).

There is evidence that Si can affect the hydraulic conductivity of roots. Gao et al. (2004) found that the water flow rate in xylem vessels of roots in maize was obviously decreased by Si. In sorghum, Hattori et al. (2007, 2008a) found that water stress increased the root hydraulic resistance to water flow, which was significantly decreased by the addition of Si. The underlying mechanism for Si-induced change in root hydraulic conductivity is not very clear.

Osmotic adjustment plays an important role in root water uptake under water-deficit conditions (Ogawa and Yamauchi 2006). Sonobe et al. (2011) investigated the effect of Si on the root water uptake in sorghum under water stress induced by polyethylene glycol. They found that the addition of Si decreased the osmotic potential of the roots and increased their water content, suggesting the involvement of osmotic adjustment in root water uptake in Si-treated plants. Their analysis also showed that the active accumulation of soluble sugars and amino acids (alanine and glutamic acid) was responsible for the osmotic adjustment. Si-induced improvement in water status and accumulation of soluble sugars in roots were also observed in rice (Ming et al. 2012a, b). Proline is one of the important compatible solutes (i.e. organic osmoprotectants compatible with cell metabolism) that are usually accumulated under stress conditions and plays an important role in osmotic adjustment (Nayyar and Walia 2003). However, de-Lacerda et al. (2003) considered the increase of proline level as an injury symptom rather than an indication of stress tolerance. Pei et al. (2010) found that the level of proline was increased in wheat leaves under water stress, while the addition of Si decreased its accumulation, suggesting that proline accumulation was a symptom of stress injury in their experimental conditions. In sorghum plants, Yin et al. (2014) observed that proline levels were significantly decreased while the sugar levels were increased by Si in the roots, and the root osmotic potential was not decreased as a result of Si addition under drought, indicating that the osmotic gradient was not involved in Si-mediated enhancement of water uptake. It can be concluded that the osmotic adjustment may be involved in Si-mediated improvement in water uptake under certain conditions. The accumulation of osmoprotectants may be regulated by transcriptional factors, as Khattab et al. (2014) observed Si-mediated enhancement of *DREB2A* and *NAC* expressions in rice under drought.

In root water uptake, aquaporins are major facilitators of water transport in plants (Maurel et al. 2008). Aquaporins can be subdivided into five subfamilies, i.e. the plasma membrane intrinsic proteins (PIPs), tonoplast intrinsic proteins (TIPs), small basic intrinsic proteins (SIPs), nodulin26-like intrinsic proteins (NIPs) and the uncharacterized X intrinsic proteins (XIPs) (Bienert et al. 2011). Among these aquaporins, PIPs and TIPs represent the core pathway of water transport between and within cells (Maurel et al. 2008). Recently, Liu et al. (2014) found that application of Si could increase the root hydraulic conductance and induce the up-regulation of PIP gene expressions in the roots of sorghum seedlings under water-deficit stress, suggesting the involvement of Si in the regulation of water uptake under water stress.

In summary, the regulative role of Si on water retention and the underlying mechanisms on its effect on water retention are quite complex. Further studies are needed to elucidate the detailed roles of Si in water uptake, transport to water loss in whole plants. For example, how Si triggers the up-regulation of gene expression of aquaporins remains unclear and warrants investigation. Si can also affect the formation of casparian bands in roots (Fleck et al. 2011; Vaculík et al. 2012). The possible impact of root anatomical change on water uptake is also worthy of investigation.

7.3.3 Silicon-Mediated Membrane Integrity, Stability and Functions

Cell membrane systems, including plasma membrane and endomembrane system, are primary targets of environmental stresses (Agarie et al. 1998). In adverse conditions, the maintenance of integrity of cell membranes is crucial for the survival of plants. Leakage of electrolytes from the cell has long been used as an indicator of membrane damage. The application of Si has been shown to decrease the leakage of electrolytes in drought-/water-stressed plants, such as rice, soybean and wheat (Agarie et al. 1998; Pei et al. 2010; Shen et al. 2010), indicating the protective role of Si against membrane damage. Ding (2006) investigated the effect of Si on the change of cell structure in drought-stressed wheat leaves. She observed that in the absence of Si, a clear plasmolysis phenomenon occurred. In the chloroplast, the thylakoids became swollen, and the matrix lamella appeared degraded; when Si was fed to the plants, this resulted in a reduction of plasmolysis and improved the chloroplast structure. The improvement of leaf cell ultrastructure was also observed in Si-treated rice plants grown under drought conditions (Ming 2012).

Fluidity is a basic and essential characteristic for the function of cell membranes. Huang and Yang (1996) suggested that proper fluidity in mitochondrial membrane was important to maintain its optimal structure. As discussed in Chap. 6, Si could help to maintain the fluidity of cell membranes at an optimal state under salt stress. However, up to now, there is little information available on the effect of Si on the membrane fluidity in plants under drought or freezing conditions.

Membrane fluidity is influenced by several factors such as interactions of proteins and lipids (Huang and Yang 1996) and lipid composition (Zhang et al. 2002). Liang et al. (2006) observed that salt stress significantly increased the ratio of phospholipids to proteins in the plasma membranes of barley leaves and added Si recovered the ratio to the level of control. This suggests that Si could regulate the ratio of phospholipids to protein and therefore fluidity of plasma membranes. However, it is unclear whether the Si-mediated decrease in the ratio of phospholipids to proteins was due to the decease of phospholipid level or increase of protein levels. In strawberry, foliar application of Si was reported to increase the amount of membrane lipids under normal conditions (Wang and Galletta 1998). Therefore, there is a possibility that the added Si increased the levels of both phospholipids and proteins, with the increase of the latter being greater. It is generally known that a decrease in fatty acid unsaturation results in a decrease in membrane fluidity (Shinitzky 1984). Wang and Galletta (1998) observed that the application of Si increased the ratios of fatty acid unsaturation in glycolipids and phospholipids in strawberry. Gong et al. (2005) reported that the application of Si increased the unsaturation of fatty acids, which was decreased under drought stress. Liang et al. (Liang et al. 2005) observed that the decrease in the ratio of unsaturated to saturated fatty acids by the addition of Si corresponded with the decrease of the tonoplast fluidity in the salt-sensitive barley cultivar. However, in the salt-tolerant cultivar, Si mediated an increase in the ratio of unsaturated to saturated fatty acids and a decrease in tonoplast fluidity. Therefore, the regulation of Si on the unsaturation fatty acids may be cultivar dependent.

Membrane composition affects its fluidity and consequently the activities of membrane proteins. In salt-stressed barley, exogenous Si is reported to increase the activities of H^+-ATPase in plasma membranes and H^+-ATPase and H^+-PPase in tonoplast, which contributed to the exclusion of Na from cytoplasm to apoplast and sequestration of Na into the vacuole, respectively (Liang et al. 2005, 2006). The increase of plasma membrane H^+-ATPase activity by Si may also contribute to increase nutrient uptake under stress conditions as discussed earlier (Kaya et al. 2006; Pei et al. 2010; Chen et al. 2011). In vitro experiments showed that Si had no effect on the plasma membrane H^+-ATPase, suggesting that the effect of Si on the membrane fluidity and enzyme activity may be secondary (Liang et al. 2006).

The level of cell wall polysaccharides in the leaves was significantly increased by Si in rice (Agarie et al. 1998). Si also influences other cell wall components, such as pectic acid, protein (Schwarz 1973), polyphenols and lignin (Raven 1983; Fleck et al. 2011). These findings suggest that Si is involved in the biosynthesis of cell wall components, therefore affecting the mechanical properties of cell walls and consequently its permeability to water. The effects of Si on the cell wall properties may in turn affect the membrane properties (Agarie et al. 1998); however, this needs to be confirmed by mechanistic experiments. Si-mediated improvement in membrane integrity and stability in stress conditions is also associated with the increased antioxidant defence (see 7.4 in this chapter).

7.4 Silicon-Mediated Antioxidative Responses

ROS-induced oxidation of functional molecules (lipids, protein, etc.) is one of the most damaging processes in living organisms (Yordanov et al. 2000; Gill and Tuteja 2010). Numerous studies have shown that Si-mediated plant tolerance against abiotic stress such as salt stress and heavy metal stress is associated with decreased oxidative damage of functional molecules in the cell (see Chaps. 5 and 6). In drought conditions, Gong et al. (2005) also observed that application of Si increased the chlorophyll and protein contents in wheat. The double bond index of fatty acids, which reflects its unsaturation, was decreased under drought, while it was significantly increased by feeding Si to the plants (Gong et al. 2005). Moreover, they found that drought stress increased the level of oxidized proteins, which was depressed in Si-fed plants. These results suggest that supply of Si can decrease oxidative damage of lipids and proteins under drought stress. Si-mediated decrease of lipid peroxidation damage under water stress has also been observed in other plants, such as chickpea, sunflower, soybean and rice (Gunes et al. 2007, 2008; Shen et al. 2010; Ming et al. 2012a). Under freezing conditions, the increased production of malondialdehyde, the end product of membrane lipid peroxidation, could be depressed by exogenous added Si, as observed in barley (Liang et al. 2008).

Si-mediated decrease in oxidative damage is attributed to its regulation on the antioxidant defence and decreased level of ROS in plants. Liang et al. (2008) found that, under freezing conditions, addition of Si increased the activities of SOD and CAT as well as contents of glutathione (GSH) and ascorbic acid in wheat. They also observed a Si-mediated decrease in H_2O_2 level in the freezing-stressed plants. In drought-stressed wheat, Gong et al. (2005) found that the application of Si increased the activities of SOD, CAT and glutathione reductase (GR), but did not change the activities of peroxidase (POD) and ascorbic peroxidase (APX), while it decreased the H_2O_2 level in the leaves. The increase of GR activity might contribute to the increased GSH level, as observed in a later study (Pei et al. 2010). Gong et al. (2005) also observed a decrease in the acid phospholipase activity in Si-treated plants under drought stress, suggesting a decrease of phospholipids deesterification damage in stressed plants. Pei et al. (2010) found that addition of Si increased the ascorbic acid concentration in the leaves of water-stressed wheat. In soybean (*Glycine max*) seedlings, it was observed that the CAT and POD activities were stimulated and the SOD activity was inhibited under drought stress, whereas their activities and H_2O_2 contents were all decreased by Si (Shen et al. 2010). Gunes et al. (2007) reported that the effects of Si on the activities of SOD and CAT were cultivar dependent. Gong et al. (2008) investigated the effects of Si on the antioxidant defence responses under the conditions of the gradually increasing drought. They found that the effects were different at different degrees of drought stress. At booting stage, the SOD activity was inhibited and POD activity was stimulated under drought stress, and they were not changed by the application of Si, whereas at filling stage when the drought intensity was stronger, the SOD activity was increased, while the POD activity and H_2O_2 level were decreased in Si-supplied plants (Gong et al. 2008).

Overall, these findings imply a rather complex mechanism of Si-mediated regulation of antioxidative enzymes' activities under drought and freezing stresses. The differences in regulative roles may be related to plant species, cultivar and stress conditions. Despite the complex regulative roles of Si in the antioxidant enzymes activities, decreased ROS level was usually observed in Si-supplied plants in stress conditions (Gong et al. 2005; Gunes et al. 2007, 2008; Liang et al. 2008; Shen et al. 2010; Ming et al. 2012a). The observed decreases in the activities of antioxidant enzymes by the addition of Si in stressed plants may be associated with the decrease of ROS level, when there was no need for plants to increase the antioxidant defence capacity (Gong et al. 2008).

References

Adatia MH, Besford RT. The effects of silicon on cucumber plants grown in recirculating nutrient solution. Ann Bot. 1986;58:343–51.

Agarie S, Hanaoka N, Ueno O, Miyazaki A, Kubota F, Agata W, Kaufman PB. Effects of silicon on tolerance to water deficit and heat stress in rice plants (*Oryza sativa* L.), monitored by electrolyte leakage. Plant Prod Sci. 1998;1:96–103.

Ahmed M, Hassen FU, Qadeer U, Aslam MA. Silicon application and drought tolerance mechanism of sorghum. Afr J Agric Res. 2011;6:594–607.

Ahsan N, Renault J, Komatsu S. Recent developments in the application of proteomics to the analysis of plant responses to heavy metals. Proteomics. 2009;9:2602–21.

Aro E-M, McCaffery S, Anderson JM. Photoinhibition and D1 protein degradation in peas acclimated to different growth irradiances. Plant Physiol. 1993;103:835–43.

Ashraf M, Harris PJC. Photosynthesis under stressful environments: an overview. Photosynthetica. 2013;51:163–90.

Ashraf M, Ahmad A, McNeilly T. Growth and photosynthetic characteristics in pearl millet under water stress and different potassium supply. Photosynthetica. 2001;39:389–94.

Barber SA. Soil nutrient bioavailability: a mechanistic approach. New York: Wiley-Interscience Publication; 1984.

Bienert GB, Bienert MD, Jahn TP, Boutry M, Chaumont F. Solanaceae XIPs are plasma membrane aquaporins that facilitate the transport of many uncharged substrates. Plant J. 2011;66:306–17.

Cattivelli L, Rizza F, Badeck FW, Mazzucotelli E, Mastrangelo AM, Francia E, Marè C, Tondelli A, Stanca AM. Drought tolerance improvement in crop plants: an integrated view from breeding to genomics. Field Crops Res. 2008;105:1–14.

Chen THH, Gusta LV, Fowler DB. Freezing injury and root development in winter cereals. Plant Physiol. 1983;73:773–7.

Chen W, Yao XQ, Cai KZ, Chen J. Silicon alleviates drought stress of rice plants by improving plant water status, photosynthesis and mineral nutrient absorption. Biol Trace Elem Res. 2011;142:67–76.

de-Lacerda CF, Cambraia J, Oliva MA, Ruiz HA, Prisco JT. Solute accumulation and distribution during shoot and leaf develipment in two sorghum genotypes under salt stress. Environ Exp Bot. 2003;49:107–20.

Deren CW. Changes in nitrogen and phosphorus concentrations of silicon-fertilized rice grown on organic soil. J Plant Nutr. 1997;20:765–71.

Detmann KC, Araújo L, Martins SCV, et al. Silicon nutrition increases grain yield, which, in turn, exerts a feed-forward stimulation of photosynthetic rates via enhanced mesophyll conductance and alters primary metabolism in rice. New Phytol. 2012;196:752–62.

Ding YF. Mechanisms of silicon-enhancement of drought tolerance in wheat seedlings. MSc thesis, Nanjing Agricultural University; 2006.

dos Santos MG, Ribeiro RV, de Oliveira RF, Machado EC, Pimentel C. The role of inorganic phosphate on photosynthesis recovery of common bean after a mild water deficit. Plant Sci. 2006;170:659–64.

Doubnerová V, Ryšlavá H. What can enzymes of C4 photosynthesis do for C3 plants under stress? Plant Sci. 2011;180:575–83.

Eneji AE, Inanaga S, Muranaka S, Li J, Hattori T, An P, Tsuji W. Growth and nutrient use in four grasses under drought stress as mediated by silicon fertilizers. J Plant Nutr. 2008;31:355–65.

Farooq M, Wahid A, Kobayashi N, Fujita D, Basra SMA. Plant drought stress: effects, mechanisms and management. Agron Sustain Dev. 2009;29:185–212.

Fleck AT, Nye T, Repenning C, Stahl F, Zahn M, Schenk MK. Silicon enhances suberization and lignification in roots of rice (*Oryza sati*va). J Exp Bot. 2011;62:2001–11.

Gao X, Zou C, Wang L, Zhang F. Silicon improves water use efficiency in maize plants. J Plant Nutr. 2004;27:1457–70.

Gao X, Zou C, Wang L, Zhang F. Silicon decreases transpiration rate and conductance from stomata of maize plants. J Plant Nutr. 2006;29:1637–47.

Gill SS, Tuteja N. Reactive oxygen species and antioxidant machinery in abiotic stress tolerance in crop plants. Plant Physiol Biochem. 2010;48:909–30.

Gong H, Chen K. The regulatory role of silicon on water relations, photosynthetic gas exchange, and carboxylation activities of wheat leaves in field drought conditions. Acta Physiol Plant. 2012;34:1589–94.

Gong HJ, Chen KM, Chen GC, Wang SM, Zhang CL. Effects of silicon on growth of wheat under drought. J Plant Nutr. 2003;26:1055–63.

Gong HJ, Zhu XY, Chen KM, Wang S, Zhang CL. Silicon alleviates oxidative damage of wheat plants in pots under drought. Plant Sci. 2005;169:313–21.

Gong HJ, Chen KM, Zhao ZG, Chen GC, Zhou WJ. Effects of silicon on defense of wheat against oxidative stress under drought at different developmental stages. Biol Plant. 2008;52:592–6.

Gunes A, Pilbeam DJ, Inal A, Bagci EG, Coban S. Influence of silicon on antioxidant mechanisms and lipid peroxidation in chickpea (*Cicer arietinum* L.) cultivars under drought stress. J Plant Interact. 2007;2:105–13.

Gunes A, Pilbeam DJ, Inal A, Coban S. Influence of silicon on sunflower cultivars under drought stress. I: Growth, antioxidant mechanisms, and lipid peroxidation. Commun Soil Sci Plant Anal. 2008;39:1885–903.

Hasanuzzaman M, Nahar K, Fujita M. Extreme temperature responses, oxidative stress and antioxidant defense in plants. In K Vahdati, C Leslie, editors. Abiotic stress – plant responses and applications in agriculture. InTech; 2013. doi: 10.5772/54833.

Hattori T, Inanaga S, Tanimoto E, Lux A, Luxova M, Sugimoto Y. Silicon-induced changes in viscoelastic properties of sorghum root cell walls. Plant Cell Physiol. 2003;44:743–9.

Hattori T, Inanaga S, Araki H, An P, Morita S, Luxová M, Lux A. Application of silicon enhanced drought tolerance in *Sorghum bicolour*. Physiol Plant. 2005;123:459–66.

Hattori T, Sonobe K, Inanaga S, An P, Tsuji W, Araki H, Eneji AE, Morita S. Short term stomatal responses to light intensity changes and osmotic stress in sorghum seedlings raised with and without silicon. Environ Exp Bot. 2007;60:177–82.

Hattori T, Sonobe K, Araki H, Inanaga S, An P, Morita S. Silicon application by sorghum through the alleviation of stress-induced increase in hydraulic resistance. J Plant Nutr. 2008a;31:1482–95.

Hattori T, Sonobe K, Inanaga S, An P, Morita S. Effects of silicon on photosynthesis of young cucumber seedlings under osmotic stress. J Plant Nutr. 2008b;31:1046–58.

Huang Y, Yang F. Physical state change of phospholipids mediated by Ma2+ modulates activity and conformation of reconstituted mitochondrial F0-F1-ATPase. In: Yang F, Wang F, editors. Membrane lipid-protein interaction andits application in medicine and agriculture. Jinan: Shandong Science Press; 1996. p. 1–17.

Huang J, Ji M, Liu Y, Zhang L, Gong D. An overview of arid and semi-arid climate change. Progressus Inquisitiones De Mutatione Climatis. 2013;9:9–14 (In Chinese with English abstract).

Kaya C, Tuna L, Higgs D. Effect of silicon on plant growth and mineral nutrition of maize grown under water-stress conditions. J Plant Nutr. 2006;29:1469–80.

Khattab HI, Emam MA, Emam MM, Helal NM, Mohamed MR. Effect of selenium and silicon on transcription factors NAC5 and DREB2A involved in drought-responsive gene expression in rice. Biol Plant. 2014;58:265–73.

Liang YC, Zhang WH, Chen Q, Ding RX. Effects of silicon on H^+-ATPase and H^+-PPase activity, fatty acid composition and fluidity of tonoplast vesicles from roots of salt-stressed barley (*Hordeum vulgare* L.). Environ Exp Bot. 2005;53:29–37.

Liang YC, Zhang WH, Chen Q, Liu YL, Ding RX. Effect of exogenous silicon (Si) on H^+-ATPase activity, phospholipids and fluidity of plasma membrane in leaves of salt-stressed barley (*Hordeum vulgare* L.). Environ Exp Bot. 2006;57:212–9.

Liang YC, Zhu J, Li ZJ, Chu GX, Ding YF, Zhang J, Sun WC. Role of silicon in enhancing resistance to freezing stress in two contrasting winter wheat cultivars. Environ Exp Bot. 2008;64:286–94.

Liu P, Yin L, Deng X, Wang S, Tanaka K, Zhang S. Aquaporin-mediated increase in root hydraulic conductance is involved in silicon-induced improved root water uptake under osmotic stress in *Sorghum bicolor* L. J Exp Bot. 2014;65:4747–56.

Lobato AKS, Coimbra GK, Neto MAM, Costa RCL, Filho BGS, Neto CFO, Luz LM, Barreto AGT, Pereira BWF, Alves GAR, Monteiro BS, Marochio CA. Protective action of silicon on water relations and photosynthetic pigments in pepper plants induced to water deficit. Res J Biol Sci. 2009;4:617–23.

Mahajan S, Tuteja N. Cold, salinity and drought stresses: an overview. Arch Biochem Biophys. 2005;444:139–58.

Matoh T, Murata S, Takahashi E. Effect of silicate application on photosynthesis of rice plants. Jpn J Soil Sci Plant Nutr. 1991;62:248–51 (in Japanese).

Maurel C, Verdoucq L, Luu DT, Santon V. plant aquaporins: membrane channels with multiple integrated functions. Annu Rev Plant Biol. 2008;59:595–624.

Millar AA, Duysen ME, Wilkerson GE. Internal water balance of barley under soil moisture stress. Plant Physiol. 1968;43:968–72.

Ming DF. Regulatory mechanisms of silicon on physiological and biochemical characteristics, ultrastructure and related gene expression of rice under water stress. PhD thesis, Zhejiang University; 2012.

Ming DF, Yuan HM, Wang YH, Gong HJ, Zhou WJ. Effects of silicon on the physiological and biochemical characteristics of roots of rice seedlings under water stress. Sci Agric Sin. 2012a;45:2510–9 (In Chinese with English abstract).

Ming DF, Pei ZF, Naeem MS, Gong HJ, Zhou WJ. Silicon alleviates PEG-induced water-deficit stress in upland rice seedlings by enhancing osmotic adjustment. J Agron Crop Sci. 2012b;198:14–26.

Nayyar H, Walia DP. Water stress induced proline accumulation in contrasting wheat genotypes as affected by calcium and abscisic acid. Biol Plant. 2003;46:275–9.

Ogawa A, Yamauchi A. Root osmotic adjustment under osmotic stress in maize seedlings 1. Transient change of growth and water relations in roots in response to osmotic stress. Plant Prod Sci. 2006;9:27–38.

Pei ZF, Ming DF, Liu D, Wan GL, Geng XX, Gong HJ, Zhou WJ. Silicon improves the tolerance to water-deficit stress induced by polyethylene glycol in wheat (*Triticum aestivum* L.) seedlings. J Plant Growth Regul. 2010;29:106–15.

Powles SB. Photoinhibition of photosynthesis induced by visible light. Annu Rev Plant Physiol Plant Mol Biol. 1984;35:15–44.

Raven JA. The transport and function of silicon in plants. Bio Rev Cambridge Philos Soc. 1983;58:179–207.

Reddy AR, Chaitanya KV, Vivekanandanb M. Drought-induced responses of photosynthesis and antioxidant metabolism in higher plants. J Plant Physiol. 2004;161:1189–202.

Schwarz K. A bound form of silicon in glycosaminoglycans and polyuronides. Proc Natl Acad Sci U S A. 1973;70:1608–12.

Shangguan ZP, Shao MA, Dyckmans J. Effects of nitrogen nutrition and water deficit on net photosynthetic rate and chlorophyll fluorescence in winter wheat. J Plant Physiol. 2000;156:46–51.

Shen X, Zhou Y, Duan L, Li Z, Eneji AE, Li J. Silicon effects on photosynthesis and antioxidant parameters of soybean seedlings under drought and ultraviolet-B radiation. J Plant Physiol. 2010;167:1248–52.

Shinitzky M. Membrane fluidity and cellular functions. In: Shinitzky M, editor. Physiology of membrane fluidity, vol. 1. Boca Raton: CRC Press; 1984. p. 1–51.

Siddique MRB, Hamid A, Islam MS. Drought stress effects on water relations of wheat. Bot Bull Acad Sinica. 2001;41:35–9.

Sistani KR, Savant NK, Reddy KC. Effect of rice hull ash silicon on rice seedling growth. J Plant Nutr. 1997;20:195–201.

Sonobe K, Hattori T, An P, Tsuji W, Eneji AE, Kobayashi S, Kawamura Y, Tanaka K, Inanaga S. Effect of silicon application on sorghum root responses to water stress. J Plant Nutr. 2011;34:71–82.

Tao H-Z, Zhao C-L, Li W-Q. Photosynthetic response to low temperature in plant. Chine J Biochem Mol Biol. 2012;28:501–8 (In Chinese with English Abstract).

Vaculík M, Landberg T, Greger M, Luxová M, Stoláriková LA. Silicon modifies root anatomy, and uptake and subcellular distribution of cadmium in young maize plants. Ann Bot. 2012;110:433–43.

Verslues PE, Agarwal M, Katiyar-Agarwal S, Zhu J, Zhu JK. Methods and concepts in quantifying resistance to drought, salt and freezing, abiotic stresses that affect plant water status. Plant J. 2006;45:523–39.

Wang SY, Galletta GJ. Foliar application of potassium silicate induces metabolic changes in strawberry plants. J Plant Nutr. 1998;21:157–67.

Wang Z, Gerstein M, Snyder M. RNA-Seq: a revolutionary tool for transcriptomics. Nat Rev Gen. 2009;10:57–63.

Whiteman PC. Control of carbon dioxide and water vapour exchange between plantand atmosphere. Dissertation, Hebrew University, Jerusalem; 1965.

Wong YC, Heits A, Ville DJ. Foliar symptoms of silicon deficiency in the sugarcane plant. Proc Cong Int Soc Sugarcane Technol. 1972;14:766–76.

Yadav SK. Cold stress tolerance mechanisms in plants. A review. Agron Sustain Dev. 2010;30:515–27.

Yin LN, Wang SW, Liu P, Wang WH, Cao D, Deng XP, Zhang SQ. Silicon-mediated changes in polyamine and 1-aminocyclopropane-1-carboxylic acid are involved in silicon-induced drought resistance in *Sorghum bicolor* L. Plant Physiol Biochem. 2014;80:268–77.

Yordanov I, Velikova V, Tsonev T. Plant responses to drought, acclimation, and stress tolerance. Photosynthetica. 2000;38:171–86.

Yoshida S. Chemical aspect of silicon in physiology of the rice plant. Bull Natl Agric Sci B. 1965;15:1–58.

Zhang W, Chen Q, Liu Y. Relationship between H^+-ATPase activity and fluidity of tonoplast in barley roots under NaCl stress. Acta Bot Sin. 2002;44:292–6.

Zhu JK. Salt and drought stress signal transduction in plants. Annu Rev Plant Biol. 2002;53:247–73.

Zhu J, Liang Y-C, Ding Y-F, Li Z-J. Effect of silicon on photosynthesis and its related physiological parameters in two winter wheat cultivars under cold stress. Sci Agric Sin. 2006;39:1780–8 (In Chinese with English abstract).

Chapter 8
Silicon-Mediated Tolerance to Other Abiotic Stresses

Abstract Silicon (Si) has been reported to mitigate some other abiotic stresses such as boron (B) toxicity, potassium (K) and phosphorus (P) deficiency, iron (Fe) deficiency, excess of nitrogen and phosphorus, UV radiation and flooding. In this chapter, current knowledge concerning the roles Si may play under such abiotic stresses is overviewed. More recently, rapid progress has been made at both the physiological and molecular levels in dissecting the mechanisms involved in Si-induced root mobilization of Fe from the rhizosphere and acquisition of root apoplastic Fe, thereby mitigating Fe-deficiency chlorosis in strategy 1 plants.

Keywords Boron toxicity • Excess of nitrogen and phosphorus • Flooding • Iron-deficiency chlorosis • Potassium and phosphorus deficiency • UV radiation

8.1 Boron Toxicity

Although boron (B) is an essential micronutrient for higher plants, there is only a narrow level between critical deficient and toxic concentrations of B in plant tissues, particularly in cereal species (Nable et al. 1990; Schnurbusch et al. 2010). Boron toxicity is a significant agronomic problem in a number of cereal-growing regions across the world, mostly in arid and semiarid climate conditions such as West Asia, Northern Africa and Southern Australia (Schnurbusch et al. 2010). Soil B excess is often accompanied by excessive salinity, the conditions conducive to multiple stresses such as osmotic stress and chlorine (Cl), sodium (Na) and B toxicity (Alpaslan and Gunes 2001). Boron is removed more slowly than salt ions during leaching (Yermiyahu et al. 2008), thereby remaining at excessive concentrations even after reclamation of saline soils (Ben-Gal and Shani 2002). Besides natural occurrence in the soil, B-contaminated irrigation water and mining pollution can also be a source of high soil B (Nable et al. 1997).

The mechanisms of B tolerance and B toxicity are not well understood. Within species, large genotypic differences exist in the capacity to tolerate high external B concentrations (Paull et al. 1992; Nable et al. 1997). These differences are based primarily on restrictions in B uptake by the roots, thereby restricting translocation of B to the shoot, rather than tolerance to high B in tissues (Nable et al. 1997). In wheat and barley, for instance, root B concentration is reduced in tolerant

Y. Liang et al., *Silicon in Agriculture*, DOI 10.1007/978-94-017-9978-2_8

genotypes due to exudation of B from roots to soil (Reid 2007) via an efflux B transporter belonging to the BOR family (Miwa et al. 2007; Sutton et al. 2007). Recently, Schnurbusch et al. (2010) have proposed that tolerance to high soil B is mediated by reduced expression of *HvNIP2;1* to limit B uptake. This gene encoding HvNIP2;1, an aquaporin from the NIP subfamily, was previously described as a silicon (Si) influx transporter in barley (see Chap. 4). Moreover, differential expression of B toxicity tolerance among the genotypes, despite similarly high total concentration of B in leaves, seems to be related to a better redistribution of B by efflux transporters from sensitive symplastic compartments into the leaf apoplast (Reid and Fitzpatrick 2009). In general, species with high B demand may have also a higher capacity to sequester B in the cell walls (Broadley et al. 2011).

Silicon and B are elements with many similar chemical properties; in aqueous solutions both exist as weak, undissociated acids and can complex readily with polyhydroxy compounds (Brown et al. 1999; Kinrade et al. 1999). Plants take up Si as uncharged monosilicic acid (pK_a of 9.8) and B as boric acid (pK_a of 9.2), either passively or actively, which closely depends on the external concentration. Due to these similarities, interactions between Si and B are possible, but there is little information available regarding this aspect.

Dicots and graminaceous species differ not only in their capacity for Si shoot accumulation but also in their B demand, which is inversely related (Table 8.1). In barley, genotypical differences in restriction of uptake by roots and transport of B into the leaves are closely correlated with similar restrictions in uptake and transport of Si (Nable et al. 1990). Liang and Shen (1994) reported that under conditions of adequate or excess B concentrations, Si supply led to a decrease in B uptake, whereas under conditions of B deficiency, Si supply enhanced B uptake in oilseed rape. In *Lilium longiflorum*, Si supply increased the range between critical deficiency and toxicity for B (Polster and Schwenk 1992). However, in cucumber, Si supply had no effect on total B concentration of leaves, but influenced distribution of B between symplastic and apoplastic compartments, especially at high B supply (Wiese et al. 2007). In fact, more B was bound to the cell wall, and less B was in the cell sap of leaves of Si-treated plants, which was further correlated with a lower B concentration of the apoplastic fluid. Amendment of Si to high B soil effectively

Table 8.1 Shoot concentrations of Si and B in different graminaceous and dicotyledonous species

Plant species	Si concentration (mg g^{-1} dw)	B concentration (μg g^{-1} dw)
Rice	41	7
Barley	30	8
Wheat	26	7
Maize	23	10
Soybean	12	29
Sunflower	7	38
Grapevine	5	42
Tomato	3	60

mitigated B toxicity in spinach, wheat and barley by decreasing root-to-shoot translocation of B and in addition by preventing reactive oxygen species (ROS) membrane damages due to modulation of the activities of antioxidative enzymes (Karabal et al. 2003; Gunes et al. 2007a, b, c; Inal et al. 2009). Also, Si alleviates the concomitant adverse effects of salinity/sodicity (Na ions) and B toxicity on barley and grapevine (*Vitis vinifera*) rootstocks (Gunes et al. 2007a; Soylemezoglu et al. 2009).

8.2 Potassium and Phosphorus Deficiency

Potassium (K) deficiency reduces mechanical stability, nutritional quality, drought tolerance and crop resistance to pathogens (Liebersbach et al. 2004; Pettigrew 2008). Potassium deficiency also leads to oxidative stress, as evidenced by accumulation of ROS and membrane lipid peroxidation (Cakmak 1994, 2005). On the other hand, the ameliorative effect of Si on plants suffering from abiotic stresses often occurs through counteracting oxidative stress by modulating antioxidant enzymes (Liang et al. 2007). It has been reported that application of Si fertilizers improved growth, yield and quality as well as the overall nutritional status of different crops (see Chap. 10). Yet, a direct role of Si in mediating K deficiency has not been uncovered. The preliminary study of Miao et al. (2010) demonstrates that the application of Si- to K-deficient soybean (*Glycine max*) plants improved both internal K status and plant growth. In addition, Si also alleviated K-deficiency-induced membrane lipid peroxidation and oxidative stress by modulating antioxidant enzymes.

Soil-available phosphorus (P) is often low, especially in acid soils, because of its poor solubility, sorption and slow diffusion (Marschner 1995). Over 50 % of the cultivated land across the world is P deficient (Vance 2001). Most of the phosphate used in mineral fertilizers is derived from rock phosphate as a non-renewable resource; globally these resources are predicted for the next 50–100 years, assuming an increased use of P fertilizers to meet food requirements of a fast-growing world population (Cordell et al. 2009; Gilbert 2009).

The root exudation of carboxylates (mainly citrate and malate) and phenolic compounds (e.g. flavonoids) has been considered as the major root response to mobilize sparingly soluble P in the rhizosphere (for review see: Neumann and Römheld 2007; Cesco et al. 2010). However, the underlying mechanism involved in biosynthesis and subsequent exudation of organic acids in response to P deficiency and/or Al excess in acid soils remains unclear (e.g. Ryan et al. 1997; Li et al. 2000; Kochian et al. 2004; Neumann and Römheld 2007). The enhanced activity of phosphoenolpyruvate carboxylase (PEPC) appears to be a key Pi-independent metabolic bypass reaction of organic acid biosynthesis in P-deficient plants, to achieve a more economic internal P utilization under conditions of P limitation (Theodoru and Plaxton 1993; Plaxton 1998). However, this mechanism has not been proven in many plant species. For instance, unlike in white lupine, P deficiency did not enhance exudation of carboxylates by wheat roots despite an increased PEPC activity (Neumann and

Römheld 1999). Root P uptake is mediated by high-affinity plasma membrane (PM)-associated Pi transporters. In general, the expression of Pi transporter genes (*PHTs*, formerly known as *PTs*) is upregulated under P deprivation (e.g. Karthikeyan et al. 2002; Tittarelli et al. 2007), although some environmental factors may alter their expression (Catarecha et al. 2007; Glassop et al. 2005; Kostic et al. 2015).

Beneficial effects of Si under P-deficiency stress have been recorded on various graminaceous species such as wheat (*Triticum aestivum*), rice (*Oryza sativa*), maize (*Zea mays*), barley (*Hordeum vulgare*) and pasture grasses grown in both soil and hydroponic conditions (Rothbuhr and Scott 1957; Ma and Takahashi 1989, 1990a, b, 1991; Owino-Gerroh and Gascho 2004; Ma 2004; de Melo et al. 2007; Eneji et al. 2008; Yang et al. 2008; Abro et al. 2009). In a long-term field experiment conducted at the Rothamsted Experimental Station, when P fertilizers were not applied, yield of barley from a field fertilized with Si was higher than in a field without Si amendment (Fisher 1929). Such beneficial effects of Si have been attributed through the varied and even contradictory explanations, including competitive sportive interaction and exchange of these two elements (Smyth and Sanchez 1980), an enhancement of soil P availability by increasing soil pH (Roy et al. 1971; Owino-Gerroh and Gascho 2004) and decrease of metal uptake (e.g. Mn, Fe, Al, Cd), thus indirectly improving P utilization by plants (Ma and Takahashi 1990a; Liang et al. 2005). Different experimental materials and conditions could be an important reason for such varied explanations in the literature.

Phosphorus sorption is the major cause of P deficiency in acid soils. Phosphate ($H_2PO_4^-$) anions are specifically adsorbed onto hydrated Fe and Al oxides by replacing OH groups from the coordination sphere of the metals (Hingston et al. 1967; Obihara and Russell 1972). The chemical competition between $H_2PO_4^-$ and silicate ($H_3SiO_4^-$) anions for the sorption sites has been demonstrated in soils resulting in displacement of Si by P, and *vice versa* (Obihara and Russell 1972; Smyth and Sanchez 1980). Freshwater eutrophication, as a possible way of internal P loading under aerobic conditions, has been attributed to Si-induced mobilization of P from anion exchange sites in sediments (Tuominen et al. 1998; Tallberg et al. 2009). Although theoretically possible, the ability of Si to compete P for the sorption sites is highly pH dependent, since pK_a value of orthosilicic acid is much higher than that of orthophosphoric acid (9.8 and 2.1, respectively). Therefore, it is unlikely that such interaction between $H_3SiO_4^-$ and $H_2PO_4^-$ may occur in agriculture soils. Indeed, the experiments of Ma and Takahashi (1990b, 1991) did not confirm that increasing amount of applied Si could affect either fixation capacity or availability of soil P. Moreover, the uptake of P was also not affected by the Si supply at a low P level in both soil and solution culture (Ma and Takahashi 1990a, b, 1991). Therefore, Ma (2004) proposed that the larger beneficial effect of Si on plant growth under P-deficiency stress was attributed to the enhanced availability of internal P through the decrease of excess Fe and Mn uptake.

On the contrary, the recent experiments on acid soils showed that supply of different Si sources increased readily available P fraction followed by decreasing of Al- and Fe-bound fractions in wheat rhizosphere (Kostic et al. 2015). Moreover, application of Si in acid soil resulted in enhancement of the expressions of

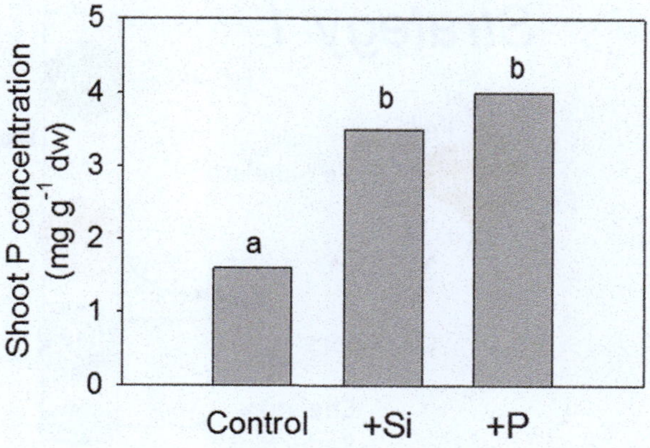

Fig. 8.1 Effect of Si supply on shoot P concentration of wheat plants grown in the acid soil (Based on Kostic-Kravljanac 2015)

P-deficiency-related genes encoding phosphoenolpyruvate carboxylase (PEPC), the organic anion efflux transport proteins, i.e. multidrug and toxic compound extrusion 1 (MATE1) and Al-activated malate transporter 1 (ALMT1), and high-affinity Pi transporters (PHT1.1 and PHT1.2) in wheat roots (L. Kostic and M. Nikolic, unpublished). Consequently, the leaf P concentration in Si-treated wheat plants without P application achieved the range of that in P-fertilized plants (Fig. 8.1).

8.3 Iron Deficiency

Despite being the fourth most abundant element in the Earth's crust, iron (Fe) deficiency is one of the major limiting factors for crop production in calcareous soils all over the world (Vose 1982). The overall processes of Fe acquisition by roots have been described in terms of two basic strategies to cope with Fe deficiency (Fig. 8.2; for review see Römheld and Nikolic 2007). Strategy 1 plant species (dicots and nongraminaceous species) respond to Fe deficit by undergoing both morphological and physiological changes. Morphological changes include an increase of the root surface characterized by the formation of lateral roots, root hairs in the apical zone, and the development of transfer cells (Schmidt 1999). All the components of strategy 1 root uptake machinery are upregulated proteins, i.e. PM-bound Fe^{III} chelate reductases [ferric reductase oxidase (FRO) family; Jeong and Connolly 2009], Fe^{II} PM transporter [iron-regulated transporter (IRT); Vert et al. 2002] and PM proton pump of the HA family P-type H^+-ATPase, which mediates acidification of the rhizosphere (Santi and Schmidt 2009). This acidification serves not only to make rhizosphere iron more available but also to facilitate the required FRO activity (Wu et al. 2005). Many strategy 1 species also enhance root exudation of low-molecular-weight compounds (e.g. carboxylates, phenolics and flavins), which can increase Fe availability due to chelation and potential reduction of Fe^{III}. Instead of the reduction-based Fe acquisition, graminaceous species (strategy 2 plants) developed the chelation-based mechanism of Fe uptake to cope with Fe-deficient conditions. This

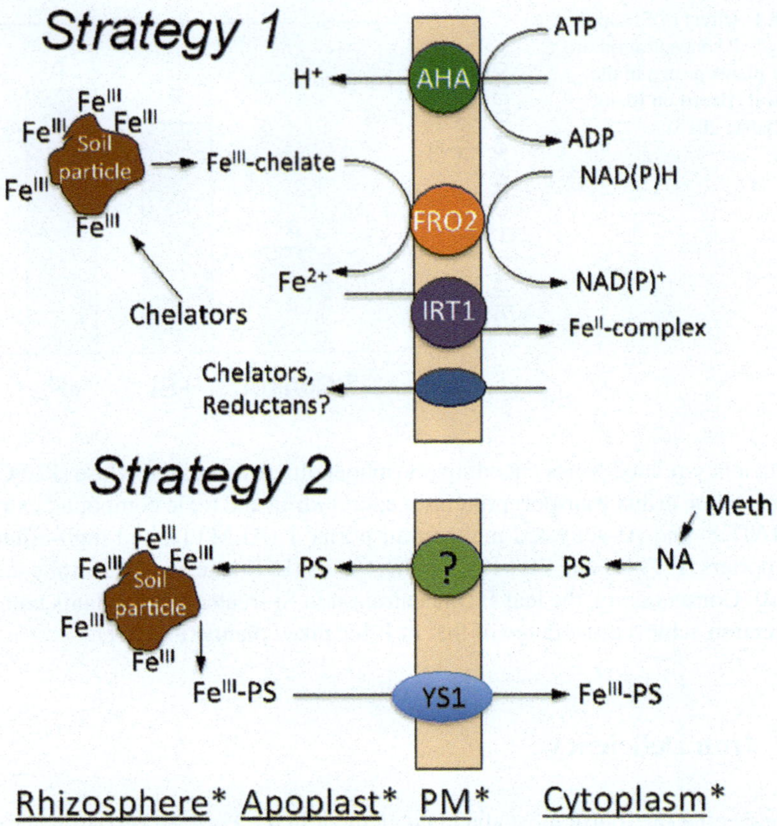

Fig. 8.2 Strategies for acquisition of Fe in response to Fe deficiency in strategy 1 (dicots and nongraminaceous species) and strategy 2 (graminaceous species) plants

includes the root biosynthesis and release of natural Fe(III) chelators such as the mugineic acid (MA) family of phytosiderophores (PSs) that chelate sparingly soluble Fe from the rhizosphere. The PS–Fe(III) complex is subsequently taken up by specific transporters belonging to yellow stripe-like (YSL) family of proteins, which also contribute to shoot transport and seed deposition of Fe (Curie et al. 2009). Nicotianamine (NA) is not only a precursor of PS biosynthesis (strategy 2 plants), but being a strong chelator of Fe(II), it plays a major role in symplastic homeostasis and phloem transport of Fe in plants (Haydon and Cobbett 2007).

Rapid progress has been made predominantly in elucidating the mechanisms of how Si mediates metal excess (see Chap. 5), whereas limited information is available on the relevance of Si nutrition under lack of Fe and other micronutrients. This is partly due to the fact that root responses to Fe deficiency have so far been studied and characterized mainly in nutrient solution experiments in which Si was omitted. Recently it has been reported that the addition of Si to the nutrient solution is able to mitigate Fe-deficiency chlorosis in strategy 1 plants such as cucumber, pumpkin

Fig. 8.3 Root application of Si mitigates symptoms of Fe-deficiency chlorosis in tomato, Si-excluding, Fe-strategy 1 species (Photograph by Miroslav Nikolic)

and soybean (Bityutskii et al. 2010; Pavlovic et al. 2011, 2013; Gonzalo et al. 2013). Interestingly, Si also prevented Fe chlorosis in tomato (*Lycopersicon esculentum*), which is known as a Si-excluder species (Fig. 8.3). However, Si supply had no effect on the nutritional status of zinc (Zn) and Mn, although it diminished leaf necrosis symptoms (Bityutskii et al. 2014), which is most probably due to an indirect effect of Si on enhancing antioxidant defence capacity in plant tissues.

Pavlovic et al. (2013) are the first to demonstrate Si-induced increase of the root apoplastic Fe pool, together with the enhanced expression levels of the proteins (FRO2, IRT1 and AHA1) involved in reduction-based Fe uptake (Fig. 8.4). Moreover, in Fe-deficient cucumber roots, Si influenced the genes involved in the carboxylate, shikimate and phenylpropanoid metabolism, thus resulting in enhanced accumulation of Fe-chelating compounds (organic acids and phenolics) for improved Fe mobilization from the rhizosphere and reutilization of root apoplastic Fe (Pavlovic et al. 2013). Application of Si also facilitated mobility and xylem translocation of Fe towards shoot, along with tissue accumulation of Fe-mobilizing compounds such as citrate (xylem sap, root and shoot tissues) or catechins (in roots) (Pavlovic et al. 2013; Bityutskii et al. 2014). Very recently, Nikolic's group has demonstrated that Si enhanced remobilization of labelled ^{57}Fe from old to younger leaves of cucumber, accompanied by Si-induced expression of genes encoding both NA biosynthesis and YSL transporters responsible for symplastic Fe unloading in leaves and phloem transport of the Fe–NA complex (Pavlovic et al. 2014). Based on these new findings, the alleviating effect of Si appears to be more indirect, by affecting activation of Fe-deficiency-associated genes responsible for enhanced root acquisition and tissue mobilization of Fe.

The alleviating effect of Si in strategy 2 plants is still under debate. Bityutskii et al. (2010) reported that addition of Si to the nutrient solution did not effectively mitigate Fe-deficiency chlorosis in maize and barley (strategy 2). In contrast, Nikolic's group has obtained opposite experimental evidence for maize plants (Fig. 8.5), even though this information is preliminary.

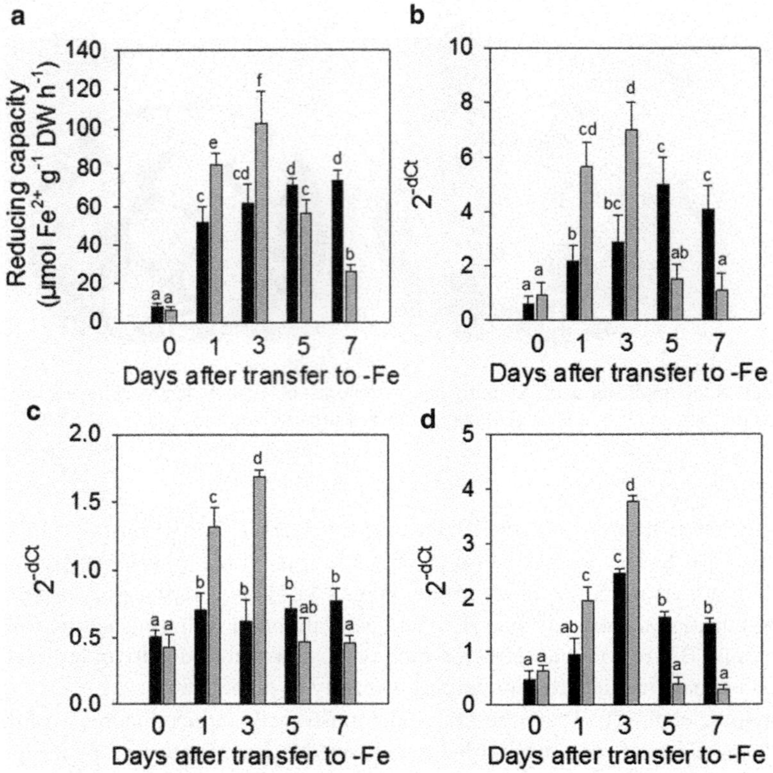

Fig. 8.4 Effect of Si supply on Fe-deficiency responses of cucumber roots during 7 days of Fe deprivation (From Pavlovic et al. 2013). FRO activity (**a**); *FRO2* relative expression (**b**); *IRT1* relative expression (**c**); *AHA1* relative expression (**d**). –Si (*black bars*); +Si (*grey bars*). Plants were pre-cultured in complete nutrient solution (10 µM Fe) –Si or + Si (1.5 mM) for 7 days and then transferred to Fe-free nutrient solution –Si or + Si (1.5 mM)

Although still in progress, these findings provide not only new evidence for the manifold beneficial role of Si in plant nutrition but, in perspective, can also be of practical importance in the development of new sustainable strategies for controlling Fe chlorosis in calcareous soils, which in general are low in available Si (Liang et al. 1994).

8.4 Excess of Nitrogen and Phosphorus

Overall, supply of Si is, in particular, of importance for the cultivation systems with dense planting and high nitrogen (N) application. To attain high yields, modern crop cultivars require large amounts of N fertilizers. However, excess N causes lodging, mutual shading and susceptibility to biotic stresses. Lodging is a major constraint to

Fig. 8.5 Root application of Si mitigates symptoms of Fe-deficiency chlorosis in maize, Si-accumulating, Fe-strategy 2 species (Photograph by Dragana Nikolic)

rice production, especially in high-yielding cultivars, because it causes loss in grain yield and quality and has some indirect effects such as hindering harvesting operation (Fallah 2012). Deposition of silica in rice increases the thickness of the culm wall and the size of the vascular bundle, preventing lodging during typhoons in Japan (Ma et al. 2001). A lodging-resistant wheat cultivar has been shown to have a higher Si content in the culm epidermis than a lodging-sensitive one (Gartner and Paris-Pireyre 1984). Supply of Si counteracts the negative effects of an increasing nitrogen supply on stalk stability and lodging susceptibility of rice (Sadanandan and Varghes 1968; Idris et al. 1975; Balasta et al. 1989, also see Fig. 8.6). Also in dicotyledonous species, such as cucumber, Si increases the rigidity of mature leaves, which are held more horizontally, increases their chlorophyll content and delays their senescence (Adatia and Besford 1986). Leaf erectness is an important factor affecting light interception in dense plant stands. In rice, leaf erectness decreases with increasing N supply, while Si increases leaf erectness (Fig. 8.7) and thus to a large extent counteracts the negative effects of high N supply on light interception (Marschner 1995). The occurrence of blast disease is significantly inhibited by Si application in over-fertilized rice field (Ohyama 1985). These aspects are discussed in more detail in Chap. 9. Excessive application of N fertilizers also results in increase of protein content in rice, and thus adequate supply of Si is important in producing low-protein rice (Morimiya 1996).

Excess P stress can hardly occur in natural soils but was observed in some greenhouse soils where P fertilizer had been heavily applied or in hydroponic culture where a high P concentration was supplied (Ma 2004). It is well known that excess

Fig. 8.6 Beneficial effect of Si fertilization on resistance against rice lodging in a calcareous paddy soil amended with heavy nitrogen (Photography by Yongchao Liang)

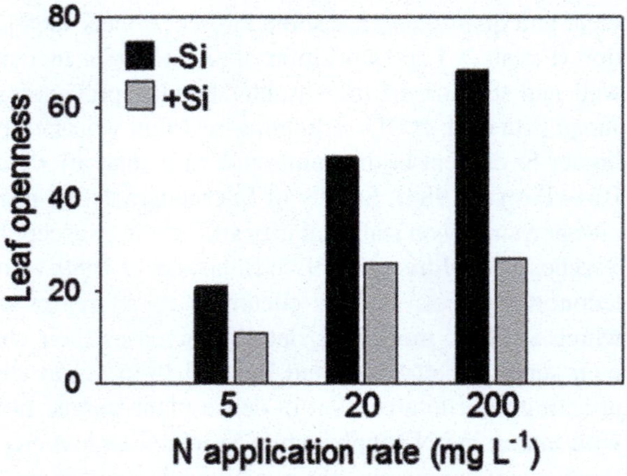

Fig. 8.7 Effect of Si supply of 3 mM on the leaf erectness (expressed as leaf openness; angle between the culm and the lamina) in rice plants at flowering grown at different *N* levels. Means of three rice cultivars recalculated from data of Yoshida (1969)

P might lead to P-induced Zn deficiency showing typical symptoms like leaf chlorosis and necrosis (Cakmak and Marschner 1987). In cucumber, application of Si alleviated the symptoms caused by P excess, resulting in an increase in the proportion of water-soluble to total Zn and an avoidance of toxic P accumulation (Marschner et al. 1990). Also, Marschner et al. (1990) demonstrated that a reduction of P supply as well as an increase in Zn supply prevented this disorder. They postulated that Si supply might increase Zn availability within the leaves and in particular in the leaf apoplast. However, an increase in the ratio of water-soluble to total Zn content by Si application has not been confirmed in the experiments of Rogalla (2001). Therefore, Si-mediated increase in internal Zn availability seems to be achieved only under very specific conditions (Wiese et al. 2007). Another explanation is that Si deposited in the endodermal cells of roots (e.g. Lux et al. 2003) may form apoplastic barriers against the radial movement of P across the root, thus decreasing the excessive uptake of P when the P concentration in the medium is high (Ma 2004). The Si-induced decrease of P uptake has been observed not only in rice (Ma and Takahashi 1990a) but also in some Si non-accumulating plants such as tomato, soybean, strawberry (*Fragaria vesca*) and cucumber (*Cucumis sativus*) (Ma et al. 2001 and references therein).

8.5 UV Radiation

The depletion of the stratospheric ozone layer is leading to more solar ultraviolet B (UV-B) radiation reaching the Earth (Madronich et al. 1998). In general, UV-B negatively affects plant cells, causing generation of ROS such as superoxide anion radicals (O_2^-), hydrogen peroxide (H_2O_2), hydroxyl radicals (·OH) and singlet oxygen (O_2). However, some plant species were unaffected, while in several species UV-B showed even stimulating effects on plant growth (Kakani et al. 2003). Plant sensitivity to UV radiation stress was influenced by water regime and nutrient status (Balakumar et al. 1993). Drought stress has been considered an important environmental factor to induce plant sensitivity to UV radiation (Alexieva et al. 2001). Under the exposure to high UV radiation, plants increase biosynthesis of phenolic compounds to absorb UV by the presence of carbonyl and hydroxyl groups, hence directly protecting their internal tissues from injury (Rozema et al. 1997; Winkel-Shirley 2002; Treutter 2005).

It is well documented that enhanced Si supply lowers stress induced by UV-B radiation in different plant species, including rice, soybean and wheat (Goto et al. 2003; Li et al. 2004; Fang et al. 2011; Shen et al. 2010; Yao et al. 2011). In leaves, Si is deposited in the space immediately beneath the cuticle layer, forming a cuticle–Si double layer in leaf blades (Hodson and Sangster 1988; Ma and Yamaji 2006; Currie and Perry 2007; Schaller et al. 2012). Such Si double layer may act like a glass layer and decrease the further transmission of UV radiation from the epidermis (Gatto et al. 1998). It still remains unclear whether the protective role of Si is associated with the plant's ability to increase formation of Si double layer in response to the UV-B or only a side effect of species-specific variations of Si accumulated in plants. Also, the role of Si in the metabolic response to UV radiation stress such as biosynthesis of low-molecular-weight UV-absorbing compounds (e.g. phenolic acids and flavonoids) is still not clear. For instance, Goto et al. (2003) reported that Si application increased Si deposits in rice leaves, but decreased cinnamyl alcohol hydrogenase activity and accumulation of ferulic and p-coumaric

Table 8.2 Effect of Si supply on chlorophyll content, net photosynthetic rate (*Ps*), lipid peroxidation, catalase (CAT) activity and H_2O_2 concentration in leaves of soybean exposed to high UV-B radiation

Treatment		Chlorophyll (mg g^{-1} fw)	*Ps* (μmol CO_2 m^{-2} s^{-1})	Lipid peroxidation (MDA nmol g^{-1} fw)	H_2O_2 (μmol g^{-1} fw)	CAT (u g^{-1} fw min^{-1})
Control	−Si	1.6	8.7	0.4	220	24
	+Si	1.8	9.2	0.5	240	26
UV	−Si	0.2	2.6	0.9	490	98
	+Si	1.4	3.0	0.6	360	36

Based on Shen et al. (2010)

acids in leaf blades. Thus, Si-treated rice plants had lower UV absorbance around 320 nm via carbonyl groups or conjugated carbonyls of phenolic acids in the most cell types of leaf blades. It is possible that Si acts indirectly by enhancement of antioxidative enzyme activities and thereby contributes to higher ROS consumption. Furthermore, the experiment of Shen et al. (2010) showed that application of Si decreased lipid peroxidation, activities of the antioxidative enzymes and generation of H_2O_2 in soybean seedlings subjected to UV-B radiation stress (Table 8.2). Therefore, UV absorption due to the optical properties of accumulated Si might be an advantage with respect to the energy demand for producing UV-absorbing substances, as the alternative protection measure against UV radiation (Schaller et al. 2013).

8.6 Flooding

Despite the fact that rice normally grows under anaerobic soil conditions, the exact role of Si in alleviating flooding stress is still unclear. Wetland plants possess diverse anatomical, morphological and physiological features that maintain aerobic metabolism and growth and thereby survival in waterlogged soils under anaerobic and reductive conditions. Rice, like other wetland species, is adapted to a low-oxygen environment by the internal aeration of root via the aerenchyma, which provides a low-resistance pathway for the diffusion of oxygen within the root (Colmer 2003, 2006). To counteract the diffusion of oxygen from the root to the anaerobic rhizosphere, rice roots contain higher amounts of suberin and lignin in the outer root parts, forming a stronger barrier to radial oxygen loss (Kotula et al. 2009). Lignin and suberin metabolism in plants shares the phenylpropanoid pathway. Parts of this metabolic pathway were enhanced by Si supply in plants subjected to various stresses (e.g. Kidd et al. 2001; Cai et al. 2008; Führs et al. 2009; Pavlovic et al. 2013). The effects of Si on rice root anatomy (e.g. development of casparian bands in the exodermis and endodermis), and on the transcription of genes related to suberin and lignin biosynthesis, have recently been demonstrated (Fleck et al. 2011).

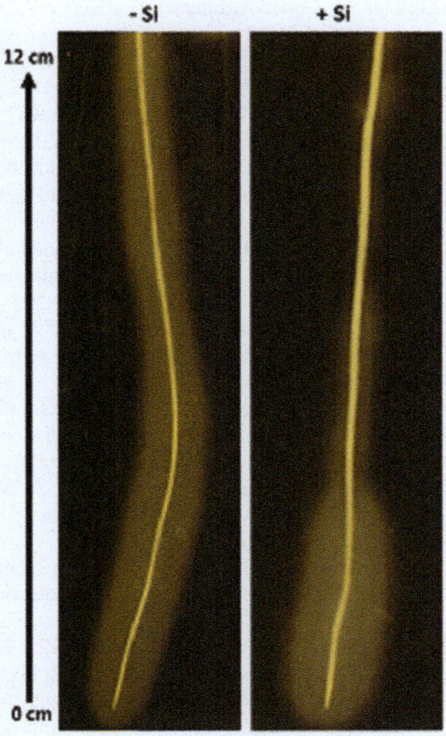

Fig. 8.8 Effect of Si supply on the oxidation power of adventitious rice roots (From Fleck et al. 2011)

In addition, under reductive conditions of flooded soils, the concentrations of readily available Fe(II) and Mn(II) are high, and plants can suffer from Fe and/or Mn toxicity (Marschner 1995). Silicon nutrition of rice plants also increased the oxidation power of submerged roots (Fig. 8.8), leading to an oxidation of Fe^{2+} and Mn^{2+} ions and their subsequent precipitation on the root surface, thereby decreasing Fe and Mn uptake (Okuda and Takahashi 1961). In plant tissues, excess Fe and Mn can induce production of ROS through nonenzymatic Fenton-like reaction (for details see Chap. 5). Indeed, under conditions of soil flooding, the application of Si decreases the intensity of ROS destruction in barley roots, however without significant changes in the activities of antioxidant enzymes (Balakhnina et al. 2012).

References

Abro SA, Qureshi R, Soomro FM, Mirbahar AA, Jakhar GS. Effects of silicon levels on growth and yield of wheat in silty loam soil. Pak J Bot. 2009;41:1385–90.

Adatia MH, Besford RT. The effects of silicon on cucumber plants grown in recirculating nutrient solution. Ann Bot. 1986;58:343–51.

Alexieva V, Sergiev I, Mapelli S, Karanov E. The effect of drought and ultraviolet radiation on growth and stress markers in pea and wheat. Plant Cell Environ. 2001;24:1337–44.

Alpaslan M, Gunes A. Interactive effects of boron and salinity stress on the growth, membrane permeability and mineral composition of tomato and cucumber plants. Plant Soil. 2001;236:123–8.

Balakhnina TI, Matichenkov VV, Wlodarczyk T, Borkowska A, Nosalewicz M, Fomina IR. Effects of silicon on growth processes and adaptive potential of barley plants under optimal soil watering and flooding. Plant Growth Regul. 2012;67:35–43.

Balakumar T, Hani Babu Vincent V, Paliwal K. On the interaction of UV-B radiation (280–315 nm) with water stress in crop plants. Physiol Plant. 1993;87:217–22.

Balasta MLFC, Perez CM, Juliano BO, Villareal CP, Lott JNA, Roxas DB. Effect of silica level on some properties of *Oriza sativa* straw and hull. Can J Bot. 1989;67:2356–63.

Ben-Gal A, Shani U. Yield, transpiration and growth of tomatoes under combined excess boron and salinity stress. Plant Soil. 2002;247:211–21.

Bityutskii NP, Yakkonen KL, Zlotina MM. Vliyanie kremniya na proyavlenie khlorosa rasteniy v usloviyakh defizita zhelesa i marganza. Agrokhimiia. 2010;2:45–51 (In Russian, with English summary).

Bityutskii N, Pavlovic J, Yakkonen K, Maksimovic V, Nikolic M. Contrasting effect of silicon on iron, zinc and manganese status and accumulation of metal-mobilizing compounds in micronutrient-deficient cucumber. Plant Physiol Biochem. 2014;74:205–11.

Broadley M, Brown P, Cakmak I, Rengel Z, Zhao F. Function of nutrients: micronutrients. In: Marschner P, editor. Marschner's mineral nutrition of higher plants. 3rd ed. Amsterdam: Elsevier; 2011. p. 233–43.

Brown PH, Bellaloui N, Hu H, Dandekar A. Transgenically enhanced sorbitol synthesis facilitates phloem boron transport and increases tolerance of tobacco to boron deficiency. Plant Physiol. 1999;119:17–20.

Cai K, Gao D, Luo S, Zeng R, Yang J, Zhu X. Physiological and cytological mechanisms of silicon-induced resistance in rice against blast disease. Physiol Plant. 2008;134:324–33.

Cakmak I. Activity of ascorbate-dependent H_2O_2-scavenging enzymes and leaf chlorosis are enhanced in magnesium- and potassium-deficient leaves, but not in phosphorus-deficient leaves. J Exp Bot. 1994;45:1259–66.

Cakmak I. The role of potassium in alleviating detrimental effects of abiotic stresses in plants. J Plant Nutr Soil Sci. 2005;168:521–30.

Cakmak I, Marschner H. Mechanism of phosphorus-induced zinc deficiency in cotton III. Changes in physiological availability of zinc in plants. Physiol Plant. 1987;70:13–20.

Catarecha P, Segura MD, Franco-Zorrilla JM, García-Ponce B, Lanza M, Solano R, Paz-Ares J, Leyva A. A mutant of the Arabidopsis phosphate transporter PHT1; 1 displays enhanced arsenic accumulation. Plant Cell. 2007;19:1123–33.

Cesco S, Neumann G, Tomasi N, Pinton R, Weisskopf L. Release of plant-borne flavonoids into the rhizosphere and their role in plant nutrition. Plant Soil. 2010;329:1–25.

Colmer TD. Long-distance transport of gases in plants: a perspective on internal aeration and radial oxygen loss from roots. Plant Cell Environ. 2003;26:17–36.

Colmer TD, Cox MCH, Voesenek LACJ. Root aeration in rice (*Oryza sativa* L.): evaluation of oxygen, carbon dioxide, and ethylene as possible regulators of root acclimatizations. New Phytol. 2006;170:767–78.

Cordell D, Drangert J-O, White S. The story of phosphorus: global food security and food for thought. Global Environ Change. 2009;19:292–305.

Curie C, Cassin G, Couch D, Divol F, Higuchi K, Le Jean M, Misson J, Schikora A, Czernic P, Mari S. Metal movement within the plant: contribution of nicotianamine and yellow stripe 1-like transporters. Ann Bot. 2009;103:1–11.

Currie HA, Perry CC. Silica in plants: biological, biochemical and chemical studies. Ann Bot. 2007;100:1383–9.

de Melo SP, Monteiro FA, Manfredini D. Silicate and phosphate combinations for marandu palisadegrass growing on an oxisol. Sci Agric (Piracicaba, Braz); 2007; 64: 275–281.

Eneji AE, Inanaga S, Muranaka S, Li J, Hattori T, An P, Tsuji W. Growth and nutrient use in four grasses under drought stress as mediated by silicon fertilizers. J Plant Nutr. 2008;31:355–65.

Fallah A. Silicon effect on lodging parameters of rice plants under hydroponic culture. Int Agric Sci. 2012;2:630–4.

Fang CX, Wang QS, Yu Y, Huang LK, Wu XC, Lin WX. Silicon and its uptaking gene Lsi1 in regulation of rice UV-B tolerance. Acta Agron Sin. 2011;37:1005–11.

Fisher RA. A preliminary note on the effect of sodium silicate in increasing the yield of barley. J Agric Sci. 1929;19:132–9.

Fleck AT, Nye T, Repenning C, Stahl F, Zahn M, Schenk MK. Silicon enhances suberization and lignification in roots of rice (*Oryza sativa*). J Exp Bot. 2011;62:2001–11.

Führs H, Götze S, Specht A, Erban A, Gallien S, Heintz D, Van Dorsselaer A, Kopka J, Braun H-P, Horst WJ. Characterization of leaf apoplastic peroxidases and metabolites in *Vigna unguiculata* in response to toxic manganese supply and silicon. J Exp Bot. 2009;60:1663–78.

Gartner S, Paris-Pireyre N. Microanalysis of silica and resistance to mechanical lodging in soft wheat. J Phys Colloque. 1984;45:511–4.

Gatto A, Escoubas L, Roche P, Commandre M. Simulation of the degradation of optical glass substrates caused by UV irradiation while coating. Optics Commun. 1998;148:347–54.

Gilbert N. The disappearing nutrient. Nature. 2009;461:716–8.

Glassop D, Smith SE, Smith FW. Cereal phosphate transporters associated with the mycorrhizal pathway of phosphate uptake into roots. Planta. 2005;222:688–98.

Gonzalo MJ, Lucena JJ, Hernández-Apaolaza L. Effect of silicon addition on soybean (*Glycine max*) and cucumber (*Cucumis sativus*) plants grown under iron deficiency. Plant Physiol Biochem. 2013;70:455–61.

Goto M, Ehara H, Karita S, Takabe K, Ogawa N, Yamada Y, Ogawa S, Yahaya MS, Morita O. Protective effect of silicon on phenolic biosynthesis and ultraviolet spectral stress in rice crop. Plant Sci. 2003;164:349–56.

Gunes A, Inal A, Bagci EG, Coban S. Silicon-mediated changes on some physiological and enzymatic parameters symptomatic of oxidative stress in barley grown in sodic-B toxic soil. J Plant Physiol. 2007a;164:807–11.

Gunes A, Inal A, Bagci EG, Coban S, Pilbeam DJ. Silicon mediates changes to some physiological and enzymatic parameters symptomatic for oxidative stress in spinach (*Spinacia oleracea* L.) grown under B toxicity. Sci Hortic. 2007b;113:113–9.

Gunes A, Inal A, Bagci EG, Coban S, Sahin O. Silicon increases boron tolerance and reduces oxidative damage of wheat grown in soil with excess boron. Biol Plant. 2007c;51:571–4.

Haydon MJ, Cobbett CS. Transporters of ligands for essential metal ions in plants. New Phytol. 2007;174:499–506.

Hingston FJ, Atkinson RJ, Posner AM, Quirk JP. Specific adsorption of anions. Nature. 1967; 215:1459–61.

Hodson MJ, Sangster AG. Observations on the distribution of mineral elements in the leaf of wheat (*Triticum aestivum* L) with particular reference to silicon. Ann Bot. 1988;62:463–71.

Idris M, Hossain MM, Choudary FA. The effect of silicon on lodging of rice in presence of added nitrogen. Plant Soil. 1975;43:691–5.

Inal A, Pilbeam DJ, Gunes A. Silicon increases tolerance to boron toxicity and reduces oxidative damage in barley. J Plant Nutr. 2009;32:112–28.

Jeong J, Connolly EL. Iron uptake mechanisms in plants: functions of the FRO family of ferric reductases. Plant Sci. 2009;176:709–14.

Kakani VG, Reddy KR, Zhao D, Sailaja K. Field crop responses to ultraviolet-B radiation: a review. Agric For Meteor. 2003;120:191–218.

Karabal E, Yücel M, Öktem HA. Antioxidant responses of tolerant and sensitive barley cultivars to boron toxicity. Plant Sci. 2003;164:925–33.

Karthikeyan AS, Varadarajan DK, Mukatira UT, D'Urzo MP, Damsz B, Raghothama KG. Regulated expression of Arabidopsis phosphate transporters. Plant Physiol. 2002;130:221–33.

Kidd PS, Llugany M, Poschenrieder C, Gunse B, Barcelo J. The role of root exudates in aluminium resistance and silicon-induced amelioration of aluminium toxicity in three varieties of maize (*Zea mays* L.). J Exp Bot. 2001;359:1339–52.

Kinrade SD, Del Nin JW, Schach AS, Sloan TA, Wilson KL, Knight CTG. Stable five- and six-coordinated silicate anions in aqueous solution. Science. 1999;285:1542–5.

Kochian LV, Hoekenga OA, Piñeros MA. How do crop plants tolerate acid soil? Mechanisms of aluminum tolerance and phosphorus efficiency. Annu Rev Plant Biol. 2004;55:459–93.

Kostic L, Nikolic N, Samardzic J, Milisavljevic M, Maksimovic V, Cakmak D, et al. Liming of anthropogenically acidified soil promotes phosphorus acquisition in the rhizosphere of wheat. Biol Fertil Soils. 2015;51:289–98.

Kostic-Kravljanac LM. Modulation of the processes in wheat rhizosphere in responses to the amendments of soils degraded by mining waste. PhD thesis, University of Belgrade; 2015 (In Serbian with English abstract).

Kotula L, Ranathunge K, Steudle E. Apoplastic barriers effectively block oxygen permeability across outer cell layers of rice roots under deoxygenated conditions: roles of apoplastic pores and of respiration. New Phytol. 2009;184:909–17.

Li XF, Ma JF, Matsumoto H. Pattern of aluminum-induced secretion of organic acids differs between rye and wheat. Plant Physiol. 2000;123:1537–43.

Li WB, Shi XH, Wang H, Zhang FS. Effects of silicon on rice leaves resistance to ultraviolet-B. Acta Bot Sin. 2004;46:691–7.

Liang YC, Shen ZG. Interaction of silicon and boron in oilseed rape plants. J Plant Nutr. 1994;17:415–25.

Liang YC, Ma TS, Li FJ, Feng YJ. Silicon availability and response of rice and wheat to silicon in calcareous soils. Commun Soil Sci Plant Anal. 1994;25:2285–97.

Liang YC, Wong JWC, Wei L. Silicon-mediated enhancement of cadmium tolerance in maize (*Zea mays* L.) grown in cadmium contaminated soil. Chemosphere. 2005;58:475–83.

Liang YC, Sun WC, Zhu YG, Christie P. Mechanisms of silicon- mediated alleviation of abiotic stresses in higher plants: a review. Environ Pollut. 2007;147:422–8.

Liebersbach H, Steingrobe B, Claassen N. Roots regulate ion transport in the rhizosphere to counteract reduced mobility in dry soil. Plant Soil. 2004;260:79–88.

Lux A, Luxová M, Abe J, Tanimoto E, Taiichiro H, Shinobu I. The dynamics of silicon deposition in the sorghum root endodermis. New Phythol. 2003;158:437–41.

Ma JF. Role of silicon in enhancing the resistance of plants to biotic and abiotic stresses. Soil Sci Plant Nutr. 2004;50:11–8.

Ma JF, Takahashi E. Effect of silicic acid on phosphorus uptake by rice plant. Soil Sci Plant Nutr. 1989;35:227–34.

Ma JF, Takahashi E. Effect of silicon on the growth and phosphorus uptake of rice. Plant Soil. 1990a;126:115–9.

Ma JF, Takahashi E. The effect of silicic acid on the growth of rice grown in a P-deficient soil. Plant Soil. 1990b;126:121–5.

Ma JF, Takahashi E. Effect of silicate on phosphate availability for rice in a P-deficient soil. Plant Soil. 1991;133:151–5.

Ma JF, Yamaji N. Silicon uptake and accumulation in higher plants. Trends Plant Sci. 2006;11:392–7.

Ma JF, Miyake Y, Takahashi E. Silicon as a beneficial element for crop plants. In: Datnoff LE, Snyder GH, Korndörfer GH, editors. Silicon in agriculture. Amsterdam: Elsevier; 2001. p. 17–39.

Madronich S, McKenzie RL, Björn LO, Caldwell MM. Changes in biologically active UV radiation reaching the Earth's surface. J Photochem Photobiol B. 1998;46:5–19.

Marschner H. Mineral nutrition of higher plants. 2nd ed. London: Academic Press; 1995.

Marschner H, Oberle H, Cakmak L, Romheld V. Growth enhancement by silicon in cucumber (*Cucumis sativus*) plants depends on imbalance in phosphorus and zinc supply. Plant Soil. 1990;124:211–9.

Miao B-H, Han X-G, Zhang W-H. The ameliorative effect of silicon on soybean seedlings grown in potassium-deficient medium. Ann Bot. 2010;105:967–73.

Miwa K, Takano J, Omori H, Seki M, Shonozaki K, Fujiwara T. Plants tolerant of high boron levels. Science. 2007;318:1417.

Morimiya Y. Role of Si in production of low protein rice and diagnosis parameters. Jpn J Soil Sci Plant Nutr. 1996;67:696–700.

Nable RO, Lance RCM, Cartwright B. Uptake of boron and silicon by barley genotypes with differing susceptibilities to boron toxicity. Ann Bot. 1990;66:83–90.

Nable RO, Banuelos GS, Paull JG. Boron toxicity. Plant Soil. 1997;193:181–98.

Neumann G, Römheld V. Root excretion of carboxylic acids and protons in phosphorus-deficient plants. Plant Soil. 1999;211:121–30.

Neumann G, Römheld V. The release of root exudates as affected by the plant's physiological status. In: Pinton R, Varanini Z, Nannipieri P, editors. The rhizosphere: biochemistry and organic substances at the soil–plant interface. Boca Raton: CRC; 2007. p. 23–72.

Obihara CH, Russell EW. Specific adsorption of silicate and phosphorus by soils. J Soil Sci. 1972;23:105–17.

Ohyama N. Amelioration of cold weather damage of rice by silicate fertilizer application. Agric Hortic. 1985;60:1385–9.

Okuda A, Takahashi E. Studies on the physiological role of silicon in crop plants. 4. Effect of silicon on the growth of barley, tomato, raddish, green onion, Chinese cabbage and their nutrients uptake. J Sci Soil Manure Jpn. 1961;32:623–6.

Owino-Gerroh C, Gascho GJ. Effect of silicon on low pH soil phosphorus sorption and on uptake and growth of maize. Commun Soil Sci Plant Anal. 2004;35:2369–78.

Paull JG, Nable RO, Ratjhen AJ. Physiological and genetic control of the tolerance of wheat to high concentrations of boron and implications for plant breeding. Plant Soil. 1992;146:251–60.

Pavlovic J, Samardzic J, Nikolic M. Interactions between iron and silicon. In: Proceedings of 17 international symposium on iron nutrition and interactions in plants; 2014 July 6–10, Gratersleben/Quedlinburg, Germany, p. 38.

Pavlovic J, Samardzic J, Ilic P, Maksimovic V, Kostic L, Stevic N, Nikolic N, Liang YC, Nikolic M. Silicon ameliorates iron deficiency chlorosis in strategy I plants: first evidence and possible mechanism(s). In Proceedings of The 5th International Conference on Silicon in Agriculture; 2011 September 13–18, Beijing, China; 2011. p. 137.

Pavlovic J, Samardzic J, Masimović V, Timotijevic G, Stevic N, Laursen KH, Hansen TH, Husted S, Schjoerring JK, Liang Y, Nikolic M. Silicon alleviates iron deficiency in cucumber by promoting mobilization of iron in the root apoplast. New Phytol. 2013;198:1096–107.

Pettigrew WT. Potassium influences on yield and quality production for maize, wheat, soybean and cotton. Physiol Plant. 2008;133:670–81.

Plaxton WC. Metabolic aspects of phosphate starvation in plants. In: Lynch JP, Deikman L, editors. Phosphorus in plant biology: regulatory roles in molecular, cellular, organismic, and ecosystem processes. Rockville: American Society of Plant Physiology; 1998. p. 229–38.

Polster J, Schwenk M. The role of boron, silicon and nucleic bases on pollen tube growth of *Lilium longiflorum* (L.). Z Naturforsch. 1992;47:102–8.

Reid R. Identification of boron transporter genes likely to be responsible for tolerance to boron toxicity in wheat and barley. Plant Cell Physiol. 2007;48:1673–8.

Reid R, Fitzpatrick K. Influence of leaf tolerance mechanisms and rain on boron toxicity in barley and wheat. Plant Physiol. 2009;151:413–20.

Rogalla H. Einfluss von Silizium auf Austauschereigenschaften des Apoplasten und indungszustand von Nährstoffen in Blättern. Dissertation, University of Hohenheim. Verlag Grauer, Stuttgart; 2001.

Römheld V, Nikolic M. Iron. In: Barker AV, Pielbeam DJ, editors. Handbook of plant nutrition. Boca Raton: CRC; 2007. p. 329–50.

Rothbuhr L, Scott F. A study of the uptake of silicon and phosphorus of wheat plants, with radiochemical methods. Biochem J. 1957;65:241–5.

Roy AC, Ali MY, Fox RL, Silva JA. Influence of calcium silicate on phosphate solubility and availability in Hawaiian latosols. Proceedings of the International Symposium on Soil Fertility Evaluation, New Delhi, Vol 1; 1971. p. 757–65.

Rozema J, Vande Staaij J, Bjorn LO, Caldwell M. UV-B as an environmental factor in plant life: stress and regulation. Trends Ecol Evol. 1997;12:22–8.

Ryan PR, Skerrett M, Findlay GP, Delhaize E, Tyerman SD. Aluminum activates an anion channel in the apical cells of wheat roots. Proc Natl Acad Sci U S A. 1997;94:6547–52.

Sadanandan AK, Varghes EJ. Studies on the silicate nutrition of rice in the laterite soil of Keral. I. Effect on growth and yield. Madras Agric J. 1968;11:261–4.

Santi S, Schmidt W. Dissecting iron deficiency-induced proton extrusion in Arabidopsis roots. New Phytol. 2009;183:1072–84.

Schaller J, Brackhage C, Dudel E. Silicon availability changes structural carbon ratio and phenol content of grasses. Environ Exp Bot. 2012;77:283–7.

Schaller J, Brackhage C, Bäucker E, Dudel EG. UV-screening of grasses by plant silica layer? J Biosci. 2013;38:413–6.

Schmidt W. Mechanisms and regulation of reduction-based iron uptake in plants. New Phytol. 1999;141:1–26.

Schnurbusch T, Hayes J, Hrmova M, Baumann U, Ramesh SA, Tyerman SD, Langridge P, Sutton T. Boron toxicity tolerance in barley through reduced expression of the multifunctional aquaporin HvNIP2;1. Plant Physiol. 2010;153:1706–15.

Shen X, Zhou Y, Duan L, Li Z, Eneji AE, Li J. Silicon effects on photosynthesis and antioxidant parameters of soybean seedlings under drought and ultraviolet-B radiation. J Plant Physiol. 2010;167:1248–52.

Smyth TJ, Sanchez PA. Effects of lime, silicate, and phosphorus applications to an Oxisol on phosphorus sorption and ion retention. Soil Sci Soc Am J. 1980;44:500–5.

Soylemezoglu G, Demir K, Inal A, Gunes A. Effect of silicon on antioxidant and stomatal response of two grapevine (*Vitis vinifera* L.) rootstocks grown in boron toxic, saline and boron toxic-saline soil. Sci Hortic. 2009;123:240–6.

Sutton T, Baumann U, Hayes J, Collins NC, Shi BJ, Schnurbusch T, Hay A, Mayo G, Pallotta M, Tester M, et al. Boron-toxicity tolerance in barley arising from efflux transporter amplification. Science. 2007;318:1446–9.

Tallberg P, Lukkari K, Räike A, Lehtoranta J, Leivuori M. Applicability of a sequential P fractionation procedure to Si in sediment. J Soil Sediment. 2009;9:594–603.

Theodoru ME, Plaxton WC. Metabolic adaptations of plant respiration to nutritional phosphate deprivation. Plant Physiol. 1993;101:339–44.

Tittarelli A, Milla L, Vargas F, Morales A, Neupert C, Meisel LA, Salvo GH, Peñaloza E, Muñoz G, Corcuera LJ, Silva H. Isolation and comparative analysis of the wheat TaPT2 promoter: identification in silicon of new putative regulatory motifs conserved between monocots and dicots. J Exp Bot. 2007;58:2573–82.

Treutter D. Significance of flavonoids in plant resistance and enhancement of their biosynthesis. Plant Biol. 2005;7:581–91.

Tuominen L, Hartikainen H, Kairesalo T, Tallberg P. Increased bioavailability of sediment phosphorous due to silicate enrichment. Water Res. 1998;32:2001–8.

Vance CP. Symbiotic nitrogen fixation and phosphorus acquisition: plant nutrition in a world of declining renewable resources. Plant Physiol. 2001;127:390–7.

Vert G, Grotz N, Dedaldechamp F, Gaymard F, Guerinot ML, Briat JF, Curie C. IRT1, an Arabidopsis transporter essential for iron uptake from the soil and for plant growth. Plant Cell. 2002;14:1223–33.

Vose PB. Iron nutrition in plants: a world overview. J Plant Nutr. 1982;5:233–49.

Wiese H, Nikolic M, Römheld V. Silicon in plant nutrition. In: Sattelmacher B, Horst WJ, editors. The apoplast of higher plants: compartment of storage, transport and reactions. Heidelberg: Springer; 2007. p. 33–47.

Winkel-Shirley B. Biosynthesis of flavonoids and effects of stress. Curr Opin Plant Biol. 2002;5:218–23.

Wu H, Lihua L, Du J, Yuan Y, Cheng X, Ling HQ. Molecular and biochemical characterization of the Fe(III)chelate reductase gene family in Arabidopsis thaliana. Plant Cell Physiol. 2005;46:1505–14.

Yang D, Zhang YL, Liu MD, Yu N. Silicon liberation characteristics of soil and its affecting factors after applying slag mucks. II. Relationship between silicon liberation and pH. Chine J Soil Sci. 2008;39:726–9 (In Chinese with English abstract).

Yao XQ, Chu JZ, Cai KZ, Liu L, Shi JD, Geng WY. Silicon improves the tolerance of wheat seedlings to ultraviolet-B stress. Biol Trace Element Res. 2011;143:507–17.

Yermiyahu U, Ben-Gal A, Keren R, Reid RJ. Combined effect of salinity and excess boron on plant growth and yield. Plant Soil. 2008;304:73–87.

Yoshida S. Effect of silica and nitrogen supply on some leaf characteristics of the rice plants. Plant Soil. 1969;31:48–56.

Chapter 9
Silicon and Plant–Pathogen Interactions

Abstract The beneficial effects of silicon (Si) at preventing plant diseases have been known for many decades, and the list of plant–pathogen interactions influenced by Si keeps expanding. However, it is quite evident that the prophylactic properties of Si will vary greatly depending on the plant and the pathogen. The recent discovery of Si-specific transporters in rice roots has been instrumental in identifying plants that possessed such transporters and were thus genetically disposed to accumulate Si in their tissues and benefit more from Si amendments. For the same reason, soil applications of Si lead to significantly more Si accumulation in plant tissues than foliar applications and are likely to yield better results. An issue that has received less attention is the specificity of Si with respect to plant pathogens. Based on the literature, there is an overwhelming abundance of reports describing the positive effects of Si against fungal pathogens compared to bacteria, virus or nematodes. Among fungal pathogens, those described as biotrophic or hemibiotrophic, such as powdery mildews and *Magnaporthe grisea,* appear to be better controlled by Si. The reasons behind this apparent specificity are still unknown but recent data suggest that Si would interfere with effector proteins released by those pathogens, thus allowing the plant to mount better defence reactions. Our recent progress at understanding Si properties will contribute to optimizing its use in the context of sustainable agriculture.

Keywords Absorption • Biotrophic pathogens • Effector proteins • Mode of action • Silicon properties

9.1 Introduction

Experimental evidence dating back to 1860 has convincingly shown that most plants, with the notable exception of horsetail, can complete their life cycle without silicon (Si) (Epstein 1999). The nutritional role of Si in plant metabolism appears to be limited, and certainly, its essentiality in plant growth has not been clearly established (Epstein 1994, 1999, 2001). Nevertheless, there has been accumulating evidence that Si absorption plays an important function in alleviating biotic and abiotic stresses. Many reports have implicated Si with improved plant growth in situations of nutrient deficiency or excess (Bloemhard and Van Moolenbroek 1994; Datnoff

Y. Liang et al., *Silicon in Agriculture*, DOI 10.1007/978-94-017-9978-2_9

et al. 2001), and more commonly, Si fertilization has been linked to increased resistance of plants to diseases, namely, in the case of powdery mildew pathogens on wheat (*Triticum aestivum*) (Leusch and Buchenauer 1989; Bélanger et al. 2003), barley (*Hordeum vulgare*) (Jiang et al. 1989), rose (*Rosa rugosa*) (Shetty et al. 2012), cucumber (*Cucumis sativus*) (Wagner 1940; Miyake and Takahashi 1983a; Adatia and Besford 1986; Menzies et al. 1991; Samuels et al. 1991a, b), muskmelon (*Cucumis melo*), zucchini squash (*Cucurbita pepo*) (Menzies et al. 1992), grape (*Vitis vinifera*) (Bowen et al. 1992) and dandelion (*Taraxacum mongolicum*) (Bélanger et al. 1995) and for other diseases such as blast (*Pyricularia grisea*) and brown spot (*Bipolaris oryzae*) on rice (*Oryza sativa*) (Datnoff et al. 1997; Rodrigues et al. 2003a, b; Rodrigues et al. 2004; Sun et al. 2010; Ning et al. 2014), Fusarium wilt (Miyake and Takahashi 1983b) and root rots (Chérif et al. 1992a, 1994).

Despite the accumulating scientific evidence of the benefits of Si in agriculture, use of Si products or amendments is still misunderstood because of the many intricacies surrounding Si properties, absorption and efficacy. In this chapter, we will review the latest scientific developments in terms of Si research and plant protection, in an effort to clarify the prophylactic role of Si.

9.2 Soil Versus Foliar Applications

One of the most controversial issues surrounding Si use and properties relates to its application. Given the difficulties in applying Si in the form of silicic acid to the root system, it has been suggested that foliar applications could overcome these limitations and confer the same benefits. Some companies have developed some Si-based products specifically for foliar applications.

The current debate regarding the efficacy of Si as foliar applications lies in two main areas: absorption and mode of action. As explained above, root transporters that carry Si to the plant's upper parts mediate Si absorption. It is well known that plants that absorb the most Si are the ones that respond the best to Si feeding, so it would be fair to conclude that Si absorption is essential for a plant to derive benefits from it. Herein lies the problem because there is no strong evidence that plants can absorb Si through the leaves, which would render foliar applications much less useful. As a matter of fact, recent studies have shown that foliar applications will not lead to significant amounts of Si being absorbed in plants compared to root applications. This was particularly true in the case of wheat where plants sprayed with a Si solution did not accumulate more Si than controls (Guével et al. 2007). Expectedly, the prophylactic effects of Si were significantly more beneficial on plants where Si was fed through the roots and that had absorbed high amounts of Si. Similar results were recently reported with soybean in trials against soybean rust (Rodrigues et al. 2009).

Despite the overwhelming evidence that root applications of Si will lead to higher absorption and better prophylactic results, there is still a movement, both commercial and scientific, to promote the benefits of foliar applications of Si. This situation has created a climate of confusion about Si and its properties because, under the premise of a similar active ingredient, we have two completely distinct phenomena at play. It is therefore important to categorize each one of them in order to better understand how foliar and root applications of Si interact with the plant.

In the case of foliar applications, the literature does indeed contain reports of disease control with Si-based solutions. In most cases, the best results have been obtained against foliar pathogens such as powdery mildews. However, very few studies have looked into the mechanisms behind the process, or even if treated plants absorbed Si. Those that did showed conclusively that plants did not absorb Si under foliar treatments (Liang et al. 2005; Guével et al. 2007; Rodrigues et al. 2009). In an interesting experiment where Hoagland's solution was used as a control, the authors showed that powdery mildew repression was similar with the latter solution as with the solution made from potassium silicate. Incidentally, salt sprays, including potassium salts such as potassium carbonates and potassium phosphates, have been reported many times to reduce disease incidence with particular emphasis on powdery mildews (Reuveni and Reuveni 1995; Bélanger and Labbé 2002; Ehret et al. 2002), an observation in line with the results obtained with Hoagland's solution. Liang et al. (2005) suggested that disease reduction caused by foliar sprays of potassium silicate was the result of an osmotic effect on spores germinating at the leaf surface. Based on this evidence, it thus appears that any foliar sprays of salts, and more specifically potassium salts, including potassium silicate, can afford a certain control of disease incidence as a result of a direct effect on the pathogen rather than one mitigated by the plant. As such, while it is acknowledged that Si foliar sprays can yield positive effects against some pathogens, these effects do not appear to be related to the unique and intrinsic properties of Si *in planta* and should therefore be discussed in a separate context.

9.3 Silicon Specificity with Plants

As discussed in Chap. 4, the protective role of Si against plant pathogens will be greatly influenced by the ability of the plant species under treatment to absorb the element. For this reason, some plant species will not respond to a Si treatment and results will often be interpreted as a failure by Si to confer protection, rather than a biological limitation. It is therefore always important to make sure that a plant is Si competent before exposing it to the element. As a general rule, all monocots are Si accumulators, and all studies done to date confirm that they do carry the influx genes for Si transport. For dicots, the picture is not as clear as most dicots are unable to accumulate Si due to the absence of the specific NIPs required for Si influx. For instance, the model plant Arabidopsis will only accumulate limited amounts of Si if fed with the element because it lacks NIPIIIs (Montpetit et al. 2012). Notable exceptions among dicots are the Cucurbitaceae that are well known to benefit from Si feeding. Not surprisingly, the first Si transporters in dicots were found in that family (Mitani and Ma 2005). More recently, Si transporters have been described in soybean, a result congruent with reports of Si accumulation and protective effect in the species (Deshmukh et al. 2013). As more genomic data become available, it will become easier to precisely classify a plant as Si accumulator or not, simply on the presence of aquaporins permeable to silicic acid.

9.4 Silicon Specificity with Pathogens

Another confusing issue about Si relates to its specificity in terms of plant pathogens that are controlled by a treatment. Assuming a starting premise with a plant that is Si competent, the prophylactic properties of Si appear to be more efficient against biotrophic or hemibiotrophic fungal pathogens. Interestingly, very few studies have ever compared the relative efficiency of Si against different plant pathogens simultaneously but the literature contains a lot more reproducible and convincing data against plant pathogens that are classified as having a biotrophic phase. For instance, powdery mildews, strict biotrophs, are particularly well controlled by Si, and the list of plants benefiting from this effect is only limited by the host range of powdery mildews and the ability of a given plant to absorb Si. Among hemibiotrophs, rice blast, caused by the fungus *Magnaporthe grisea*, and brown spot caused by *Bipolaris oryzae* (Ning et al. 2014) are arguably the most commonly reported diseases to be controlled by Si. However, even though less numerous, there are reports of efficacy against necrotrophs. More specifically, pathogens such as *Rhizoctonia solani* on rice (Rodrigues et al. 2003b) and *Pythium ultimum* on cucumber (Chérif and Bélanger 1992) have been delayed in their development when inoculated on plants treated with Si. In the latter case, the prophylactic effects were rather modest and short lasting, but measurable. These limited examples associating Si with protection against necrotrophs have been sufficient to label Si as conferring broad-spectrum disease resistance in plants (Van Bockhaven et al. 2013), but the reality remains that there is an overwhelming bias toward biotrophs and hemibiotrophs.

Interestingly, the literature contains limited examples of plant pathogens other than fungi that are controlled by Si. For instance, very few cases of control of bacterial diseases have been associated with Si, with the notable exception of *Ralstonia solanacearum* on tomato (Dannon and Wydra 2004) and bacterial blight on rice (Xue et al. 2010a, b). While tomato is considered a Si non-accumulator species, it is nonetheless noteworthy that *R. solanacearum* has a hemibiotrophic lifestyle with a complex repertoire of Type III effector proteins (see Section 9.7). Recent evidence does suggest that priming would play a role in this interaction (Ghareeb et al. 2011), as observed with fungi.

It is uncertain if this bias for pathogens with a biotrophic phase is simply the result of the plant–pathogen interactions that have been studied or the fact that Si amendments tend to be more efficient against this particular type of pathogens. As a matter of fact, very little attention has been given to the specificity, or lack thereof, of Si with regard to the pathogens it controls. In the context of optimizing the use of

Si in agriculture, it would certainly be appropriate to define with greater precision the spectrum of pathogens that are indeed more susceptible to a Si treatment. Part of this answer may come from a better understanding of the exact mode of action of Si *in planta*.

9.5 The Mode of Action of Si

Among all the intricacies related to Si properties, its mode of action remains arguably the most controversial and debated. Several review papers have proposed a commensurate number of hypotheses, and although a consensus has started to emerge, many unanswered questions persist. The initial theory concerning the mode of action of Si in plant's prophylaxis involved a mechanical barrier impeding fungal progress. Indeed, over the years, it had been generally accepted that polymerized Si in the plant cell wall and apoplast prevented pathogen penetration (Wagner 1940; Heath and Stumpf 1986; Carver et al. 1987, 1994; Ishiguro 2001). However, as early as 1965, this theory was put into doubt by Okuda and Takahashi (1965) citing Yoshi's results (Yoshi 1941) of noncorrelation between Si treatment and leaf toughness as measured by a needle-puncture method: 'From this result, it seemed that Si protected the rice plant against blast disease, but the increase in mechanical toughness of the plant tissue resulting from absorbed Si is not sufficient to explain the mechanism of protection'. Nevertheless, this theory has survived over the years and is still cited. Carver et al. (1987) upon the observation of Si accumulation in papillae consistent with findings from Kunoh and Ishizaki (1975) stated that polymerized Si at attempted sites of penetration might provide an additional mean of resistance against penetration. In other works, Kim et al. (2002) proposed the reinforcement of cell walls in rice as a mechanism for enhanced resistance provided by Si treatment. However, no evidence has ever directly linked cell wall reinforcement with penetration failure by the fungus. It should be noted that the logical association proposed between Si deposition and pathogen resistance stems from the fact that Si has been reported in several pathosystems to accumulate at infection sites (Samuels et al. 1991a, b) (Fig. 9.1). This probably derives from a higher transpiration rate at sites, where the cuticle is damaged, rather than a defensive process, as Si accumulation after pathogen penetration would be too late to prevent penetration. As a matter of fact, Chérif et al. (1992a) observed the accumulation of Si in needle-punctured leaf holes and showed the absence of such deposits when plants were grown under saturated humidity. Even though Si is effectively deposited at preferential sites of penetration, and is also continuously deposited at higher rates after penetration has occurred, the hypothesis of cell wall reinforcement by Si to explain enhanced resistance of plants against pathogenic fungi has been strongly contested in recent years.

Assuming that Si does not act exclusively as a physical barrier, the possibility that Si played an active role in the resistance process remained speculative. However, the opportunity for exploring this new hypothesis first came with the pathosystem cucumber–*Pythium* spp. Si applied at a concentration of 1.7 mM in the nutrient

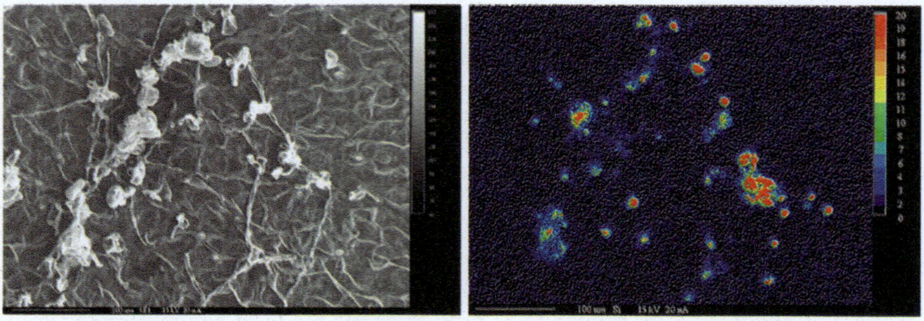

Fig. 9.1 Scanning electron (*left*) and X-ray (*right*) microanalysis showing that the accumulation of Si is coincident with *E. cichoracearum* presence on Si-treated *A. thaliana* leaves. The concentration of Si is indicated by color (see inset), where *red* represents the highest concentration of Si and *black* indicates no Si

solution was found to provide resistance to *Pythium ultimum* in cucumber (Chérif and Bélanger 1992). As in the case of the foliar interaction described by Samuels et al. (1991a, b), the plants reacted more promptly to the infection by *P. ultimum* and the number of infected cells accumulating an electron-dense, phenolic-like material was far greater (Chérif et al. 1992b). This material hindered the propagation of the parasite into the vascular system (Chérif et al. 1992b). Interestingly, none of these deposits, nor the cell walls at sites of penetration of the fungus, contained any trace of Si (Chérif et al. 1992b) as determined by SEM and scanning X-ray analysis. This excluded the possibility of Si functioning as a mechanical strengthener of cell walls to limit *P. ultimum* progress.

Corroborating evidence that Si played more than a mechanical role *in planta* was provided when the interaction between cucumber and powdery mildew was further analyzed at the chemical level (Fawe et al. 1998). The detection and analysis of antifungal compounds in infected cucumber leaves led to the determination of their nature and to the estimation of the effect of Si amendment on their appearance and accumulation (Fawe et al. 1998). Some of these metabolites, identified as flavonoids and phenolic acids, were specifically and strongly induced in a pattern typical of phytoalexins. Thus, Si was hypothesized to play an active role in disease resistance by stimulating the defence mechanisms, namely, phytoalexins, of cucumber in reply to fungal attack Fig. 9.2). This hypothesis was originally received with scepticism by the scientific community who argued that (1) monocots, for which the most important effects of Si feeding were reported, behaved differently and (2) direct genetic evidence to support the hypothesis was still lacking. Accordingly, subsequent efforts focused on the demonstration that monocots responded to Si feeding in the same manner as dicots did. Through electron microscopy studies and subsequent chemical analyses of rice plants, it was shown that rice plants treated with Si fended off rice blast infection through production of electron dense material that was fungitoxic to *Pyricularia oryzae* inside leaf cells (Rodrigues et al. 2003a). This finding was in direct contradiction with a previous publication that argued that amorphous Si deposition had prevented fungal penetration (Kim et al. 2002).

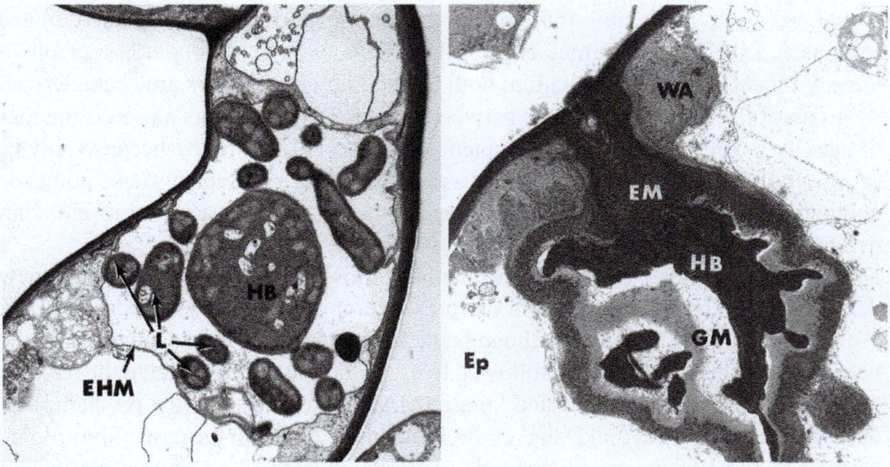

Fig. 9.2 Transmission electron micrographs of ultra-thin sections of *A. thaliana* leaves infected by *E. cichoracearum* (From Ghanmi et al. 2004). Si treatment (*right*) led to more efficient defence compared to control plants (*left*). *EHM* extrahaustorial membrane, *HB* haustorial body, *Ep* epidermis, *GM* granular material, *WA* wall apposition, *EM* electron-dense material, *L* lobes

Rodrigues et al. (2004) subsequently showed that the fungitoxic material was composed of momilactones, known as rice phytoalexins. At the same time, evidence was provided that Si amendments significantly reduced powdery mildew infection on wheat and that this phenomenon was once again attributable to a cellular reaction to fungal presence (Bélanger et al. 2003).

Since the deposited part of Si forming a physical barrier does not account for all the protection afforded by the element, it has been suggested that the soluble part of Si, silicic acid, could be involved in the prophylactic role of the element (Samuels et al. 1991a, b). Soluble Si has been reported inside the cell, in the cytosol, in chloroplastic membranes as well as in association with RNA and DNA (Aleshin et al. 1987). Also, high amounts of soluble Si have been detected in epidermal cells of barley at the beginning of their defence reaction against *Erysiphe graminis*, where it could play an important biochemical or physiological role during the penetration of the parasite (Zeyen et al. 1993). These data on the presence of Si inside the cells are scarce and debated, notably because of the poor measurement techniques available at the time. However, they suggest that Si can have a series of intracellular sites of action to explain its stimulating properties in plant disease resistance. Among others, it could alter directly or indirectly gene expression, by association with genes and modification of their configuration or by acting on the signal transduction pathway(s) activated by infection (Fawe et al. 2001). This second hypothesis is particularly interesting, as Si could represent, among other things, a natural activator of plant disease resistance. On the basis of their observations with cucumber, Fawe et al. (2001) suggested a model to explain how Si would play a role in induced resistance. According to their model, Si bioactivity was compared to that of known activators/secondary messengers of systemic acquired resistance (SAR) whereby it would act as a modulator influencing the timing and extent of plant defence responses. Like secondary messengers, the effects of Si on secondary metabolism are significant only after elicitation; both Si and known activators are characterized by a saturable effect. A difference between known SAR activators and Si is the loss of activity when Si feeding is interrupted (Samuels et al. 1991a, b), because polymerization of Si leads to its inactivation as an inducer of resistance. These points of comparison prompted the authors to propose that Si acted as a signal in inducing defence responses.

Induced resistance (IR) is a mechanism allowing plants to synthesize new defence compounds in response to the presence of a pathogen (Glazebrook 2005; Walters et al. 2013). During a pathogen attack, plants initiate active defences by the perception of an elicitor signal, followed by transduction of the signal to the nucleus by a network of mitogen-activated kinase (MAPK) cascades and the production of defence proteins (Jones and Dangl 2006; Benhamou 2009). In the case of biotrophic pathogens, most plants establish a type of IR called 'systemic acquired resistance' (SAR) that requires the activation of the salicylic acid (SA) signalling pathway (Delaney et al. 1994; Zhou et al. 1998; Glazebrook 2005) and the presence of the defence regulatory protein nonexpressor of pathogenesis-related protein1 (NPR1; Durrant and Dong 2004). Also, the IR mechanism is often associated with the augmented capacity to mobilize cellular defence responses only after a contact with pathogens (Conrath et al. 2002). This phenomenon, often called 'priming', allows the plant to respond more quickly and effectively to an attack, with minimal metabolic cost (Katz 1998; van Hulten et al. 2006). In this context, several studies have now shown that a pretreatment with Si will prime plants to better respond to pathogen infections (Fauteux et al. 2005; Chain et al. 2009; Van Bockhaven et al. 2013). However, the molecular mechanisms underlying such priming associated with Si remain poorly understood.

9.6 Transcriptomic Analyses

In an effort to better understand how Si affected the plant's metabolism, a microarray study was performed in 2006 (Fauteux et al. 2006) on both control and powdery mildew-stressed Arabidopsis plants, with or without Si application (Fig. 9.3). The expression of all but two genes was unaffected by Si in control plants, a result contradicting reports of possible direct effect of Si on plant physiology as a fertilizer. In contrast, inoculation of plants, treated or not with Si, altered the expression of a set of nearly 4,000 genes. Following functional categorization, many of the upregulated genes were defence-related, whereas a large proportion of downregulated genes were involved in primary metabolism. Regulated defence genes included R genes, stress-related transcription factors, genes involved in signal transduction, the biosynthesis of stress hormones (SA, JA, ethylene) and the metabolism of reactive oxygen species. In inoculated plants treated with Si, the magnitude of downregulation was attenuated by over 25 %, an indication of stress alleviation. Those results

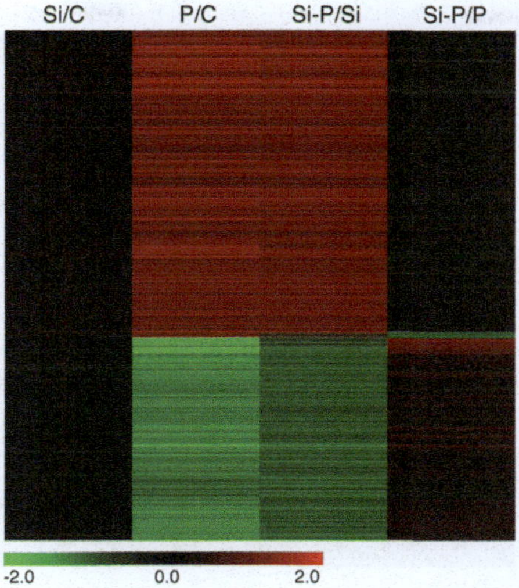

Fig. 9.3 Differential gene expression in Arabidopsis leaves following Si treatment and/or pathogen inoculation (From Fauteux et al. 2006). The columns represent the contrasts between the treatments: control (C), silicon (Si), *Erysiphe cichoracearum* (P) or a combination of both (Si-P). Each of the 3,970 differentially expressed genes ($p<0.01$, ≥ 1.5-fold change) in at least one contrast is represented by a *colored* line indicating the mean ($n=6$) relative transcript level: green corresponds to a Log_2 ratio of −2 (*down*regulation) and red corresponds to a Log_2 ratio of 2 (*up*regulation)

suggested that Si treatment had no effect on the metabolism of unstressed plants but that it had beneficial properties attributable to modulation of a more efficient response to pathogen stress.

However, following the discovery of Si transporters (Ma et al. 2006) that confirmed the inability of *Arabidopsis* to absorb large quantities of Si because of a lack of Lsi1 transporters, the previous results had to be interpreted with caution until similar studies with high Si-accumulating plants could be carried out. Incidentally, a large transcriptomic analysis (55,000 unigenes) with wheat, a high Si-accumulating plant (Chain et al. 2009), was reported a few years later with plants under both control and pathogen stress (*Blumeria graminis* f.sp. *tritici* (Bgt) (Fig. 9.4). The response to the supply of Si on control (uninfected) plants was limited to 47 genes of diverse functions, mainly related to stress, providing little evidence of regulation of a specific metabolic process. Plants reacted to inoculation with Bgt by an upregulation of many genes linked to stress and metabolic processes and a downregulation of genes linked to photosynthesis. Supplying Si to inoculated plants largely prevented disease development, a phenotypic response that translated into a nearly perfect reversal of genes regulated by the effect of Bgt alone. These results confirmed that Si played a limited role in a plant's transcriptome in the absence of stress, even in the case of a high Si-accumulating monocot such as wheat. On the other hand, the benefits of Si, in the form of biotic stress alleviation, were remarkably aligned with a counterresponse to transcriptomic changes induced by the pathogen Bgt.

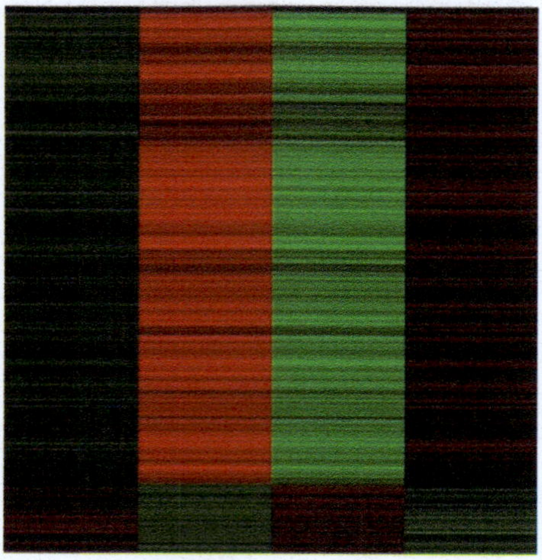

Fig. 9.4 Visual representation of differential gene expression in wheat leaves for plants having received or not Si amendment and/or *Blumeria graminis* f. sp. *tritici* inoculation (Chain et al. 2009). The columns represent contrasts between the treatments: control (Si–B–), Si alone (Si + B–), inoculation alone (Si–B+) or a combination of both (Si + B+). In total, 880 genes have been identified as differentially expressed ($p < 0.01$, ≥1.5-fold change) in at least one contrast. Each gene corresponds to a colored line indicating the mean ($n = 3$) relative transcript level: green corresponds to a $\log_2$ ratio of –2 (down-regulation), and red corresponds to a $\log_2$ ratio of 2 (upregulation)

9.7 The Search for a Better Model

While the transcriptomic analyses provided unique insights into how Si alleviated plant stress, it did not offer direct evidence of a biochemical role for Si. In the meantime, many additional reports came to support that Si did not directly induce immunity but rather primed plants for a better defence response against pathogens (Van Bockhaven et al. 2013). This property even extended to insects with studies showing that Si offered protection through a better arsenal and deployment of defence responses (Reynolds et al. 2009; Ye et al. 2013). Interestingly, this also represented a deviation from the initial hypothesis that Si acted as a mechanical barrier in plant–herbivore interactions. In spite of this mounting evidence of a positive role in priming plants, the mode of action of Si *in planta* remained unresolved.

For the most part, all studies that have shown better defence responses in presence of Si are strictly based on correlative observations. Hypotheses that Si can act as a secondary messenger, a modulator of defence responses or a priming agent have never been fully tested in the absence of a proper genetic model. In this context, Arabidopsis offers a wide range of accessions and mutants that could be helpful in assessing the role of Si. However, as convenient and versatile as Arabidopsis can be as a model plant, its usefulness for Si studies is compromised by its limited absorption of the element owing to the absence of influx transporters. As mentioned earlier, since the discovery of influx transporters in rice in 2006, it is now clear that

Arabidopsis lacks the aquaporins carrying the necessary features for Si permeability and is therefore considered a Si non-accumulator (Deshmukh et al. 2013). To circumvent this problem, Montpetit et al. (2012) proposed to increase Si absorption in Arabidopsis by the insertion of heterologous influx transporters, such as one from wheat in this specific situation.

The concept that it is possible to transform a non-accumulator plant species into an accumulating one opens up a wide array of possibilities to exploit the beneficial properties of Si. However, this presupposes that such transformed plants will display the expected phenotype in presence of Si. Vivancos et al. (2015) presented the first demonstration of this outcome whereby transformed plants were a lot more resistant to powdery mildew when fed with Si compared to control plants or transformed plants deprived of Si. This suggests that beneficial effects of Si are universal among plant species as long as a plant can absorb the element through the presence of influx transporters.

Considering the impressive body of information available on the Arabidopsis–powdery mildew interaction, Vivancos et al. (2015) took advantage of these resistant phenotypes to investigate the probable mechanisms behind the protective role of Si. As stated earlier, Si was initially described as providing a mechanical barrier impeding fungal penetration (Fauteux et al. 2005), and for a long period, this mode of action stood uncontested. However, this hypothesis was challenged by Samuels et al. (1991a, b) and Chérif et al. (1992a, b, 1994) who associated the protective role of Si with the elicitation of defence mechanisms. For these reasons, it was relevant to analyze the specific and well-described markers of resistance in the Arabidopsis–powdery mildew interaction and determine if Si was involved in their expression. Results clearly showed an increase in expression of genes encoding enzymes involved in the SA pathway directly associated with Si feeding and resistant phenotypes. At the same time, SA concentrations were also augmented, thus strengthening the hypothesis that priming occurred through that pathway. Interestingly, the production of camalexin remained unchanged, a result consistent with reports that camalexin production is useful against necrotrophs and not involved with biotrophs (Rogers et al. 1996). These results thus suggest that the response is aligned with the specific pathogen under study and that Si somehow facilitates this response but does not elicit directly the priming machinery.

Other factors militate in support of this indirect role of Si associated with the manifestation of priming. Silicic acid is an uncharged molecule for which no evidence of biochemical activity has ever been obtained. It has been argued that soluble Si could somehow be directly involved in the elicitation of defence responses, namely, as a secondary messenger, much in the same manner as salicylic acid (Fawe et al. 2001; Fauteux et al. 2005; Van Bockhaven et al. 2013). However, the presence of silicic acid in the symplastic environment and subsequent interactions with key defence molecules has simply not been corroborated by scientific data and remain speculative. Furthermore, the fact that the expression of *NPRI* and other defence-related genes was unchanged in *pad4* and *sid2* Arabidopsis mutants under Si treatment (Vivancos et al. 2015) would indicate that silicic acid does not act as a surrogate for salicylic acid as previously suggested (Fawe et al. 2001; Van Bockhaven et al. 2013).

The large array of available Arabidopsis mutants offered the unique opportunity to validate the hypothesis that priming of defence reactions explained how Si protected plants against diseases. Indeed, by using mutants able to absorb larger quantities of Si but deficient in the activation of the SA pathway, Vivancos et al. (2015) were able to test directly if the protective effect of Si became null or significantly altered, given the presumed inability of the plant to mount defence reactions against powdery mildews. It was therefore quite surprising to observe that plants transformed for high Si absorption and fed with Si displayed resistant phenotypes in spite of having lost the ability to produce defence reactions through the SA pathway. These observations strongly suggest that other factors than defence reactions are at play.

Based on these observations, it might be tempting to conclude that resistance conferred by Si on SA-deficient mutants supports the concept of a physical barrier. However, a mechanical barrier physically stopping a germinating spore would not lead to elicitation of defence mechanisms as observed here and in countless recent papers (Qin and Tian 2005; Bi et al. 2006; Kanto et al. 2007). On the other hand, these results bring a unique perspective as they show that if the priming state associated with Si feeding is altered, one can still obtain resistant phenotypes.

If these results appear contradictory at first, they provide an opportunity to consider an alternative hypothesis that would unify the modes of action behind the observed phenomena. It is well known that the prophylactic role of Si has been more documented and is more efficient against pathogens with a biotrophic phase (e.g. powdery mildews, oomycetes, rice blast). In the last few years, with the advent of high-throughput sequencing, the annotation of plant pathogen genomes has highlighted the presence and importance of effector proteins, most notably in the case of biotrophs and hemibiotrophs. Effectors will modify host cell structure, metabolism and function and interfere with signal pathways required for host invasion or for triggering host resistance (Giraldo and Valent 2013). Recent developments have located effectors in the apoplast, the extrahaustorial matrix or the cytoplasm after translocation across the plant membrane. Interestingly, amorphous Si deposition in plants is located in the apoplast and more precisely at the interface of the plasma membrane and the cell wall (Bauer et al. 2011; Zhang et al. 2013). This area is the site of intense interactions of many effectors with plant targets and sites of attempted penetration by biotrophic fungi (Bozkurt et al. 2012). As a matter of fact, the appressorium and the haustorium of powdery mildew fungi are structures of active release of effectors (Giraldo and Valent 2013); the appressorium will release effectors in the apoplastic compartment to prevent the action of plant proteases and the haustorium will release them into the cytoplasm through the extrahaustorial matrix (EHMx) to alter plant defences. Given that the apoplast and the EHMx are within the confines of Si deposition (Ghanmi et al. 2004), and based on our observations, it thus seems not only plausible but also logical that Si would interfere with effectors reaching their targets. This would thus prevent the invading fungus from inhibiting the plant defence, which results in the expression of the complete array of defence mechanisms as observed in this work and elsewhere. In addition, the intercellular space is a hostile environment for a fungal pathogen, and the latter will rely on apoplastic effectors to inhibit the release of a wide array of proteases and other plant molecules that would adversely affect its development (Win et al. 2012; Giraldo and Valent 2013). From our results, it does appear that this initial barrier is indeed quite efficient and has significantly delayed fungal infection. In consideration of the superior prophylactic role of Si against biotrophs, the heavy reliance of biotrophs on effectors to maintain their virulence and the site of Si deposition coinciding with effectors release, our results support a link between Si and effectors and certainly future efforts in testing this hypothesis.

In conclusion, the latest studies of the role of Si in plant–pathogen interactions have confirmed the association between Si and priming but have also uncovered a new phenomenon suggesting that mechanisms other than priming are involved in the way Si protects plants. These unforeseen results may be helpful in defining a unifying theory explaining the elusive and debated mode of action of Si in alleviating biotic stresses.

References

Adatia MH, Besford RT. The effects of silicon on cucumber plants grown in recirculating nutrient solution. Ann Bot. 1986;58:343–51.

Aleshin NE, Avakyan ER, Dyakunchak SA, Aleshkin EP, Baryshok VP, Voronkov MG. Role of silicon in resistance of rice to blast. Doklady Botanical sciences-Akademii nauk SSSR (USA); 1987.

Bauer P, Elbaum R, Weiss IM. Calcium and silicon mineralization in land plants: transport, structure and function. Plant Sci. 2011;180:746–56.

Bélanger RR, Labbé C. Control of powdery mildews without chemicals: prophylactic and biological alternatives for horticultural crops. In: Bélanger RR, Bushnell WR, Dik AJ, Carver TLW, editors. The powdery mildews: a comprehensive treatise. St Paul: APS Press; 2002. p. 256–67.

Bélanger RR, Bowen PA, Ehret DL, Menzies JG. Soluble silicon: its role in crop and disease management of greenhouse crops. Plant Dis. 1995;79:329–36.

Bélanger RR, Benhamou N, Menzies JG. Cytological evidence of an active role of silicon in wheat resistance to powdery mildew (*Blumeria graminis* f sp *tritici*). Phytopathology. 2003;934:402–12.

Benhamou N. La résistance chez les plantes. Principes de la stratégie défensive et applications agronomiques Éditions. Cachan: TEC & DOC Lavoisier; 2009.

Bi Y, Tian SP, Guo YR, Ge YH, Qin GZ. Sodium silicate reduces postharvest decay on hami melons: induced resistance and fungistatic effects. Plant Dis. 2006;90:279–83.

Bloemhard CMJ, Van Moolenbroek J. Management of mineral elements of roses grown in closed rockwool systems. Acta Hortic. 1994;401:481–92.

Bowen P, Menzies J, Ehret D, Samuels L, Glass AD. Soluble silicon sprays inhibit powdery mildew development on grape leaves. J Am Soc Hortic Sci. 1992;117:906–12.

Bozkurt TO, Schornack S, Banfield MJ, Kamoun S. Oomycetes, effectors, and all that jazz. Curr Opin Plant Biol. 2012;15:483–92.

Carver TLW, Zeyen RJ, Ahlstrand GG. The relationship between insoluble silicon and success or failure of attempted primary penetration by powdery mildew (*Erysiphe graminis*) germlings on barley. Physiol Mol Plant P. 1987;31:133–48.

Carver TLW, Ingerson-Morris SM, Thomas BJ, Gay AP. Light-mediated delay of primary haustorium formation by *Erysiphe graminis* f sp *avenae*. Physiol Mol Plant P. 1994;45:59–79.

Chain F, Côté-Beaulieu C, Belzile F, Menzies JG, Bélanger RR. A comprehensive transcriptomic analysis of the effect of silicon on wheat plants under control and pathogen stress conditions. Mol Plant Microbe Inter. 2009;22:1323–30.

Chérif M, Bélanger RR. Use of potassium silicate amendments in recirculating nutrient solutions to suppress *Pythium ultimum* on long English cucumber. Plant Dis. 1992;76:1008–11.

Chérif M, Benhamou N, Menzies JG, Bélanger RR. Silicon induced resistance in cucumber plants against *Pythium ultimum*. Physiol Mol Plant P. 1992a;41:411–25.

Chérif M, Benhamou N, Menzies JG, Bélanger RR. Studies of silicon distribution in wounded and *Pythium ultimum*-infected cucumber plants. Physiol Mol Plant P. 1992b;41:371–85.

Cherif M, Asselin A, Bélanger RR. Defense responses induced by soluble silicon in cucumber roots infected by *Pythium* spp. Phytopathology. 1994;84:236–42.

Conrath U, Pieterse CMJ, Mauch-Mani B. Priming in plant–pathogen interactions. Trends Plant Sci. 2002;7:210–6.

Dannon EA, Wydra K. Interaction between silicon amendments, bacterial wilt development and phenotype of *Ralstonia solanacearum* in tomato genotypes. Physiol Mol Plant Pathol. 2004;64:233–43.

Datnoff LE, Deren CW, Snyder GH. Silicon fertilization for disease management of rice in Florida. Crop Prot. 1997;16:525–31.

Datnoff LE, Heft Snyder G, Korndörfer GH, editors. Silicon in agriculture. Amsterdam: Elsevier Science BV; 2001.

Delaney TP, Uknes S, Vernooij B, Friedrich L, Weymann K, Negrotto D, Gaffney T, Gut-Rella M, Kessmann H, Ward E, et al. A central role of salicylic acid in plant disease resistance. Science. 1994;266:1247–50.

Deshmukh RK, Vivancos J, Guérin V, Sonah H, Labbé C, Belzile F, Bélanger RR. Identification and functional characterization of silicon transporters in soybean using comparative genomics of major intrinsic proteins in Arabidopsis and rice. Plant Mol Biol. 2013;83:303–15.

Durrant WE, Dong X. Systemic acquired resistance. Annu Rev Phytopathol. 2004;42:185–209.

Ehret DL, Menzies JG, Bogdanoff C, Utkhede RS, Frey B. Foliar applications of fertilizer salts inhibit powdery mildew on tomato. Can J Plant Pathol. 2002;24:437–44.

Epstein E. The anomaly of silicon in plant biology. Proc Natl Acad Sci U S A. 1994;91:11–7.

Epstein E. Silicon. Annu Rev Plant Physiol Plant Mol Biol. 1999;50:641–64.

Epstein E. Silicon in plants: facts vs. concepts. In: Datnoff LE, Snyder GH, Korndöfer GH, editors. Silicon in agriculture. Amsterdam: Elsevier Science; 2001. p. 1–15.

Fauteux F, Rémus-Borel W, Menzies JG, Bélanger RR. Silicon and plant disease resistance against pathogenic fungi. FEMS Microbiol Lett. 2005;249:1–6.

Fauteux F, Chain F, Belzile F, Menzies JG, Bélanger RR. The protective role of silicon in the Arabidopsis-powdery mildew pathosystem. Proc Natl Acad Sci U S A. 2006;103:17554–9.

Fawe A, Abou-Zaid M, Menzies JG, Bélanger RR. Silicon-mediated accumulation of flavonoid phytoalexins in cucumber. Phytopathology. 1998;88:396–401.

Fawe A, Menzies JG, Chérif M, Bélanger RR. Silicon and disease resistance in dicotyledons. In: Datnoff LE, Snyder GH, Korndöfer GH, editors. Silicon in agriculture. Amsterdam: Elsevier; 2001. p. 159–70.

Ghanmi D, McNally DJ, Benhamou N, Menzies JG, Bélanger RR. Powdery mildew of Arabidopsis thaliana: a pathosystem for exploring the role of silicon in plant-microbe interactions. Physiol Mol Plant Pathol. 2004;64:189–99.

Ghareeb H, Bozsó Z, Ott PG, Repenning C, Stahl F, Wydra K. Transcriptome of silicon-induced resistance against Ralstonia solanacearum in the silicon non accumulator tomato implicates priming effect. Physiol Mol Plant Pathol. 2011;75:83–9.

Giraldo MC, Valent B. Filamentous plant pathogen effectors in action. Nat Rev Microbiol. 2013;11:800–14.

Glazebrook J. Contrasting mechanisms of defense against biotrophic and necrotrophic pathogens. Annu Rev Phytopathol. 2005;43:205–27.

Guével MH, Menzies JG, Bélanger RR. Effect of root and foliar applications of soluble silicon on powdery mildew control and growth of wheat plants. Eur J Plant Pathol. 2007;119:429–36.

Heath MC, Stumpf MA. Ultrastructural observations of penetration sites of the cowpea rust fungus in untreated and silicon-depleted French bean cells. Physiol Mol Plant P. 1986;29:27–39.

Ishiguro K. Review of research in Japan on the roles of silicon in conferring resistance against rice blast. Stud Plant Sci. 2001;8:277–91.

Jiang D, Zeyen RJ, Russo V. Silicon enhances resistance of barley to powdery mildew (*Erysiphe graminis* f sp *hordei*). Phytopathology. 1989;79:119–25.

Jones JDG, Dangl JL. The plant immune system. Nature. 2006;444:323–9.

Kanto T, Maekawa K, Aino M. Suppression of conidial germination and appressorial formation by silicate treatment in powdery mildew of strawberry. J Gen Plant Pathol. 2007;73:1–7.

Katz VA. A benzothiadiazole primes parsley cells for augmented elicitation of defense responses. Plant Physiol. 1998;117:1333–9.

Kim SG, Kim KW, Park EW, Choi D. Silicon-induced cell wall fortification of rice leaves: a possible cellular mechanism of enhanced host resistance to blast. Phytopathology. 2002;92:1095–103.

Kunoh H, Ishizaki H. Silicon levels near penetration sites of fungi on wheat, barley, cucumber and morning glory leaves. Phys Plant Pathol. 1975;5:283–7.

Leusch HJ, Buchenauer H. Effect of soil treatments with silica-rich lime fertilizers and sodium trisilicate on the incidence of wheat by *Elysiphe graminis* and *Septoria nodorum* depending on the form of N-fertilizer. J Plant Dis Prot. 1989;96:154–72.

Liang YC, Sun WC, Si J, Römheld V. Effects of foliar and root applied silicon on the enhancement of induced resistance to powdery mildew in *Cucumis sativus*. Plant Pathol. 2005;54:678–85.

Ma JF, Tamai K, Yamaji N, Mitani N, Konishi S, Katsuhara M, Ishiguro M, Murata Y, Yano M. A silicon transporter in rice. Nature. 2006;440:688–91.

Menzies JG, Ehret DL, Glass ADM, Samuels AL. The influence of silicon on cytological interactions between *Sphaerotheca fuliginea* and *Cucumis sativus*. Physiol Mol Plant P. 1991;39:403–14.

Menzies J, Bowen P, Ehret D, Glass AD. Foliar applications of potassium silicate reduce severity of powdery mildew on cucumber, muskmelon, and zucchini squash. J Am Soc Hortic Sci. 1992;117:902–5.

Mitani N, Ma JF. Uptake system of silicon in different plant species. J Exp Bot. 2005; 56:1255–61.

Miyake Y, Takahashi E. Effect of silicon on the growth of solution-cultured cucumber plant. Soil Sci Plant Nutr. 1983a;29:71–83.

Miyake Y, Takahashi E. Effect of silicon on the growth of cucumber plant in soil culture. Soil Sci Plant Nutr. 1983b;29:463–71.

Montpetit J, Vivancos J, Mitani-Ueno N, Yamaji N, Remus-Borel W, Belzile F, Ma JF, Belanger RR. Cloning, functional characterization and heterologous expression of TaLsi1, a wheat silicon transporter gene. Plant Mol Biol. 2012;79:35–46.

Ning DF, Song AL, Fan FL, Li ZJ, Liang YC. Effects of slag-based silicon fertilizer on rice growth and brown-spot resistance. Plos One. 2014;9(7), e102681. doi:10.1371/journal. pone.0102681.

Okuda A, Takahashi E. The role of silicon. The mineral nutrition of the rice plant. Madison: Johns Hopskins Press; 1965. p. 146–56.

Qin GZ, Tian SP. Enhancement of biocontrol activity of *Cryptococcus laurentii* by silicon and the possible mechanisms involved. Phytopathology. 2005;95:69–75.

Reuveni M, Reuveni R. Efficacy of foliar sprays of phosphates in controlling powdery mildews in field-grown nectarine, mango trees and grapevines. Crop Prot. 1995;14:311–4.

Reynolds OL, Keeping MG, Meyer JH. Silicon-augmented resistance of plants to herbivorous insects: a review. Ann Appl Biol. 2009;155:171–86.

Rodrigues FÁ, Benhamou N, Datnoff LE, Jones JB, Bélanger RR. Ultrastructural and cytochemical aspects of silicon-mediated rice blast resistance. Phytopathology. 2003a;93:535–46.

Rodrigues FÁ, Valeb FXR, Korndörfer GH, Prabhud AS, Datnoff LE, et al. Influence of silicon on sheath blight of rice in Brazil. Crop Prot. 2003b;22:23–9.

Rodrigues FÁ, McNally DJ, Datnoff LE, Jones JB, Labbé C, Benhamou N, Bélanger RR. Silicon enhances the accumulation of diterpenoid phytoalexins in rice: a potential mechanism for blast resistance. Phytopathology. 2004;94:177–83.

Rodrigues FA, Duarte HSS, Domiciano GP, Souza CA, Korndörfer GH, Zambolim L. Foliar application of potassium silicate reduces the intensity of soybean rust. Aust Plant Pathol. 2009;38:366–72.

Rogers EE, Glazebrook J, Ausubel FM. Mode of action of the *Arabidopsis thaliana* phytoalexin camalexin and its role in Arabidopsis-pathogen interactions. Mol Plant Microbe Inter. 1996;9:748–57.

Samuels AL, Glass ADM, Ehret DL, Menzies JG. Distribution of silicon in cucumber leaves during infection by powdery mildew fungus (*Sphaerotheca fuliginea*). Can J Bot. 1991a;69:140–6.

Samuels AL, Glass ADM, Ehret DL, Menzies JG. Mobility and deposition of silicon in cucumber plants. Plant Cell Environ. 1991b;14:485–92.

Shetty R, Jensen B, Shetty NP, Hansen M, Hansen CW, Starkey KR, Jørgensen HJL. Silicon induced resistance against powdery mildew of roses caused by *Podosphaera pannosa*. Plant Pathol. 2012;61:120–31.

Sun WC, Zhang J, Fan QH, Xue GF, Li ZJ, Liang YC. Silicon-enhanced resistance to rice blast is attributed to silicon-mediated defence resistance and its role as physical barrier. Eur J Plant Pathol. 2010;128:39–49.

Van Bockhaven J, de Vleesschauwer D, Höfte M. Towards establishing broad-spectrum disease resistance in plants: silicon leads the way. J Exp Bot. 2013;64:1281–93.

Van Hulten M, Pelser M, van Loon LC, Pieterse CMJ, Ton J. Costs and benefits of priming for defense in Arabidopsis. Proc Natl Acad Sci U S A. 2006;103:5602–7.

Vivancos J, Labbé C, Menzies JG, Bélanger RR. Silicon-mediated resistance of Arabidopsis against powdery mildew involves mechanisms other than the SA-dependent defense pathway. Mol Plant Pathol. 2015. doi:10.1111/mpp.12213.

Wagner F. The importance of silicic acid for the growth of some cultivated plants, their metabolism, and their susceptibility to true mildews. Phytopathol Z. 1940;12:427–79.

Walters DR, Ratsep J, Havis ND. Controlling crop diseases using induced resistance: challenges for the future. J Exp Bot. 2013;64:1263–80.

Win J, Chaparro-Garcia A, Belhaj K, Saunders DGO, Yoshida K, Dong S, Schornack S, Zipfel C, Robatzek S, Hogenhout SA, Kamoun S. Effector biology of plant-associated organisms: concepts and perspectives. Cold Spring Harbor Symp Quant Biol. 2012;77:235–47.

Xue GF, Sun WC, Al S, Li ZJ, Fan FL, Liang YC. Influence of silicon on rice growth, resistance to bacterial blight and activity of pathogenesis-related proteins. Sci Agric Sin. 2010a;43:690–7 (In Chinese with English abstract).

Xue GF, Al S, Sun WC, Li ZJ, Fan FL, Liang YC. Influences of silicon on activities of antioxidant enzymes in rice leaves infected by *Xoo* strain in relation to bacterial blight resistance. Plant Nutr Fertil Sci. 2010b;16:591–7 (In Chinese with English abstract).

Ye M, Song Y, Long J, Wang R, Baerson SR, Pan Z, Zhu-Salzman K, Xie J, Cai K, Luo S, Zeng R. Priming of jasmonate-mediated antiherbivore defense responses in rice by silicon. Proc Natl Acad Sci U S A. 2013;110:3631–9.

Yoshi H. Studies on the nature of rice blast resistance. Kyusu Imp Univ Sci Fakultato Terkultura Bull. 1941;9:277–307.

Zeyen RJ, Ahlstrand GG, Carver TLW. X-ray microanalysis of frozen-hydrated, freeze-dried, and critical point dried leaf specimens: determination of soluble and insoluble chemical elements at *Erysiphe graminis* epidermal cell papilla sites in barley isolines containing Ml-o and ml-o alleles. Can J Bot. 1993;71:284–96.

Zhang C, Wang L, Zhang W, Zhang F. Do lignification and silicification of the cell wall precede silicon deposition in the silica cell of the rice (*Oryza sativa* L.) leaf epidermis? Plant Soil. 2013;372:137–49.

Zhou N, Tootle TL, Tsui F, Klessig DF, Glazebrook J. PAD4 functions upstream from salicylic acid to control defense responses in Arabidopsis. Plant Cell. 1998;10:1021–30.

Chapter 10
Silicon and Insect Pest Resistance

Abstract In general, silicon (Si) is involved in plant resistance against insect pest damage via two major defence mechanisms: physical defence and induced biochemical (chemical) defence. Si deposited as opaline phytoliths within plants may act as an antiherbivore defence through increasing hardness and abrasiveness of tissues and wearing of herbivore mouthparts. In turn, this would create a feeding deterrent, which may reduce the palatability and digestibility of leaves, thereby potentially impacting on herbivore performance (physical defence). On the other hand, increasing evidence shows that biochemical changes and induced resistance, because of Si uptake by the plant, are also responsible for the alleviation of insect pest damage. More recently, it has been confirmed that Si plays a positive role in priming plants for a better defence response against pest infestation (e.g. rice leaf roller). The primed state seems to be mediated via jasmonic acid signalling pathways.

Keywords Induced biochemical defence • Insect pest • Physical defence • Silicon

10.1 General

The earliest report on Si-mediated resistance to maize Hessian fly, *Mayetiola destructor*, can be dated back to the 1920s (McColloch and Salmon 1923). Later, Ponnaiya (1951) found that there was a close relationship between sorghum (*Sorghum bicolor*) resistance to central shoot fly, *Atherigona indica infuscate*, and silicon (Si) concentration in shoots. The first report on Si-increased plant resistance to an insect pest was associated with the rice stem borer, *Chilo simplex* (Sasamoto 1953; also see Reynolds et al. 2009). According to the report by Sasamoto (1953), the application of a slag-based silicate fertilizer to rice reduced *C. simplex* damage and increased growth impetus of the plant, which was likely attributable to increased strength of the rice stem following Si accumulation (Sasamoto 1955). Si has since been well documented to enhance the resistance against various insect herbivores in many crops of agricultural importance (see Table 10.1) and many other plant species. It has been well documented that Si has a clear prophylactic effect against a wide range of insect-feeding guilds, including lepidopteran borers, folivores, phloem-feeding insects, xylem-feeding insects and other plant feeders (Table 10.1; also see Reynolds et al. 2009).

Y. Liang et al., *Silicon in Agriculture*, DOI 10.1007/978-94-017-9978-2_10

Table 10.1 Selected reports in the literature on the prophylactic effects of Si on insect herbivores in crops

Insect species	Crop	Reference
African striped rice stem borer (*Chilo zacconius*)	Rice	Ukwungwu and Odebiyi (1985)
Asiatic rice borer (*Chilo suppressalis*)	Rice	Sasamoto (1958), Nakano et al. (1961), Djamin and Pathak (1967), Dravé and Laugé (1978), and Hou and Han (2010)
Yellow rice borer (*Scirpophaga incertulas*)	Rice	Panda et al. (1975)
Brown planthopper (*Nilaparvata lugens*)	Rice	Yoshihara et al. (1979)
Small brown planthopper (*Laodelphax striatellus*)	Rice	Liu et al. (2007)
Green leafhopper (*Nephotettix bipunctatus cincticeps*)	Rice	Savant et al. (1997)
White-backed planthopper (*Sogatella furcifera*)	Rice	Salim and Saxena (1992) and Savant et al. (1997)
Leaf spider (*Tetranychus* spp.)	Rice	Savant et al. (1997)
Leaf roller (*Cnaphalocrocis medinalis*)	Rice	Hanifa et al. (1974), Sudhakar et al. (1991), Ramachandran and Khan (1991), Nakata et al. (2008), and Ye et al. (2013)
Green rice caterpillar (*Naranga aenescens*)	Rice	Nakata et al. (2008)
Hessian fly (*Phytophaga destructor*)	Wheat	Miller et al. (1960)
Hessian fly (*Mayetiola destructor*)	Wheat	McColloch and Salmon (1923)
Greenbug (*Schizaphis graminum*)	Wheat	Basagli et al. (2003), Gomes et al. (2005), Goussain et al. (2005), and Costa and Moraes (2006)
Aphid (*Sitobion avenae*)	Wheat	Dias et al. (2014)
Hessian fly (*Mayetiola destructor*)	Maize	McColloch and Salmon (1923)
Stem borer (*Sesamia calamistis*)	Maize	Setamou et al. (1993)
Asian corn borer (*Ostrinia furnacalis*)	Maize	Horng and Chu (1990)
Fall armyworm (*Spodoptera frugiperda*)	Maize	Goussain et al. (2002)
Central shoot fly (*Atherigona indica infuscate*)	Sorghum	Ponnaiya (1951)
Stalk borer (*Eldana saccharina*)	Sugarcane	Rao (1967), Meyer and Keeping (2005), Keeping and Meyer (2002), Keeping and Meyer (2006), Keeping and Kvedaras (2008), and Kvedaras et al. (2007, 2009)
Stem borer (*Diatraea saccharina*)	Sugarcane	Pan et al. 1979, Elawad et al. (1985), and Anderson and Sosa (2001)
Spittlebug (*Mahanarva fimbriolata*)	Sugarcane	Korndörfer et al. (2011)
Whitefly (*Bemisia tabaci*)	Cucumber	Correa et al. (2005)
Weevil (*Cylas formicarius*)	Sweet potato	Singh et al. (1993)

Interestingly, almost all case studies have shown a positive relationship between pest damage control and accumulation of Si in plant organs (Table 10.1), with the notable exception of greenbugs on wheat (*Triticum aestivum*) and barley (*Hordeum vulgare*) (Lanning 1966a, b). However, the mechanisms underpinning the role of Si against insects still remain poorly understood. Nakata et al. (2008) demonstrated that pest damage, as indicated by chewing damage by rice leaf roller, *Cnaphalocrocis medinalis*, and rice green caterpillar, *Naranga aenescens*, developed more clearly in the leaves of the *lsi1* mutant, defective in active Si uptake. Since this mutant accumulates less Si compared to the wild type, the authors concluded that Si accumulation in plants was closely associated with insect pest damage. So far, only two review papers have dealt with Si effects on insect pest control (see Laing et al. 2006; Reynolds et al. 2009). Reynolds et al. (2009) provided a comprehensive overview of the then current knowledge of Si-augmented resistance to insect pest damage and pointed out a need for further research on the possible involvement of Si in upregulation of defence-related genes such as those related to biosynthesis of jasmonic acid (JA) or salicylic acid (SA) following insect attack. As expected, over the past years, rapid progress has been made in interactions between Si and JA in defence against insect herbivores involving priming of JA-mediated defence responses by Si and the promotion of Si accumulation by JA (Ye et al. 2013; see below). This updated knowledge of Si interactions with insect pests is overviewed in this chapter.

10.2 Nutritional and Physical Defences

10.2.1 Nutritional Defences

Elevated levels of Si in plant tissues may increase the bulk density of the insect diet so that insects are unable to ingest sufficient quantities of nutrients (carbohydrates and nitrogen) and water (Smith et al. 1971; Panda and Kush 1995; Massey et al. 2006). For example, Massey et al. (2006) showed that *S. exempta* feeding on high-Si plants suffered reduced digestion efficiency in three (out of five) grass species and increased its rate of consumption in two (out of the five) grass species. Subsequently, Massey and Hartley (2009) found that the conversion efficiency of ingested food into the body mass of *S. exempta* was significantly reduced by high amount of Si contained in plants and so was the amount of N absorbed from the diet. Increased consumption of plant tissues was recorded for *S. eridania* (Peterson et al. 1988) and *S. gregaria* (Massey et al. 2006) as a consequence of poor dietary quality. However, Si may also alter the palatability of plant tissues and deter herbivore feeding, as reported by Massey et al. (2006) for *S. gregaria* and *S. exempta* feeding on grass species. Salim and Saxena (1992) also reported decreased consumption of high-Si rice materials by white-backed planthopper, *S. furcifera.* Decreased stalk damage and larval weight gain as a result of decreased consumption were recorded for stem borer *E. saccharina* when fed with Si-amended sugarcane (*Saccharum officinarum*) compared with the control sugarcane (Keeping and Meyer 2006; Kvedaras et al. 2007).

When Asian corn borer *O. furnacalis* was exposed to the artificial corn diet containing Si ranging from 0.1, 3, 5 or 10 %, mortality of larvae and pupae increased, pre-oviposition period was prolonged and the oviposition period was shortened with increasing Si concentration (Horng and Chu 1990). A correlation analysis showed that pupal weight, fecundity, net reproductive rate and intrinsic rate of increase were all negatively correlated with Si content in the corn diet, while the duration of larval and pupal stages, longevity of adults and mean generation times were not significantly affected (Horng and Chu 1990). Later, several studies also showed that Si fertilization altered survival, reproduction and host plant preferences of chewing (Goussain et al. 2002; Massey et al. 2006; Kvedaras et al. 2009) and sucking insects (Basagli et al. 2003; Correa et al. 2005; Korndörfer et al. 2011). Costa and Moraes (2006) showed that the application of Si significantly reduced the number of nymphs, the population growth rate, the post-reproductive period and the longevity of the greenbug. A significant reduction of the number of aphids was observed during aphids' colonization as a result of Si application. More recently, Si fertilization was reported to induce resistance of wheat plants to apterous morphs via antibiosis as shown by reduced fecundity, reproductive period, longevity, intrinsic rate of increase and net reproductive rate; however, alates were unaffected (Dias et al. 2014). Plants treated with Si had fewer alate aphids in both the vegetative and reproductive phases compared with Si-untreated plants.

Interestingly, insect performance can be promoted by high levels of N in the diet but may be counteracted by Si. According to Chu and Horng (1991), the Asian corn borer (*O. furnacalis*) preference was decreased under free-choice conditions for potted corn plants treated with Si, but was increased for the plants receiving high-N treatment. Later work by Meyer and Keeping (2005) showed that Si fertilization could mitigate the promotional effects of applied N on populations of the stalk borer, *E. saccharina*, in sugarcane. It suggests that Si fertilization would enable growers to resume applying the recommended rates of N and avoid N limitation of crop yield by counteracting the increased *E. saccharina* infestations at high rates of N (Meyer and Keeping 2005). This may be one of the reasons why Si fertilization can increase crop performance and yield.

10.2.2 Physical Defences

It has long been recognized that Si deposited as opaline phytoliths in the cell walls of plant leaf and stem epidermis may act, directly or indirectly, as an antiherbivore defence. As a mechanism of physical defence, Si increases hardness and abrasiveness of the plant tissues, which concomitantly accelerates the wear on herbivore mouthparts; hence, Si acts as a feeding deterrent, reducing the palatability and digestibility of leaves, thereby potentially impacting on herbivore performance (Kaufman et al. 1985; Salim and Saxena 1992; Ma et al. 2001; Kvedaras et al. 2007; Massey and Hartley 2009; Reynolds et al. 2009 and references therein; Dias et al. 2014). However, this hypothesis that Si increases plant abrasiveness and wearing of herbivore mouthparts has been correlated only with fossil record and remained correlative, although the evolution of high-crowned teeth (Ungulata), continuously growing teeth (Rodentia and Lagomorpha) and enlarged mandibles (Lepidoptera and Orthoptera) has all been linked to a grass-based diet (Isely 1944; Stebbins 1981; Jernvall and Fortelius 2002). Indeed, there is some evidence showing that increases in Si concentration can deter feeding by some mammals (Gali-Muhtasib et al. 1992) and stem-boring insects (Djamin and Pathak 1967; Moore 1984). However, direct evidence was not provided until Massey et al. (2006) first demonstrated a direct effect of Si on the abrasiveness of grasses and the adverse impact of Si on herbivore preference and performance. They showed that the addition of Si resulted in increases in leaf abrasiveness in four of the five grass species studied and Si supply also deterred feeding by two folivores and reduced their growth rates and digestion efficiency. Furthermore, Si significantly reduced the growth rates, digestion efficiency and pupal mass of insect folivores, but did not affect the growth performance nor did it act as a feeding deterrent on the phloem feeder tested.

It has been widely reported that the hardness of plant parts affected mandibular wear in leaf beetle (*P. versicolora*) (Raupp 1985; King et al. 1998), bee (*M. rotundata*) (Kokko et al. 1993; Schaber et al. 1993), noctuid lepidopteran larvae (*S. exigua*) (Korth et al. 2006), weevil (*P. callosus*) (Barnes and Giliomee 1992), stalk borer (Kvedaras et al. 2009) and even in the stylet of a true bug (*D. hesperus*) (Roitberg et al. 2005). Mandibular wear is often attributed to Si deposited within plant tissues (Kvedaras et al. 2009). Insects feeding on high-Si plants suffered higher mandibular wear, as has been reported for stem borer *C. suppressalis* (Sasamoto 1958; Djamin and Pathak 1967) and rice leaf roller larvae *C. medinalis* (Hanifa et al. 1974; Ramachandran and Khan 1991) feeding on rice and for fall armyworm (*S. frugiperda*) feeding on corn plants (Goussain et al. 2002). However, these earlier studies and reports of Si effect on mandibular wear of arthropods should be explained with caution due to the flaws in their experimental design. According to Kvedaras et al. (2009), some earlier studies dealing with the relationship between mandibular wear of lepidopteran larvae and the high- and low-Si diet plants failed to describe the research method (Djamin and Pathak 1967) or relied only on visual comparisons (Sasamoto 1958; Hanifa et al. 1974; Goussain et al. 2002) at selected extremes of mandibular wear, with no statistical analysis. Furthermore, some of the comparisons between larvae feeding on rice cultivars with high and low Si (Djamin and Pathak 1967; Hanifa et al. 1974; Ramachandran and Khan 1991) demonstrated only mere correlation and not causality. Dravé and Laugé (1978) used a quantitative, statistical approach to compare larval mandibular wear, but they only tested the effect of exogenous Si, rather than *in planta*, thereby ignoring the natural context in which the insect would be affected by Si when feeding on the plant (Kvedaras et al. 2009). Furthermore, although Zouhourian-Saghiri et al. (1983) quantified mandibular wear by using plant material with contrasting Si levels, they did not exclude the role of other abrasive materials including cellulose and lignin in promoting wear in *L. migratoria*. More recently, by using the novel method of Smith et al. (2007), Kvedaras et al. (2009) first accurately and quantitatively measured the mandibular wear of a sugarcane stalk borer *E. saccharina* feeding on live plants with experimentally elevated Si content. Their results showed that although there was a trend for increased wear in larvae on sugarcane treated with Si, the effects of Si, cultivar or feeding site (leaf bud, root band and internode) on mandibular wear of *E. saccharina* were not statistically different. Redmond and Potter (2007) also showed that no excessive wearing was found on the mandibular teeth in black cutworm, *Agrotis ipsilon*, and root-feeding masked chafer grubs, *Cyclocephala* spp., feeding on cultivated creeping bentgrass amended with calcium silicate fertilizer compared with the Si-untreated controls. However, Massey and Hartley (2009) observed a significantly and rapidly increased mandibular wear in *S. exempta* that developed on two out of three grass species fertilized with Si, compared with the Si-untreated controls. Therefore, it seems that more quantitative experimentally sound work should be done to confirm the adverse effect of Si in plant tissues on the mandibular wear of insects.

By measuring both the percentage of total chlorophyll liberated from high- and low-Si grass leaves by mechanical grinding and the chlorophyll content in the faeces of locust (*S. gregaria*), Hunt et al. (2008) demonstrated that high-Si grasses released less chlorophyll after grinding and retained more after passing through the gut of the locust, illustrating that Si levels are correlated with increased mechanical protection. These results suggest that Si may defend grasses at least in part through reducing mechanical breakdown of the leaf and that mechanical protection of resources in chlorenchyma cells is a newly reported and potentially important mechanism by which Si protects grasses.

10.3 Induced Biochemical Defences

As early as 1958, by conducting a laboratory choice study, Sasamoto showed that larvae of the rice stem borer, *C. suppressalis*, preferred Si-untreated rice stalks as diet over Si-treated ones and the larvae feeding on the Si-treated rice stalks showed increased mandibular wear, suggesting that host choice by an insect depended not only on the physical properties of the food but also on its chemical properties. Notably, larvae preferred water extracted from the stem of the rice plant grown on N-rich manure to that extracted from the stem of rice plants fertilized with Si (Sasamoto 1958). Goussain et al. (2005) showed a clear adverse effect on the development of greenbug (*S. graminum*) feeding on Si-treated wheat plants. Although the penetration of aphid stylet was not impeded by Si in wheat plants, the stylet was withdrawn more often resulting in a reduction of probing time, suggesting that chemical changes and induced resistance, because of Si uptake by the plant, were likely responsible for the reduction of aphid performance, as opposed to a physical impediment (as the stylet eventually did reach the phloem). Gomes et al. (2005) investigated the effect of Si and previous infestations with aphids on the induction of resistance of wheat to the greenbug and showed that Si alone or together with pre-infestation negatively affected the greenbug's preference and population increase rate and concomitantly induced a significant increase in the activities of peroxidase (POX), polyphenol oxidase (PPO) and phenylalanine ammonia-lyase (PAL) in wheat. The interactions of Si with whitefly (*B. tabaci*) in cucumber (*Cucumis sativus*) (Correa et al. 2005) and greenbug in wheat (Gomes et al. 2005) also suggest that soluble Si is important in induced resistance to insect herbivores, as has been shown for Si-mediated resistance induction to fungal pathogens in various crops (for review, see Fauteux et al. 2005; see also Chap. 9). More recent studies have also provided evidence that soluble Si is involved in induced chemical or biochemical defences to insect herbivore attack through the enhanced production of defensive enzymes or possibly the enhanced release of plant volatiles (for review, see Reynolds et al. 2009; Kvedaras et al. 2009). More recently, Ye et al. (2013) have explored the role jasmonic acid (JA) may play in Si-enhanced resistance by silencing the expression of allene oxide synthase (*OsAOS*; active in JA biosynthesis) and *CORONATINE INSENSITIVE1* (*OsCOI1*; active in JA perception) genes in transgenic rice plants via RNAi and examining resulting changes in Si accumulation and defence responses against the infestation by caterpillars of *Cnaphalocrocis medinalis* (rice leaf roller). Rice resistance to *C. medinalis* larvae increased by Si pretreatment in wild-type plants but not in *OsAOS* and *OsCOI1* RNAi lines. Compared with untreated controls, defence responses in wild-type Si-treated plants were enhanced upon *C. medinalis* attack, including higher levels of JA accumulation, increased levels of transcripts encoding defence marker genes and elevated activities of POD, PPO and trypsin protease inhibitor. Additionally, compared to wild-type plants, Si deposition and Si cell expansion were reduced in leaves of *OsAOS* and *OsCOI1* RNAi plants, and steady-state transcript levels of the Si transporters *OsLsi1*, *OsLsi2* and *OsLsi6* were reduced as well in Si-pretreated plants upon *C. medinalis* attack. These results clearly demonstrate a strong link between Si and JA in defence against insect herbivores involving priming of JA-mediated defence responses by Si and the promotion of Si accumulation by JA.

10.4 Conclusions and Perspectives

Taken together, in addition to physical defence, Si plays a positive role in priming plants for a better defence response against pest infestation (e.g. rice leaf roller; see Ye et al. 2013). The primed state might be mediated via SA, JA, ethylene and/or ROS signalling pathways. However, convincing evidence to support such an induced biochemical resistance against pest infestation (and fungal and bacterial diseases; see Chap. 9) is still not sufficient. The further research needs should focus on addressing the following key questions: (a) Which defence is the main mechanism for Si-augmented resistance to insect herbivores, physical barrier (defence) or induced biochemical defence or both? (b) Is JA or SA signalling pathway a universal mechanism for Si-augmented resistance to infestation of all categories of pests? In order to answer these questions, more work is needed to provide direct and convincing evidence for unravelling the mechanisms involved, particularly at the post-transcriptional and proteomic level.

References

Anderson DL, Sosa Jr O. Effect of silicon on expression of resistance to sugarcane borer (*Diatraea saccharina*). Proc J Am Soc Sug Cane Technol. 2001;21:43–50.

Barnes BN, Giliomee JH. Fruit-feeding behaviour of banded fruit weevil, *Phlyctinus callosus* (Schonherr) (Col., Curculionidae), in apple orchards. J Appl Entomol. 1992;113:407–15.

Basagli MA, Moraes JC, Carvalho GA, Ecole CC, Goncalves-Gervasio R d CR. Effects of sodium silicate application on the resistance of wheat plants to the green-aphid *Schizaphis graminum* (Rond.) (Hemiptera: Aphididae). Neotrop Entomol. 2003;32:659–63.

Chu YI, Horng SB. Infestation and reproduction of Asia corn borer on slag-treated corn plants. Chine J Entomol. 1991;11:19–24.

Correa RSB, Moraes JC, Auad AM, Carvalho GA. Silicon and acibenzolar-S-methyl as resistance inducers in cucumber, against the whitefly *Bemisia tabaci* (Gennadius) (Hemiptera: Aleyrodidae) biotype B. Neotrop Entomol. 2005;34:429–33.

Costa RR, Moraes JC. Effects of silicon acid and of acibenzolar-S-methyl on *Schizaphis graminum* (Rondani) (Hemiptera: Aphididae) in wheat plants. Neotrop Entomol. 2006;35:834–9.

Dias PAS, Sampaio MV, Rodrigues MP, Korndorfer AP, Oliveira RS, Ferreira SE, Korndörfer GH. Induction of resistance by silicon in wheat plants to alate and apterous morphs of *Sitobion avenae* (Hemiptera: Aphididae). Environ Entomol. 2014;43:949–56.

Djamin A, Pathak MD. Role of silica in resistance to Asiatic rice borer, *Chilo suppressalis* Walker, in rice varieties. J Econ Entomol. 1967;60:347–51.

Dravé EH, Laugé G. Etude de laction de la silice surlusure des mandibules de la pyrale du riz: *Chilo suppressalis* (F. Walker) (Lep. Pyralidae Crambinae). Bulletin de la Societe Entomologique de France. 1978;83:159–62.

Elawad SH, Allen JR, Gascho GJ. Influence of UV-B radiation and soluble silicates on the growth and nutrient concentration of sugarcane. Proc Soil Crop Sci Soc Florida. 1985;44:134–41.

Fauteux F, Rémus-Borel W, Menzies JG, Bélanger RR. Silicon and plant disease resistance against pathogenic fungi. FEMS Microbiol Lett. 2005;249:1–6.

Gali-Muhtasib HU, Smith CC, Higgins JJ. The effect of silica in grasses on the feeding-behavior of the prairie vole *Microtus ochrogaster*. Ecology. 1992;73:1724–9.

Gomes FB, Moraes JC, Santos CD, Goussain MM. Resistance induction in wheat plants by silicon and aphids. Sci Agric. 2005;62:547–51.

Goussain MM, Moraes JC, Carvalho JG, Nogueira NL, Rossi ML. Efeito da aplicacao de silicio em plantas de milho no desenvolvimento biologico da lagarta-do-cartucho *Spodoptera frugiperda* (J.E. Smith) (Lepidoptera: Noctuidae). Neotrop Entomol. 2002;31:305–10.

Goussain MM, Prado E, Moraes JC. Effect of silicon applied to wheat plants on the biology and probing behaviour of the greenbug *Schizaphis graminum* (Rond.) (Hemiptera: Aphididae). Neotrop Entomol. 2005;34:807–13.

Hanifa AM, Subramaniam TR, Ponnaiya BWX. Role of silica in resistance to the leafroller, *Cnaphalocrocis medinalis* Guenee, in rice. Indian J Exp Biol. 1974;12:463–5.

Horng SB, Chu YI. Development and reproduction of Asian corn borer (*Ostrinia furnacalis* Guenee) fed on artificial diet containing silica. Chine J Entomol. 1990;10:325–35.

Hou ML, Han YQ. Silicon-mediated rice plant resistance to the Asiatic rice borer (Lepidoptera: Crambidae): effects of silicon amendment and rice varietal resistance. J Econ Entomol. 2010;103:1412–9.

Hunt JW, Dean AP, Webster RE, Johnson GN, Ennos AR. A novel mechanism by which silica defends grasses against herbivory. Ann Bot. 2008;102:653–6.

Isely FB. Correlation between mandibular morphology and food specificity in grasshoppers. Ann Entomol Soc Am. 1944;37:47–67.

Jernvall J, Fortelius M. Common mammals drive the evolutionary increase of hypsodonty in the Neogene. Nature. 2002;417:538–40.

Kaufman PB, Dayanandan P, Franklin CI, Takeoka Y. Structure and function of silica bodies in the epidermal system of grass shoots. Ann Bot. 1985;55:487–507.

Keeping MG, Kvedaras OL. Silicon as a plant defence against insect herbivory: response to Massey, Ennos and Hartley. J Anim Ecol. 2008;77:631–3.

Keeping MG, Meyer JH. Calcium silicate enhances resistance of sugarcane to the African stalk-borer, *Eldana saccharina* Walker (Lepidoptera: Pyralidae). Agric For Entomol. 2002;4:265–74.

Keeping MG, Meyer JH. Silicon-mediated resistance of sugarcane to *Eldana saccharina* Walker (Lepidoptera: Pyralidae): effects of silicon source and cultivar. J Appl Entomol. 2006;130:410–20.

King BH, Crowe ML, Blackmore MD. Effects of leaf age on oviposition and on offspring fitness in the imported willow leaf beetle *Plagiodera versicolora* (Coleoptera: Chrysomelidae). J Insect Behav. 1998;11:23–36.

Kokko EG, Schaber BD, Entz T. Quantification of mandibular wear of female alfalfa leafcutter bees, *Megachile Rotundata* (F.) (Hymenoptera: Megachilidae), by image analysis. Can Entomol. 1993;125:93–9.

Korndörfer AP, Grisoto E, Vendramin JD. Induction of insect plant resistance to the spittlebug *Mahanarvafimbriolata* Stål (Hemiptera: Cercopidae) in sugarcane by silicon application. Neotrop Entomol. 2011;40:387–92.

Korth KL, Doege SJ, Park SH, Goggin FL, Wang Q, Gomez SK, Liu G, Jia L, Nakata PA. *Medicago truncatula* mutants demonstrate the role of plant calcium oxalate crystals as an effective defense against chewing insects. Plant Physiol. 2006;141:188–95.

Kvedaras OL, Keeping MG, Goebel FR, Byrne MJ. Larval performance of the pyralid borer *Eldanasaccharina* Walker and stalk damage in sugarcane: influence of plant silicon, cultivar and feeding site. Intl J Pest Manage. 2007;53:183–94.

Kvedaras OL, Byrne MJ, Coombes NE, Keeping MG. Influence of plant silicon and sugarcane cultivar on mandibular wear in the stalk borer *Eldana saccharina*. Agric For Entomol. 2009;11:301–6.

Laing MD, Gatarayiha MC, Adandonon A. Silicon use for pest control in agriculture: a review. Proc S Afr Sugar Technol Ass. 2006;80:278–86.

Lanning FC. Relation of silicon in wheat to disease and pest resistance. J Agric Food Chem. 1966a;14:350–2.

Lanning FC. Relation of silicon in barley to disease, cold and pest resistance. J Agric Food Chem. 1966b;14:636–8.

Liu F, Song Y, Bao S, Lu H, Zhu S, Liang G. Resistance to small brown planthopper and its mechanism in rice varieties. Acta Phytophylacica Sinica. 2007;34:449–54 (In Chinese with English abstract).

Ma JF, Miyake Y, Takahashi E. Silicon as a beneficial element for crop plants. In: Datnoff LE, Snyder GH, Korndörfer GH, editors. Silicon in agriculture. Amsterdam: Elsevier; 2001. p. 17–39.

Massey FP, Hartley SE. Physical defences wear you down: progressive and irreversible impacts of silica on insect herbivores. J Anim Ecol. 2009;78:281–91.

Massey FP, Ennos AR, Hartley SE. Silica in grasses as a defence against insect herbivores: contrasting effects on folivores and a phloem feeder. J Anim Ecol. 2006;75:595–603.

McColloch JW, Salmon SC. The resistance of wheat to the hessian fly–a progress report. J Econ Entomol. 1923;16:293–8.

Meyer JH, Keeping MG. Impact of silicon in alleviating biotic stress in sugarcane in South Africa. Sugarc Intl. 2005;23:14–8.

Miller BS, Robinson RJ, Johnson JA, Jones ET, Ponnaiya BWX. Studies on the relation between silica in wheat plants and resistance to Hessian fly. J Econ Entomol. 1960;53:995–9.

Moore D. The role of silica in protecting Italian ryegrass (*Lolium multiflorum*) from attack by dipterous stem-boring larvae (*Oscinella frit* and other related species). Ann Appl Biol. 1984;104:161–6.

Nakano K, Abe G, Taketa N, Hirano C. Silicon as an insect resistant component of host plant, found in the relation between the rice stem-borer and rice plant. Jpn J Appl Entomol Zool. 1961;5:17–27.

Nakata Y, Ueno M, Kihara J, Ichii M, Taketa S, Arase S. Rice blast disease and susceptibility to pests in a silicon uptake-deficient mutant lsi1 of rice. Crop Prot. 2008;27:865–8.

Pan YC, Kuroshio K, Hokama Y. The effect of bagasse furnace ash on the growth of plant cane. Sugarc J. 1979;42:14–6.

Panda N, Kush GS. Host plant resistance to insects. Wallingford: CAB International; 1995.

Panda N, Pradhan B, Samalo AP, Rao PSP. Note on the relationship of some biochemical factors with the resistance in rice varieties to yellow rice borer. Indian J Agric Sci. 1975;45:499–501.

Peterson SS, Scriber JM, Coor JG. Silica, cellulose and their interactive effects on the feeding performance of the southern armyworm *Spodoptera eridania* (Cramer) (Lepidoptera: Noctuidae). J Kansas Entomol Soc. 1988;61:169–77.

Ponnaiya BWX. Studies on the genus Sorghum. II. The cause of resistance in sorghum to the insect pest *Atherigona indica* M. Madras Univ J. 1951;21:203–17.

Ramachandran R, Khan ZR. Mechanisms of resistance in wild rice *Oryza brachyantha* to rice leaf-folder *Cnaphalocrocis medinalis* (Guenée) (Lepidoptera: Pyralidae). J Chem Ecol. 1991;17:41–65.

Rao SDV. Hardness of sugarcane varieties in relation to shoot borer infestation. Andhra Agric J. 1967;14:99–105.

Raupp MJ. Effects of leaf toughness on mandibular wear of the leaf beetle, *Plagiodera versicolora*. Ecol Entomol. 1985;10:73–9.

Redmond CT, Potter DA. Silicon fertilization does not enhance creeping bentgrass resistance to cutworms and white grubs. USGA Turfgrass Environ Res [Online]. 2007;6:1–7.

Reynolds OL, Keeping MG, Meyer JH. Silicon-augmented resistance of plants to herbivorous insects: a review. Ann Appl Biol. 2009;155:171–86.

Roitberg BD, Gillespie DR, Quiring DMJ, Alma CR, Jenner WH, Perry J, Peterson JH, Salomon M, VanLaerhoven S. The cost of being an omnivore: mandible wear from plant feeding in a true bug. Naturwissenschaften. 2005;92:431–4.

Salim M, Saxena RC. Iron, silica, and aluminium stresses and varietal resistance in rice: effects on whitebacked planthopper. Crop Sci. 1992;32:212–9.

Sasamoto K. Studies on the relation between insect pests and silica content in rice plant (II). On the injury of the second generation larvae of rice stem borer. Oyo Kontyu. 1953;9:108–10.

Sasamoto K. Studies on the relation between insect pests and silica content in rice plant (III). On the relation between some physical properties of silicified rice plant and injuries by rice stem borer, rice plant skipper and rice stem maggot. Oyo Kontyu. 1955;11:66–9.

Sasamoto K. Studies on the relation between silica content of the rice plant and insect pests. IV. On the injury of silicated rice plant caused by the rice-stem-borer and its feeding behaviour. Jpn J Appl Entomol Zool. 1958;2:88–92.

Savant NK, Snyder GH, Datnoff LE. Silicon management and sustainable rice production. Adv Agron. 1997;58:151–99.

Schaber BD, Kokko EG, Entz T, Richards KW. The effect of spring burning of seed alfalfa fields on leaf characteristics and on mandibular wear of female leaf cutter bees [*Megachile rotundata* (F.)] (Hymenoptera: Megachilidae). Can Entomol. 1993;125:881–6.

Setamou MF, Schulthess F, Bosque-Perez NA, Thomas-odjo A. Effect of N and Si on the bionomics of *Sesamia calamistis* Hampson (Lepidoptera: Noctuidae). Bull Entomol Res. 1993;83:405–11.

Singh B, Yazdani SS, Singh R. Relationship between biochemical constituents of sweet potato cultivars and resistance to weevil (*Cylas formicarius* Fab.) damage. J Entomol Res. 1993;17:283–8.

Smith GS, Nelson AB, Boggino EJA. Digestibility of forages *in-vitro* as affected by content of silica. J Anim Sci. 1971;33:466–7.

Smith MT, Kvedaras OL, Keeping MG. A novel method to determine larval mandibular wear of the African stalk borer, *Eldana saccharina* Walker (Lepidoptera: Pyralidae). Afr Entom. 2007;15:204–8.

Stebbins GL. Coevolution of grasses and herbivores. Annu Missouri Bot Gard. 1981;68:75–86.

Sudhakar GK, Singh R, Mishra SB. Susceptibility of rice varieties of different durations to rice leaf folder, *Cnaphalocrocis medinalis* Guen evaluated under varied land situations. J Entomol Res. 1991;15:79–87.

Ukwungwu MN, Odebiyi JA. Resistance of some rice varieties to the African striped borer, *Chilo zacconius* Bleszynski. Insect Sci Appl. 1985;6:163–6.

Ye M, Song Y, Long J, Wang R, Baerson SR, Pan Z, Zhu-Salzman K, Xie J, Cai K, Luo S, Zeng R. Priming of jasmonate-mediated antiherbivore defense responses in rice by silicon. Proc Natl Acad Sci U S A. 2013;110:3631–9.

Yoshihara TK, Sogawa MD, Pathak BO, Sakamura S. Soluble silicic acid as a sucking inhibitory substance in rice against the rice brown planthopper (Delphacidae: Homoptera). Entomologia Experimentalis et Applicata. 1979;26:314–22.

Zouhourian-Saghiri L, Kobilinsky A, Gillon Y, Gagnepain C. Lois d'usure mandibulaire chez *Locusta migratoria* (Orthopt. Acrididae) son utilization pour la datation des ailes. Ann Entomol Soc France. 1983;19:335–52.

Chapter 11
Effect of Silicon on Crop Growth, Yield and Quality

Abstract Silicon (Si) has widely been reported to increase the growth and biomass, yield and quality of a broad range of crops including monocotyledonous crops such as rice, wheat, maize, barley, millet, sorghum and sugarcane that actively take up and accumulate high amounts of Si in their organs and some dicotyledonous crops such as cotton and some vegetable and fruit crops. The yield increment, however, may be attributable not only to the beneficial effects of Si including growth promotion, lodging resistance and biotic and abiotic stress resistance but also to some indirect effects such as pH adjustment and acquisition of macro- and micronutrients contained in the silicate fertilizers, especially when slags or Si-containing mineral ores are used as sources of silicate fertilizer.

Keywords Growth • Photosynthesis • Nutrient uptake • Quality • Silicon • Yield

11.1 General

It has been well documented that silicon (Si) is effective in enhancing the growth and yield of many crops of agricultural and horticultural importance (for review, see Lian 1976; Elawad and Green 1979; Savant et al. 1997, 1999; Wang et al. 2001; Singh et al. 2005; Guntzer et al. 2012). The major crops that are widely reported to positively respond to Si fertilization include some monocotyledonous crops such as rice (*Oryza sativa*), wheat (*Triticum aestivum*), maize (*Zea mays*), barley (*Hordeum vulgare*), millet (*Setaria italica*), sorghum (*Sorghum bicolor*) and sugarcane (*Saccharum officinarum*) that actively absorb and accumulate high amount of Si in their organs and some dicotyledonous crops such as cotton (*Gossypium*), soybean (*Glycine max*) and some vegetable and fruit crops that are also able to accumulate Si through specific transporters (see Chap. 4). Table 11.1 shows the effect of large-scale field application of blast furnace slag-based silicate fertilizer on crop yield and economic benefit indicated by ratio of added revenue to Si fertilizer cost in north-eastern China during 2005–2006 (Y. Liang et al., unpublished). The data from Table 11.1 clearly show that although the average yield increase percent for all the crops tested except maize due to Si fertilizer application was above 10 %, the ratio of benefit to cost differed greatly with crop species mainly due to the per unit area crop yield and the price of the products. Although the average yield increase percent for

Y. Liang et al., *Silicon in Agriculture*, DOI 10.1007/978-94-017-9978-2_11

Table 11.1 The effect of field application of blast furnace slag-based silicate fertilizer on crop yield and benefit/cost ratio during 2005–2006 in northeastern China (Y. Liang et al., unpublished)

Crops tested	Yield increase percent range	Average yield increase (%)	Benefit/ cost ratio	No. of trials
Rice	3.5–28.5	10.3**	4.4	50
Maize	5.6–10.4	7.7*	3.1	44
Cucumber	9.35–25.6	13.7**	42.9	40
Tomato	8.7–15.9	12.0**	35.7	35
Soybean	7.5–13.6	11.0**	1.7	32

* and ** indicate significantly different at $P<0.05$ and $P<0.01$, respectively, compared with Si-untreated controls

soybean (11 %) was statistically significant compared with Si-untreated control application of Si fertilizer was still less acceptable by local farmers because of its lower economic benefit (lower ratio of benefit to Si fertilizer cost). The application of Si fertilizer to greenhouse-grown cucumber (*Cucumis sativus*) and tomato (*Lycopersicon esculentum*) led to extremely high economic benefits and thus was welcomed commercially (Table 11.1).

11.1.1 Si-Improved Growth and Yield of Monocotyledonous Crops

11.1.1.1 Rice

Field application of slag-based silicate fertilizers in rice paddy field started in Japan in the early 1950s and in South Korea in the 1960s–1970s. This practice contributed greatly to rice production sustainability and food security in these countries (Savant et al. 1997; Park 2001; Ma and Takahashi 2002; also see Chap. 1). The use of silicate fertilizers is also a rather common agricultural practice to increase the growth and yield of rice in Southeast Asia including China mainland, Chinese Taiwan, Thailand, the Philippines, Ceylon, Vietnam, Sri Lanka, Pakistan, India and Indonesia and in Central, South and North America, including Florida, Colombia and Brazil, and in West Africa, including Nigeria (Lian 1976; Snyder et al. 1986; Yamauchi and Winslow 1989; Datnoff et al. 1992; Winslow 1992; Liang et al. 1994; Savant et al. 1997; Winslow et al. 1997; Alvarez and Datnoff 2001a; Correa-Victoria et al. 2001; Korndörfer and Lepsch 2001; Prabhu et al. 2001). Lian (1976) and Elawad and Green (1979) reviewed the rice yield responses to Si fertilization, mostly in temperate regions such as Japan, Korea and Chinese Taiwan, while Savant et al. (1997) summarized the positive effects of Si fertilization on the growth and yield of rice grown on highly weathered soils such as inceptisols, alfisols, ultisols and oxisols in the subtropical to tropical zones. In these areas more beneficial effects of Si on rice could be expected due to low Si bioavailability. In addition to lowland rice, beneficial effects of Si fertilization on the yields of upland rice have also been reported in the Philippines, China, Colombia, Brazil and West Africa (IRRI 1965, 1966; Yamauchi and Winslow 1989; Datnoff et al. 1992; Winslow 1992; Liang et al. 1994; Savant et al. 1997; Correa-Victoria et al. 2001; Prabhu et al. 2001).

Si fertilization is reported to increase the growth of rice plants during the entire growth stage, benefiting rice plants not only in the nursery (at seedling stage) but also in the field after transplanting (IRRI 1965, 1966; Liang et al. 1994; Savant et al. 1997). It was reported that the application of various slags to nursery plants led to an increase in the number of leaves and dry matter yield of rice (Savant et al. 1997 and references therein). The application of black to grey ash of rice hulls improved the growth of rice seedlings and biomass (Sawant et al. 1994; Savant et al. 1997). After transplanting, Si fertilization increased the number of tillers and panicles (IRRI 1965, 1966; Liang et al. 1994). Ma et al. (1989) reported that the dry weight of straw and grain in rice plants not receiving Si during the reproductive stage decreased by 20 % and 50 %, respectively, compared with those of plants receiving Si throughout the growth period. When Si was supplied at the reproductive stage, the dry weight of rice straw and rice grain was increased by 243 % and 30 %, respectively. Regardless of whether Si was supplied or not, its effect was limited to the vegetative and ripening stages. The percentage of filled spikelets was also affected by the addition of Si during the reproductive stage, while the 1,000-grain weight was not influenced. These results suggest that the supply of Si during the reproductive stage is crucially important for the growth of rice plants.

More recently, Wang et al. (2014) reported the beneficial effects of steel slag applied at various rates on rice growth, yield and soil properties in a subtropical paddy field of China (Table 11.2). They showed that while steel-slag amendment had no positive effects on rice growth characteristics other than the shoot/root ratio ($P<0.05$), it significantly increased the grain yield and percentage of ripened grain (Table 11.2). Further research showed that the yield of the first early rice crop was significantly higher when slag was applied at a rate of 8 Mg ha^{-1} than when lower rates were applied ($P<0.05$, Table 11.3, Wang et al. 2015), and the yield of the late rice crop with slag added at a rate of 4 and 8 Mg ha^{-1} was also significantly higher than in the control ($P<0.05$); however, no significant rice yield response was observed when steel slag was applied at a rate of 2 Mg ha^{-1} only ($P>0.05$). Furthermore, no significant residual effects of slag application were observed on the yields of both the second early paddy and the vegetable crops, suggesting that application of slag is needed for the third crop season. On the other hand, slag effects on rice yield may be caused partly by its indirect effect, as could be seen from Table 11.2 that soil organic carbon (C), total nitrogen (N), total phosphorus (P), total iron (Fe), total manganese (Mn), available P_2O_5, available SiO_2 and ferric Fe contents increased significantly with the application levels of steel slag. Agarie et al. (1992) investigated the effects of Si and light intensity on the growth and dry mass of three rice cultivars and found that Si applications promoted seedling growth and dry mass accumulation. The effect of Si on growth improvement was dependent on light exposition (canopy) and the culvivars tested; the shade-resistant cultivar, cv. Koshihikari, responded the best to Si amendment in shade conditions. The beneficial

Table 11.2 Rice growth and yield characteristics and soil properties of plots amended with different rates of steel slag

	Application level (mg ha^{-1})			
	0	2	4	8
Growth characteristics				
Total dry matter (Mg ha^{-1})	17.20±0.45	17.32±0.27	17.46±0.29	17.56±0.25
Shoot biomass (Mg ha^{-1})	14.94±0.30	14.95±0.22	15.06±0.49	15.06±0.52
Root biomass (Mg ha^{-1})	2.31±0.19	2.36±0.16	2.37±0.09	2.50±0.16
Shoot/root ratio	6.42±0.04[a]	6.30±0.16[a,b]	6.17±0.07[b]	6.08±0.07[c]
Yield characteristics				
Grain yield (Mg ha^{-1})	8.09±0.15[b]	8.22±0.13[b]	8.33±0.12[b]	8.43±0.09[2]
1,000-grain weight (g)	24.37±0.55	24.19±0.19	24.29±0.75	24.37±0.49
Ripened grains (%)	0.85±0.00[b]	0.85±0.01[b]	0.87±0.01[a]	0.87±0.01[a]
Soil properties				
pH	6.48±0.13	6.81±0.12[b]	6.93±0.37	7.16±0.66
Bull density (g cm^{-3})	1.05±0.10	1.04±0.07	0.95±0.03	0.97±0.04
Organic C (g kg^{-1})	18.10±0.28[b]	18.07±0.16[b]	18.53±0.09[2,b]	18.89±0.35[a]
Total N (g kg^{-1})	1.17±0.08[b]	1.43±0.12[a]	1.40±0.12[a]	1.40±0.04[a]
Total P (g kg^{-1})	1.06±0.04[b]	1.39±0.09[a]	1.17±0.13[a,b]	1.19±0.04[a,b]
Total Fe (g kg^{-1})	23.35±0.58[b]	26.57±159[a,b]	27.84±1.76[a]	27.79±1.10[a]
Total Mn (mg $^{-1}$)	216.60±1.81[b]	244.81±6.82[b]	293.39±18.17[a]	314.23±25.26[a]
Available P_2O_5 (mg kg^{-1})	53.12±1.24[d]	65.05±2.76[c]	79.85±7.62[b]	96.37±3.03[a]
Available SiO_2 (mg kg^{-1})	254.09±5.12[c]	485.34±750[b,c]	764.00±15.35[b]	1232.37±226.04[a]
Ferric Fe (g kg^{-1})	3.18±0.18[c]	4.66±0.49[b,c]	5.75±0.77[b]	8.76±1.53[a]

From Wang et al. (2014)
Different letters in a single row indicate statistical differences ($P<0.05$)

Table 11.3 Yields of crops amended with various rates of steel slag

	Application level (Mg ha^{-1})			
Crop yield (Mg ha^{-1})	0	2	4	8
First early rice crop	8.09±0.15[a]	822±0.13[ab]	8.33±0.12[ab]	8.43±0.09[b]
Late rice crop	7.46±0.16[a]	7.49±0.12[a]	8.08±0.34[b]	8.14±0.28[b]
Vegetable crop	26.9±2.8[a]	26.9±1.9[a]	27.0±32[a]	27.0±42[a]
Second early rice crop	7.87±0.09[a]	7.78±0.11[a]	7.93±0.12[a]	7.84±0.09[a]

From Wang et al. (2015)
Different letters within a row indicate statistical differences ($P<0.05$)

effects of Si may be attributed to an increase in water use efficiency (WUE) and maintenance of photosynthetic activity. At the cell level, the growth promotion by the addition of Si in rice has been linked to enhanced cell elongation, but not cell division in the epidermal cells (Hossain et al. 2002). Isa et al. (2010) found that Si-enhanced rice growth was independent of silica deposition and suggested an important physiological role of Si in the cell wall, although they did not report any evidence.

The yield responses to Si fertilization depend largely upon plant-available Si content in soil, plant-available Si content of the fertilizers used, soil pH, N application levels and other environmental factors such as climate conditions (Liang et al. 1994; Savant et al. 1997; Park 2001; Wang et al. 2001, 2014, 2015). According to the review by Savant et al. (1997), rice yield responses to Si fertilization differed greatly from one experiment to another, ranging from 4.6 to 48 %. In 16 provinces of China, yield increment caused by Si fertilization ranged from 0 to 400 % with an average of 10 %, depending on the severity of Si deficiency in the soils tested (Wang et al. 2001). Large-scale field experiments (about 76 million ha) at 50 trial sites across northeast China show that amendment of furnace slag-based calcium silicate fertilizers to paddy soils derived from bleaching-meadow soil, meadow soil, bleaching soil, chernozem and black soil showed a significantly positive yield increase ranging from 3.5 to 28.5 % with an average of 10.3 % (Y. Liang et al. unpublished; see Table 11.1). Interestingly, a yield increase of as high as 76 % due to the application of Si at 3 t/ha in a form of furnace phosphorus slag was reported for upland rice grown on an inceptisol with pH 4.7 and Al saturation of 18 % at the Santa Rosa site of Colombia (Correa-Victoria et al. 2001). However, it should be stressed that the yield increment was attributable not only to the beneficial effects of Si including growth promotion, lodging resistance, disease and pest resistance and abiotic stress resistance but also to some indirect effects such as pH adjustment and acquisition of macro- and micronutrients contained in the Si fertilizers (e.g. slag-based Si fertilizer and potash feldspar or potash-rich minerals manufactured by calcination or hydrothermal chemical reaction method). Another important issue that needs to be emphasized is associated with residual effects or after-effect of slag-based Si fertilizers and feldspar- or potash-rich mineral-based Si fertilizers because these categories of fertilizers are characterized by their slow-releasing properties. Thus, yearly applications may not be needed and the subsequent application rates can be considerably reduced (Datnoff et al. 1997; Correa-Victoria et al. 2001). For example, Si applications in a rice–sugarcane (*Saccharum officinarum*) rotation in southern Florida resulted in increased yields of both rice and sugarcane that was grown immediately after rice (Alvarez and Datnoff 2001b). Correa-Victoria et al. (2001) also reported that the application of slag in Colombia increased upland rice yield by 100 % due to its residual effect.

11.1.1.2 Wheat

Wheat is another staple food crop that is widely reported to positively respond to silicate fertilizers (Wang et al. 2001) and that also has an active uptake and accumulation of Si in plant organs (Rafi and Epstein 1999; Montpetit et al. 2012). Zhu and Chen (1963) reported an obvious yield increase in wheat ranging from 6 to 12 % following the application of steel-slag-based calcium silicates in northern China. Subsequent field trials conducted across China also showed significantly positive yield responses to application of silicate fertilizers (ranging from 5 to 12 %) (Wang et al. 2001). In other instances, consecutive four-year field trials indicated that the application of a mixture of powdered sodium metasilicate and sodium disilicate increased wheat yield by 4.1–9.3 % on a calcareous paddy soil (Liang et al. 1994). Liu et al. (2011) also conducted long-term field trials with slow-released potassium silicate and showed that wheat yield was increased by 13.8 % on the average.

Xia et al. (1999) and Cui et al. (1999) observed a ca. 10 % yield increase when Si was applied by foliar spray and as a soil amendment. Yu and Gao (2012) investigated the effects of Si on the yield of two wheat cultivars and found that appropriate Si levels could increase the grain yield of wheat, and the increase resulted from the increase in spike number and grain number per spike. By contrast, Segalin et al. (2013) tested the effect of foliar applications of Si on wheat yield and quality and did not observe any improvement in five cultivars. This may be related to the fact that foliar applications did not lead to Si accumulation in wheat as reported by Guével et al. (2007).

11.1.1.3 Sugarcane

Sugarcane is a Si-accumulating plant species and the second most Si-responsive crop after rice. Samuels (1969) reported that the aboveground parts of 12-month-old sugarcane plants contained 379 kg ha^{-1} of Si, compared to 362 kg ha^{-1} of K and 140 kg ha^{-1} of N. As a result, Si deficiency in soils could be a yield-declining factor in sugarcane, resulting in symptoms such as twisted leaves and leaf freckling (Wang et al. 2001). It has been well documented that Si nutrition has a definite agronomic role in sugarcane crop cultivation, especially on highly weathered tropical soils such as oxisols, ultisols, entisols and histosols (organic soils) (for review, see Savant et al. 1999; Meyer and Keeping 2001). Earlier field trials conducted in Hawaii, Mauritius, Puerto Rico, Florida, South Africa, Brazil and Australia demonstrated that the use of silicate slag as a source of Si for sugarcane increased yield by 10–50 % on Si-low soils (Ayres 1966; Clements 1965; Fox et al. 1967; Samuels 1969; Cheong and Halais 1970; Haysom and Chapman 1975; Gascho 1976; Elawad et al. 1982; Anderson et al. 1991; Alvarez and Datnoff 2001b; Meyer and Keeping 2001; Berthelsen et al. 2001). Similar results were obtained in Asia including China, Chinese Taiwan, Indonesia, Malaysia and Pakistan (Savant et al. 1999 and the references therein; Wang et al. 2001; Ashraf et al. 2009). The review of the literature by Savant et al. (1999) unveiled several levels of Si-caused yield increases in cane ranging from 10 to 50 % and from 5 to 35 %. Some recent trials in China also gave positive cane yield responses to application of Si fertilizer. For example, Jiang et al. (2011) found that the application of Si (720 kg SiO_2 ha^{-1}) increased the sugarcane and sugar yields by 9.0 % and 9.7 %, respectively. Huang et al. (2011) also found that sugarcane yield was significantly increased by application of Si.

The observed positive sugarcane responses to Si fertilization have been attributable to a number of factors including prevention of aluminium and manganese toxicities in highly weathered acid soils, improved water use efficiency, protection from fungal and insect pest damage, improved phosphorus nutrition, improved mechanical strength and improved photosynthesis through better use of sunlight (Anderson et al. 1991; see also Chaps. 5, 7, 9 and 10). For instance, Zeng et al.

(2007) found that Si application could increase the chlorophyll content and activity of nitrate reductase of lower (old) leaves of sugarcane, prolong the leaf functional period and increase the population leaf area index. Si also increased the export of photosynthetic products from middle and lower leaves at night.

Application of Si can influence the availability of nutrients in the soil and nutrient concentrations in sugarcane plants. Huang et al. (2011) observed that the application of Si increased the levels of soil-available Si and exchangeable calcium (Ca) and magnesium (Mg), but decreased the soil organic matter and available P content. Huang et al. (1992) found that Si promoted the translocation of 11 nutrients including N, P, potassium (K), Mg, Ca, Mn, zinc (Zn), copper (Cu), Fe, molybdenum (Mo) and boron (B) to the growing parts of the plant, which could enhance the growth of sugarcane, as well as sugar synthesis and accumulation. Huang et al. (1997) also observed increased concentrations of N and K in leaves following Si application, but the effects on P concentrations were different in different trial sites. The authors suggested that Si could improve the P status in the leaves, depending on the soil P level. These studies suggest that some of the beneficial effects of Si on sugarcane growth are indirect and related to Si-mediated improvement in availability of nutrients in the soil and nutrient uptake by plants.

Si may play a role in the synthesis, storage and retention of sucrose in sugarcane plants. Ji et al. (1992) observed that Si application decreased the activities of acid invertase in mature leaves, but increased the activities of neutral invertase at elongation and maturity stages. Si-mediated changes facilitated sucrose accumulation in the cane. Pawar et al. (2003) observed that foliar applications of Si increased the sucrose synthase and sucrose phosphate synthase activities in the leaves, although the mean commercial sugar content was not changed.

The improvement of sugarcane yields by Si may also be attributed to its induced resistance to various biotic and abiotic stresses, as has been discussed in previous chapters. The cane yield responses to Si fertilization are more significant under environmental stress than under normal conditions. Also, the cane yield response is genotype dependent. For instance, cane yield was increased with Si addition by 59 % and 28 % in the salt-sensitive and salt-tolerant genotype, respectively, compared with the controls (Ashraf et al. 2009).

11.1.1.4 Maize

Maize is also one of the cereal crops that actively take up and accumulate Si into its organs (Liang et al. 2006; Mitani et al. 2009; see Chap. 4). Maize growth and yield are also highly responsive to Si fertilization (Yuan et al. 1996; Li et al. 1999; Wang et al. 2001 and references therein; Liu et al. 2011; also see Table 11.1). As early as the 1960s, Zhu and Chen (1963) conducted field trials on maize with steel slag in Liaoning province of northern China and reported a yield increment ranging from 8.5 to 10.2 %. By conducting 8 field trials on maize in northern China, Li et al. (1999) showed that the yield of summer maize receiving Si fertilization was increased by 473–900 kg ha^{-1} over the control ($P<0.01$). The application of Si fertilizer significantly increased concentrations of N, P, Zn and Mn in maize plants.

Thus, the yield response to Si may be related to improved uptake of these nutrients (Li et al. 1999). Yuan et al. (1996) agreed that the positive maize yield responses to Si fertilization could be attributable to the increased ear numbers and grain size. Consecutive field trials indicated that, on the average, the application of Si fertilizer resulted in maize yield increase by 7.3 % (Liu et al. 2011) and by 7.7 % (Y. Liang et al., unpublished, Table 11.1).

Maize yield responses to Si fertilization may be impacted by climate and plant-available Si in soils as well. According to Li et al. (1999), the application of Si resulted in a maize yield increase by 10 % in 1997 due to a severe drought stress during maize-growing season, while yield increment of 5 % was observed in 1998 when no drought stress occurred. It seems to suggest that the beneficial effects of Si on plant growth and yield are particularly distinct under drought stress conditions.

11.1.2 Dicotyledonous Crops

11.1.2.1 Cucumber

Cucumber is a typical intermediate type of plant species that also actively takes up and accumulates Si into its organs (Liang et al. 2006; Nikolic et al. 2007). Beneficial effects of Si on cucumber, especially under biotic and abiotic stress, have been most widely reported (Miyake and Takahashi 1983a, b; Adatia and Besford 1986; Marschner et al. 1990; Chérif and Bélanger 1992; Wang et al. 2007; Pavlovic et al. 2013; Liu et al. 2014; also see Chaps. 5, 8 and 9). Four-year field trials show that, on the average, the application of slag-based silicate fertilizers to greenhouse cucumber increased the yield by 13.7 % (Y. Liang et al., unpublished, Table 11.1). The beneficial effect of Si on the seed germination of cucumber has also been reported. For example, Li and Ma (2002) reported that when the available Si in soil was in the range of 55 to 203 mg kg^{-1}, the activities of both protease and lipase and respiration rate were obviously increased during seed germination. The seed vigour was also increased. These results clearly show that suitable Si level could enhance the seed germination of cucumber.

During seedling growth, Li and Ma (2002) observed Si-mediated increases of the photosynthetic rate, root activities and nitrate reductase activity. In a pot trial, Wang et al. (2007) observed that applications of Si up to 125 mg kg^{-1} improved leaf chlorophyll levels, photosynthetic rate and water use efficiency. Similar results have recently been observed in a hydroponic experiment (Liu et al. 2014).

11.1.2.2 Tomato

Although tomato (*Solanum lycopersicum*) is a typical Si-excluder plant species as compared to rice, a typical Si accumulator (Nikolic et al. 2007), Si fertilization has been reported to increase the growth and yield (Liang et al. 1993; Liu 1997; Liu et al. 2011, Table 11.1). Liang et al. (1993) showed that adding 50 μg Si L^{-1} to a nutrient solution increased tomato yield by 62 %. Furthermore, field trials indicated that Si fertilization increased tomato yield by up to 15–30 % due to increased fruit numbers and sizes (Liang et al. 1993; see Table 11.1). Tomato fruits became ripened four days earlier with higher commercially produced tomato yield due to Si fertilization compared with the control treatment (Liang et al. 1993). Liu (1997) conducted several field trials to compare the effects of Si and Ca fertilizer on tomato growth, yield and quality. The results showed that the application of Si fertilizer significantly increased tomato resistance to diseases, fruit size and consequently yield. In addition, the combined application of Si–Ca fertilizer improved the taste of tomato fruit due to increased sugar content, which was not observed if only Ca fertilizer was added without Si fertilization. Cao et al. (2013) investigated the effects of different levels of Si on growth and H_2O and CO_2 exchange of tomato grown hydroponically and found that 0.6 mM (T1) and 1.2 mM Si (T2) significantly increased the plant height and dry mass of roots, stem and leaves. The contents of photosynthetic pigments (such as chlorophyll a and b and carotenoid) are also increased by applied Si at both T1 and T2 levels. However, a higher concentration of Si at 1.8 mM Si (T3) did not improve the plant growth or contents of photosynthetic pigments. The leaf net photosynthetic rates were also increased at both T1 and T2 Si levels as compared to the control, but it was slightly decreased by applied Si at T3 level. The leaf transpiration rate is decreased at all the three levels of Si tested. Si application also increased the instantaneous water use efficiency.

The form of Si influences its effect on tomato growth. Xue et al. (2012) observed that foliar application of both inorganic and organic Si could enhance the growth of tomato seedlings and increase single tomato fruit weight, with the promotion being more obvious by application of inorganic Si.

11.1.2.3 Others

Recently, Si fertilizers have been applied to many other crops of agricultural and horticultural importance (for review, see Wang et al. 2001; Korndörfer and Lepsch 2001). For example, field trials with slag-based silicate fertilizers showed that soybean responded positively with its averaged yield increase of 11 % (Y. Liang et al. unpublished; see Table 11.1). Long-term field trials in 26 provinces of China demonstrated that on the average, the application of a slow-released potassium silicate formulation increased the yield of potato (*Solanum tuberosum*) by 12.3 %, peanut (*Arachis hypogaea*) by 6.7 %, radish (*Raphanus sativus*) by 11.2 %, soybean by 5.1 %, green bean (*Phaseolus vulgaris*) by 6.0 %, sugar beet (*Beta vulgaris*) by 4.7 %, cabbage (*Brassica oleracea*) by 15.2 %, chilli pepper (*Capsicum annuum*) by 8.4 %, pumpkin (*Cucurbita maxima*) by 11.7 %, peach (*Prunus persica*) by 18.1 %, grapevine (*Vitis vinifera*) by 6.5 %, banana (*Musa* sp.) by 4.8 %, citrus species by 12.3 %, longan (*Dimocarpus longan* Lour) by 10.7 %, tea (*Camelia sinensis*) by 11.0 %, ginseng (*Panax ginseng*) by 3.2 % and papaya (*Carica papaya*) by 9.7 % (Liu et al. 2011). Balakhnina et al. (2012) observed that the application of Si stimulated the growth and biomass production of both shoot and roots in barley. In citrus, the application of Si fertilizers enhanced the growth by 30–80 %, promoted fruit maturation by 2–4 weeks and improved fruit quantity (Meena et al. 2014, and reference therein). Li and Ma (2003) found that, within a suitable range of Si application rates, the growth of cotton seedlings was promoted. Meanwhile, the uptake of P, Zn and B in seedlings was enhanced, while the uptake of N, K, Mn, Ca and Mg was decreased. These results suggest that Si could improve the nutritional metabolism of cotton plants and therefore the growth. Si application also affected the floricultural quality of gerbera. Kamenidou et al. (2010) found that Si-applied gerbera plants had thicker flower peduncles and increased flower diameters and height. Furthermore, the gerbera plants treated with Si flowered earlier than the controls.

Crop yield responses to Si fertilizers are more evident under various forms of abiotic and biotic stresses than under normal conditions.

11.2 Crop Quality

Si fertilization is proven not only to enhance crop growth and yield but also to improve crop quality. Si fertilizers are mainly reported to improve the quality of rice grain, sugarcane, vegetables and fruits. For instance, brown rice rate, milled rice rate and head rice rate coupled with fatty acid content were significantly higher in Si-treated rice than in the Si-untreated rice, while chalky grain rate and chalkiness were lower (Zhang et al. 2007; Shang et al. 2009). Yu and Gao (2012) investigated the effects of Si on grain quality in two wheat cultivars and found that Si application did not affect the 1,000-grain weight and protein content of grain in the two cultivars tested; on the other hand, in cv. Longmai 26, Si application enhanced the content of wet gluten, flour water absorption and paste fracture time, and the extension quality of paste was also improved. Sugarcane juice quality characteristics like Brix (% soluble solids in juice), Pol (% sucrose in juice), commercial cane sugar (CCS) and sugar recovery in both salt-sensitive and salt-tolerant sugarcane genotypes were also significantly improved by Si (Ashraf et al. 2009). The addition of Si to the hydroponic solution enhanced fruit firmness, total soluble solids and vitamin C content in tomato fruits (Liang et al. 1993; Stamatakis et al. 2003; Xue et al. 2012), while fertilization of strawberry with Si resulted in an increased tissue consistency and durability of fruits during post harvest (Babini et al. 2012). In apple, the application of Si increased the content of soluble solid and vitamin C and reduced the level of titratable acid in the fruit, but did not affect the fruit hardness (Su et al. 2011). Shi et al. (2010) found that Si applications increased the contents of total soluble solids, sugar and acids, but decreased the level of nitrate in grape. Wang et al. (2007) observed a yield increase in cucumber in the range of 5.1 to 10.2 %, depending on the application rate of Si. Liu et al. (2014) observed not only an increase in single cucumber fruit weight but also an improvement in cucumber quality. They found that the levels of sugar and vitamin C were significantly increased, while the level of NO_3^--N was decreased. The role of Si in increasing sugar concentration in the cucumber fruit is still unclear, but may be related to Si-promoted photosynthesis, as previously suggested (Li and Ma 2002). The decrease of NO_3^{-}-N level in cucumber may be due to Si-mediated increase in N use efficiency, as observed in rice (Detmann et al. 2012). Further study is needed to clarify the mechanisms for these Si-mediated changes in fruit quality.

In addition, supplementation of Si is reported to improve the flower quality traits of gerbera (Savvas et al. 2002; Kamenidou et al. 2010), zinnia (Kamenidou et al. 2010) and rose (Voogt and Sonneveld 2001) under greenhouse conditions.

References

Adatia MH, Besford RT. The effects of silicon on cucumber plants grown in recirculating nutrient solution. Ann Bot. 1986;58:343–51.

Agarie S, Agata W, Kubota F, Kaufman PB. Physiological roles of silicon in photosynthesis and dry matter production in rice plants I. Effects of silicon and shading treatments. Jpn J Crop Sci. 1992;61:200–6.

Alvarez J, Datnoff LE. The economic potential of silicon for integrated management and sustainable rice production. Crop Prot. 2001a;20:43–8.

Alvarez J, Datnoff LE. The economics of silicon for integrated management and sustainable production of rice and sugarcane. In: Datnoff LE, Snyder GH, Korndorfer GH, editors. Silicon in agriculture. Amsterdam: Elsevier; 2001b. p. 221–39.

Anderson DL, Snyder GH, Martin FG. Multyear response of sugarcane to calcium silicate slag on Everglades Histosols. Agron J. 1991;83:870–4.

Ashraf M, Rahmatullah, Ahmad R, Afzal M, Tahir MA, Kanwal S, Maqsood MA. Potassium and silicon improve yield and juice quality in sugarcane (*Saccharum officinarum* L.) under salt stress. J Agron Crop Sci. 2009;195:284–91.

Ayres AS. Calcium silicate slag as a growth stimulant for sugarcane on low-silicon soils. Soil Sci. 1966;101:216–27.

Babini E, Marconi S, Cozzolino S, Ritota M, Taglienti A, Sequi P, Valentini M. Bio-available silicon fertilization effects on strawberry shelf-life. Acta Horticult. 2012;934:815–8.

Balakhnina TI, Matichenkov VV, Wlodarczyk T, Borkowska A, Nosalewicz M, Fomina IR. Effects of silicon on growth processes and adaptive potential of barley plants under optimal soil watering and flooding. Plant Growth Regul. 2012;67:35–43.

Berthelsen S, Noble AD, Garside AL. Silicon research down under: past, present, and future. In: Datnoff LE, Snyder GH, Korndörfer GH, editors. Silicon in agriculture. Amsterdam: Elsevier; 2001. p. 241–55.

Cao BL, Xu K, Shi J, Xin GF, Liu CY, Li X. Effects of silicon on growth, photosynthesis and transpiration of tomato. Plant Nutr Fertil Sci. 2013;19:354–60 (in Chinese with English abstract).

Cheong YWY, Halais P. Needs of sugar cane for silicon when growing in highly weathered latosols. Exp Agric. 1970;6:99–106.

Chérif M, Bélanger RR. Use of potassium silicate amendments in recirculating nutrient solutions to suppress *Pythium ultimum* on long English cucumber. Plant Dis. 1992;76:1008–11.

Clements HF. The roles of calcium silicate slags in sugarcane growth. Repts Hawaiian Sugarc Technol. 1965;25:103–26.

Correa-Victoria FJ, Datnoff LE, Okada K, Friesen DK, Sanz JI, Snyder GH. Effects of silicon fertilization on disease development and yields of rice in Colombia. In: Datnoff LE, Snyder GH, Korndorfer GH, editors. Silicon in agriculture. Amsterdam: Elsevier Science; 2001. p. 313–22.

Cui DJ, Wang YF, Liu YJ, Zhang XS, Song HW. Study on effects of silicon and potassium in winter wheat. Chine J Soil Sci. 1999;30(3):121–2 (in Chinese).

Datnoff LE, Snyder GH, Deren CW. Influence of silicon fertilizer grades on blast and brown spot development and on rice yields. Plant Dis. 1992;76:1011–3.

Datnoff LE, Deren CW, Snyder GH. Silicon fertilization for disease management of rice in Florida. Crop Prot. 1997;16:525–31.

Detmann KC, Araújo L, Martins SCV, Sanglard LMVP, Reis JV, Detmann E, et al. Silicon nutrition increases grain yield, which, in turn, exerts a feed-forward stimulation of photosynthetic rates via enhanced mesophyll conductance and alters primary metabolism in rice. New Phytol. 2012;196:752–62.

Elawad SH, Green Jr VE. Silicon and the rice plant environment: a review of recent research. Riv Riso. 1979;28:235–53.

Elawad SH, Gascho GJ, Street JJ. Response of sugarcane to silicate source and rate. I. Growth and yield. Agric J. 1982;74:481–4.

Fox RL, Silva JA, Younge OR, Plucknett DL, Sherman GD. Soil and plant silicon and silicate response by sugarcane. Soil Sci Soc Am Proc. 1967;31:775–9.

Gascho GJ. Silicon status of Florida sugarcane. Proc Soil Crop Sci Soc Florida. 1976;36:188–91.

Guével MH, Menzies JG, Bélanger RR. Effect of root and foliar applications of soluble silicon on powdery mildew control and growth of wheat plants. Eur J Plant Pathol. 2007;119:429–36.

Guntzer F, Keller C, Meunier JD. Benefits of plant silicon for crops: a review. Agron Sustain Dev. 2012;32:201–13.

Haysom MBC, Chapman LS. Some aspects of the calcium silicate trials at Mackay. Proc Qld Soc Sugarc Technol. 1975;42:117–22.

Hossain MT, Mori R, Soga K, Wakabayashi K, Kamisaka S, Fujii S, Yamamoto R, Hoson T. Growth promotion and an increase in cell wall extensibility by silicon in rice and some other Poaceae seedlings. J Plant Res. 2002;115:23–7.

Huang X, Ji M, Zhang L, Zhou Y. Study on behavior mechanism of increasing production and sugar of cane caused by silicon. IV Effect of silicon on distribution of nutritive elements in sugarcane. Sugarc Canesugar. 1992;4(6–8):24 (In Chinese with English abstract).

Huang X, Zhang Z, Ke Y, Xiao C, Peng Z, Wu L, Li Y, Jian H, Cen J, Zhang Y, Zhong S. Effects of silicate fertilizer on nutrition of leaves, yield and sugar of sugarcanes. Trop Subtrop Soil Sci. 1997;6:242–6 (In Chinese with English abstract).

Huang HR, Xu L, Bokhtiar SM, Manoj KS, Li YR, Yang LT. Effect of calcium silicate fertilizer on soil characteristics, sugarcane nutrients and its yield parameters. J South Agric. 2011;42:756–9.

IRRI, International Rice Research Institute. Annual report 1965. Los Banos: Laguna; 1965.

Isa M, Bai S, Yokoyama T, Ma JF, Ishibashi Y, Yuasa T, Iwaya-Inoue M. Silicon enhances growth independent of silica deposition in a low-silica rice mutant, *lsi1*. Plant Soil. 2010;331:361–75.

Ji M, Huang X, Yu J. Study on behavior mechanism of increasing production and sugar of cane caused by silicon. VI Effect of silicon on activities of inbertases in blades of cane. Sugarc Canesugar. 1992;4(6–8):24 (In Chinese with English abstract).

Jiang ZP, Liao Q, Wei GP, Tan YM, Chen GF, Liu B, Wang YN. Effects of optimum fertilization model between silicon and phosphorous on yield and quality of sugarcane in lateristic red earth area of Guangxi. J Anhui Agric Sci. 2011;39:22233–6 (In Chinese with English abstract).

Kamenidou S, Cavins TJ, Marek S. Silicon supplements affect floricultural quality traits and elemental nutrient concentrations of greenhouse produced gerbera. Sci Hortic. 2010;123:390–4.

Korndörfer GH, Lepsch I. Effect of silicon on plant growth and crop yield. In: Datnoff LE, Snyder GH, Korndorfer GH, editors. Silicon in agriculture. Amsterdam: Elsevier; 2001. p. 133–47.

Li QF, Ma CC. Effect of available silicon in soil on cucumber seed germination and seedling growth metabolism. Acta Hortic Sinica. 2002;29:433–7 (in Chinese with English abstract).

Li QF, Ma CC. Effect of available silicon in soil on nutritive metabolism of cotton seedling. Sci Agric Sin. 2003;36:726–30 (in Chinese with English abstract).

Li CH, Chu TD, Liu XB, Yang Q. Silicon nutrition effects and its study and application development in China. Proceedings of Symposium of Plant Nutrition, Shaanxi Science and Technology Press, China; 1999. pp. 329–33. (In Chinese)

Lian S. Silica fertilization of rice. In: The fertility of paddy soils and fertilizer applications for rice. Taipei: Food Fertilizer Technology Center; 1976. p. 197–220.

Liang YC, Chen XH, Ma TS, Qian ZJ, Liu LR. Effect of Si on the growth, yield and quality of tomato. Jiangsu Agric Sci. 1993;4:48–50 (in Chinese).

Liang YC, Ma TS, Li FJ, Feng YJ. Silicon availability and response of rice and wheat to silicon in calcareous soils. Commun Soil Sci Plant Anal. 1994;25:2285–97.

Liang YC, Hua HX, Zhu YG, Zhang J, Cheng CM, Römheld V. Importance of plant species and external silicon concentration to active silicon uptake and transport. New Phytol. 2006;172:63–72.

Liu HY. Preliminary report on effect of Si fertilizer on tomato growth. J Guizhou Agric Coll. 1997;16:76–7 (In Chinese).

Liu JM, Han C, Sheng XB, Liu SK, Qi X. Potassium-containing silicate fertilizer: its manufacturing technology and agronomic effects. Oral presentation at 5th International Conference on Silicon in Agriculture; September 13–18, Beijing; 2011.

Liu H, Li J-M, Zheng G, Du Q-J, Pan T-H, Chang Y-B. Effects of silicon on plant growth and fruit quality of cucumber. Acta Agriculturae Boreali-occidentalis Sinica. 2014;23(8):117–21 (in Chinese with English abstract).

Ma JF, Takahashi E. Soil, fertilizer, and plant silicon research in Japan. Amsterdam: Elsevier; 2002.

Ma JF, Nishimura K, Takahashi E. Effect of silicon on the growth of rice plant at different growth stages. Soil Sci Plant Nutr. 1989;35:347–56.

Marschner H, Oberle H, Cakmak L, Romheld V. Growth enhancement by silicon in cucumber (*Cucumis sativus*) plants depends on imbalance in phosphorus and zinc supply. Plant Soil. 1990;124:211–9.

Meena VD, Dotaniya ML, Coumar V, Rajendiran S, Ajay, Kundu S, Rao AS. A case for silicon fertilization to improve crop yields in tropical soils. Proc Natl Acad Sci, India, Sect. B Biol. Sci.; 2014; 84: 505–518.

Meyer JH, Keeping MG. Past, present and future research of the role for sugarcane in southern Africa. In: Datnoff LE, Snyder GH, Korndörfer GH, editors. Silicon in agriculture. Amsterdam: Elsevier; 2001. p. 257–75.

Mitani N, Yamaji N, Ma JF. Identification of maize silicon influx transporters. Plant Cell Physiol. 2009;50:5–12.

Miyake Y, Takahashi E. Effect of silicon on the growth of solution-cultured cucumber plant. Soil Sci Plant Nutr. 1983a;29:71–83.

Miyake Y, Takahashi E. Effect of silicon on the growth of cucumber plant in soil culture. Soil Sci Plant Nutr. 1983b;29:463–71.

Montpetit J, Vivancos J, Mitani-Ueno N, Yamaji N, Remus-Borel W, Belzile F, Ma JF, Belanger RR. Cloning, functional characterization and heterologous expression of TaLsi1, a wheat silicon transporter gene. Plant Mol Biol. 2012;79:35–46.

Nikolic M, Nikolic N, Liang Y, Kirkby EA, Römheld V. Germanium-68 as an adequate tracer for silicon transport in plants. Characterization of silicon uptake in different crop species. Plant Physiol. 2007;143:495–503.

Park CS. Past and future advances in silicon research in the Republic of Korea. In: Datnoff LE, Snyder GH, Korndörfer GH, editors. Silicon in agriculture. Amsterdam: Elsevier; 2001. p. 359–71.

Pavlovic J, Samardzic J, Masimović V, Timotijevic G, Stevic N, Laursen KH, Hansen TH, Husted S, Schjoerring JK, Liang Y, Nikolic M. Silicon alleviates iron deficiency in cucumber by promoting mobilization of iron in the root apoplast. New Phytol. 2013;198:1096–107.

Pawar MW, Joshi SS, Amodkar VT. Effect of foliar application of phosphorus and micronutrients on enzyme activities and juice quality in sugar cane. Sugarc Technol. 2003;5(3):161–5.

Prabhu AS, Filho MPB, Filippi MC, Datnoff LE, Snyder GH. Silicon from rice disease control perspective in Brazil. In: Datnoff LE, Snyder GH, Korndörfer GH, editors. Silicon in agriculture. Amsterdam: Elsevier; 2001. p. 293–311.

Rafi MM, Epstein E. Silicon absorption by wheat (*Triticum aestivum* L.). Plant Soil. 1999;211:223–30.

Samuels G. Silicon and sugar. Sugar y Azucar. 1969;66(4):25–9.

Savant NK, Snyder GH, Datnoff LE. Silicon management and sustainable rice production. Adv Agron. 1997;58:151–99.

Savant NK, Korndorfer GH, Datnoff LE, Snyder GH. Silicon nutrition and sugarcane production. A review. J Plant Nutr. 1999;22:1853–903.

Savvas D, Manos G, Kotsiras A, Souvaliotis S. Effects of silicon and nutrient-induced salinity on yield, flower quality and nutrient uptake of gerbera grown in a closed hydroponic system. J Appl Bot. 2002;76:153–8.

Sawant AS, Patil VH, Savant NK. Rice hull ash applied to seedbed reduces deadhearts in transplanted rice. Intl Rice Res Notes. 1994;19(4):21–2.

Segalin SR, Huth C, Rosa TD, Pahins DB, Mertz LM, Nunes UR, Martin TN. Foliar application of silicon and the effect on wheat seed yield and quality. J Seed Sci. 2013;35:86–91.

Shang QY, Zhang WZ, Han YD, Rong R, Xu H, Xu ZJ, Chen WF. Effect of silicon fertilizer application on yield and grain quality of japonica rice from Northeast China. China J Rice Sci. 2009;23:661–4 (in Chinese with English abstract).

Shi Y-Z, Rong J-F, Su L, Li C, Feng Y. Study on effects of silicon on grapevine physiology and fruit quality. Jilin Agric. 2010;11:98–100 (in Chinese).

Singh KK, Kalyan S, Singh RS, Raghavendra S, Chandel RS. Silicon nutrition in rice – a review. Agric Rev. 2005;26(3):223–8.

Snyder GH, Jones DB, Gascho GJ. Silicon fertilization of rice on Everglades Histosols. Soil Sci Soc Am J. 1986;50:1259–63.

Stamatakis A, Papadantonakis N, Lydakis-Simantiris N, Kefalas P, Savvas D. Effects of silicon and salinity on fruit yield and quality of tomato grown hydroponically. Acta Hortic. 2003;609:141–7.

Su XW, Wei SC, Jiang YM, Huang YY. Effects of silicon on quality of apple fruit and Mn content in plants on acid soils. Shandong Agric Sci. 2011;6:59–61 (In Chinese with English abstract).

Voogt W, Sonneveld C. Silicon in horticultural crops grown in soilless culture. In: Datnoff LE, Snyder GH, Korndörfer GH, editors. Silicon in agriculture. Amsterdam: Elsevier; 2001. p. 115–31.

Wang HL, Li CH, Liang YC. Agricultural utilization of silicon in China. In: Datnoff LE, Snyder GH, Korndörfer GH, editors. Silicon in agriculture. Amsterdam: Elsevier; 2001. p. 343–52.

Wang XY, Zhang YL, Yu N, Meng QL, Li J. Effect of silicon fertilizer on cucumber photosynthesis and yield in protected field. J Changjiang Veget. 2007;2:45–7 (in Chinese with English abstract).

Wang W, Lai DYF, Li S, Kim PJ, Zeng C, Li P, Liang Y. Steel slag amendment reduces methane emission and increases rice productivity in subtropical paddy fields in China. Wetland Ecol Manage. 2014;22:683–91.

Wang W, Sardans J, Lai DYF, Wang C, Zeng C, Tong C, Liang Y, Penuelas J. Effects of steel slag application on greenhouse gas emissions and crop yield over multiple growing seasons in a subtropical paddy field in China. Field Crops Res. 2015;171:146–56.

Winslow MD. Silicon, disease resistance and yield of rice genotypes under upland cultural conditions. Crop Sci. 1992;32:1208–13.

Winslow MD, Okada K, Correa-Victoria F. Silicon deficiency and the adaptation of tropical rice ecotypes. Plant Soil. 1997;188:239–48.

Xia SY, Huang SH, Wang QS. Study on yield increase effect of foliar application of silicon in wheat and rice. Soils Fertil. 1999;1:36–8 (In Chinese).

Xue G, Zhang G, Sun Y, Liao S, Chen Y. Influences of spraying two different forms of silicon on plant growth and quality of tomato in solar greenhouse. Chine Agric Sci Bull. 2012;28(16):272–6.

Yamauchi M, Winslow MD. Effect of silica and magnesium on yield of upland rice in the humid tropics. Plant Soil. 1989;113:265–9.

Yu LH, Gao JL. Effects of silicon on yield and grain quality of wheat. J Triticeae Crops. 2012;32:469–73 (in Chinese with English abstract).

Yuan ZG, Zhang QR, Yin WR. Effect of combined application of Si, Zn and Mn on growth rate of wheat and corn. Soil Fertil. 1996;1:56–48 (In Chinese).

Zeng XL, Liang JN, Tan ZW. Effects of silicate on some photosynthetic characteristics of sugarcane leaves. J Huazhong Agric Univ. 2007;26:330–4 (in Chinese with English abstract).

Zhang GL, Dai QG, Wang JW, Zhang HC, Huo ZY, Ling L, et al. Effects of silicon fertilizer rate on yield and quality of japonica rice Wuyujing. Chin J Rice Sci. 2007;21:299–303.

Zhu Q, Chen EF. The properties of iron and steel slags and their effects on crops grown on different types of soils. Acta Pedol Sin. 1963;11:70–83.

Chapter 12
Silicon Sources for Agriculture

Abstract This chapter overviews the major silicate fertilizers commercially available and describes briefly their manufacturing processes. Silicon (Si) sources for agricultural use range from chemical products to natural minerals to by-products of steel and iron industries. All these products are shown to be effective in improving crop growth and yield. Slag-based silicate fertilizers are cost-effective, but more attention should be paid to their potential environmental risks which may arise from the heavy metals contained in the fertilizers. Soluble potassium or sodium silicates are completely water soluble and can be used as foliar fertilizers, but are usually too expensive for soil application. Slow-releasing potassium-containing or potassium-rich silicates that are manufactured using feldspar as raw materials are not only cost-effective and agriculturally effective but also environmentally friendly.

Keywords Potassium silicate • Slag-based silicate • Slow-releasing potassium silicate • Sodium silicate

12.1 Slag-Based Silicate Fertilizer

Slag-based silicate fertilizers refer to calcium silicate fertilizers that are processed using slags, by-products or industrial waste materials from the processing of iron and steel-making industries. Although slags from the processing of iron and steel-making industries have been extensively used first by monasteries in Europe since the Middle Ages (fifth to fifteenth centuries) (see Chap. 1), they are usually treated as liming or phosphorus-containing amendments because silicon (Si) has not been recognized as an essential element for plant growth. The first patents for using Si slag as a fertilizer were obtained by Zippicotte in the USA in 1881 (Zippicotte 1881). Based on the distinct beneficial agronomic effects of slags on paddy rice (*Oryza sativa*) and degraded paddy soils in field trials conducted across Japan after World War II, slags were first approved officially as silicate fertilizers by the Ministry of Agriculture, Forestry and Fisheries of Japan in 1955 and has since been commercially available on the market. Slags that contain a sufficient amount of Si available to the plant, but free of toxic components, were recognized as 'silicate fertilizer', and an official standard of silicate fertilizer was established (Ma and Takahashi 2002).

Y. Liang et al., *Silicon in Agriculture*, DOI 10.1007/978-94-017-9978-2_12

Slags that are used to process silicate fertilizers include blast furnace slags, silicomanganese slags, ferronickel slags, stainless-steel slags, carbon-steel slags, manganese slags and converter slags (Gascho 2001); another common source is the electric furnace production of elemental phosphorus such as phosphorus slag.

Slag is a by-product of the production of metals or alloy steels by smelting the mineral ore with limestone and cokes in a blast furnace or electric arc furnace at high temperature (e.g. 1,300 °C), followed by cooling (by either air or water) the material floated on the surface. During the smelting process, silicate components in the ore react with limestone, and Si contained in the ore is often released from its tightly bound state, thus leading to the separation of calcium silicate and, consequently, the increased solubility of Si compounds. Meanwhile, Fe and other metals in the ore are reduced and separated, and slag remains as a by-product (Ma and Takahashi 2002). The main components of slag are calcium silicate, magnesium (Mg), aluminium (Al) and iron (Fe), with trace amount of manganese (Mn), nickel (Ni) and chromium (Cr) included. According to the official standard of silicate fertilizer set up in Japan in 1955 (Ma and Takahashi 2002), the plant-available Si content (0.5 N HCl-soluble SiO_2) and alkali component (mainly CaO) content in slag should be more than 20 % and 35 %, respectively, and maximum allowable limits for toxic components Ni, Cr and titanium (Ti) are 0.4 %, 4 % and 1.5 %, respectively. This standard was revised in 1987; slags containing more than 10 % soluble SiO_2 (0.5 N HCl-soluble SiO_2) were also recognized as silicate fertilizers. Listed in Table 12.1 is the composition of 10 slags used as calcium silicate fertilizers in Japan (Ma and Takahashi 2002). It can be seen clearly that, owing to the differences in manufacturing processes of iron and alloy industries, the Si solubility, plant-available Si content and value of the by-product differ considerably with the sources of slag used (Gascho 2001; Ma and Takahashi 2002). The main slag used in Japan has been blast furnace slag followed by silicomanganese slag (Gascho 2001).

Table 12.1 Composition of slags used as calcium silicate fertilizer (%)

	CaO	SiO_2	MgO	MnO	Al_2O_3	Fe_2O_3
Pig iron slags	35–45	30–41	3–7	0.3–1.7	12–20	0.3–1.7
Steel mill slags	37–65	9–22	0.6–1.5	0.5–10.0	0.1–7.5	1.5–3.5
Stainless-steel slags	43–48	23–28	10–15	0.5–1.0	1–4	0.5–1.2
Ferromanganese Slags	33–37	25–30	4–7	5–14	4–7	0.5–1.0
Silicomanganese Slags	30–45	30–40	1–4	6–10	3–12	3–10
Ferronickel slags	8–12	40–50	20–26	–	4–6	4–5
Nickel slags	17–20	40–45	20–25	–	2.5–4	14–20
Ferrochrome slags	47–53	27–32	9–12	–	9–12	0.2–0.5
Magnesium slags	50–55	29–33	8–12	0.1–0.3	1–3	2–4
Dephosphorylated Slags	48–50	40–45	0.2–0.4	–	1–4	0.1–0.3

From Ma and Takahashi (2002)

Table 12.2 Heavy metal content of the waste materials

	Heavy metal content/mgkg^{-1}						
Material	Cu	Zn	Cr	As	Cd	Pb	Hg
Ca silicate	2.8	6.3	7.8	35.5	0.24	1.16	0.145
BF slag 1	8.2	6.6	34.7	96.0	1.52	0.72	0.117
BF slag 2	13.6	124.8	42.5	57.8	1.00	1.01	0.120
BF slag 3	12.3	5.8	36.8	64.4	0.70	0.69	0.122
Steel slag	18.2	58.9	1,472.2	213.6	0.99	0.38	1.363
Processing mud	25.5	65.7	1,211.2	366.5	1.37	49.9	0.290
Fly ash	64.7	135.7	91.3	0.008	0.56	25.2	0.026

From Haynes et al. (2013)
BF, air-cooled blast furnace; processing mud, bauxite-processing mud

One of the most important issues to address is the potential toxicity and environmental risk problems that may be caused by the toxic heavy metals contained in these slags (Gascho 2001). Table 12.2 shows the content of major heavy metals in some of the Australian slags. Steel slag and processing mud contain elevated concentrations of Cr and arsenic (As), and processing mud also contains high lead (Pb) concentrations. However, these materials were still considered to be safe because research (Zhou and Haynes 2011) showed that TCLP-extractable Cr (USEPA 1992) in steel slag (0.002 mg L^{-1}) and red mud (1.093 mg L^{-1}) was well below the regulatory limit of 5.0 mg L^{-1} and TCLP-Pb in red mud (0.016 mg L^{-1}) was also low (Haynes et al. 2013). Compared to the maximum allowable limits (MALs) for Cr (4 %) set up in Japan (Ma and Takahashi 2002), the total Cr content of all these slags (Table 12.2) is far below the MAL. However, in consideration of the potential environmental risks to be caused by the recycled use of waste materials in farmland, more strict maximum allowable limits (MALs) have been set up for soil amendments or soil conditioners. For example, the MALs for Cr, mercury (Hg), As, Cd and Pb in China are 50, 5, 10, 10 and 50 mg kg^{-1}, respectively.

The major problem in early development of calcium silicate fertilizers in the USA using a by-product of the production of elemental P during the 1970s is related to the radon activity of some calcium silicate slags, depending on the location of the phosphate rock mines (Gascho 2001). Further research by Snyder et al. (1986) using a source of calcium silicate slag with lower radon activities demonstrated the value of the slag for rice and sugarcane (*Saccharum officinarum*) production in the high organic mucks of the Everglades and in some associated sands in South Florida.

12.2 Fused Magnesium Phosphate

Fused magnesium phosphate fertilizer containing phosphorus (P), Mg, Ca and Si was manufactured in Japan in the 1950s (Ma and Takahashi 2002). This Si-containing fertilizer is manufactured by melting rock phosphate together with serpentinite, followed by quick cooling and pulverization. Fused magnesium phosphate appears as a furnace phosphate fertilizer but contains 16–26 % soluble SiO_2. Fertilizers containing more than 20 % acid-soluble SiO_2 are recognized as silicate fertilizer.

12.3 Slow-Releasing Potassium Silicate Fertilizer

Potassium silicate appeared as a slow-releasing K fertilizer in 1978 in Japan and in the 1990s in China. In Japan, coal fly ash from coal power plant is calcined together with potassium carbonate or potassium hydroxide and magnesium hydroxide at about 900 °C in a furnace. The main minerals are K_2O (Al, Fe)O_3 SiO_2 and K_2O MgO SiO_2 (Ma and Takahashi 2002 and references therein). According to the official standard of commercial fertilizer in Japan, the content of citrate-soluble potassium (K_2O), 0.5 M HCl-soluble silicate (SiO_2) and citrate-soluble magnesium (MgO) in the fertilizer must be greater than 20 %, 25 % and 3.0 %, respectively, and the content of nonreactive water-soluble potassium (K_2O) must be less than 3 %. Potassium in the potassium silicate fertilizer is slow released. Around 2000, the annual demand of this fertilizer reached 50,000 tons, over 90 % of which was used for rice production in Japan.

In China, slow-released potassium silicate fertilizer is also available commercially. However, slow-released potassium silicate fertilizer is a patented product manufactured with potash feldspar or potash-rich minerals by calcination or hydrothermal chemical reaction method at 130–250 °C (Han 2005; Liu et al. 2011). The hydrothermal chemical reaction technique involves using additives consisting of boron/magnesium mine powder or slight-burnt boron/magnesium mine powder. The chemical reaction is given in Eq. (12.1):

$$6KAlSi_3O_8 + 19Ca(OH)_2 + H_2O \rightarrow 3Ca_3Al_2(SiO_4)_2(OH)_4 + Ca_{10}Si_{12}O_{31}(OH)_6 8H_2O + 6KOH \quad (12.1)$$

This slow-released potassium silicate contains 22–25 % of citrate-soluble SiO_2, 5.0–7.0 % of citrate-soluble K_2O, 35–40 % of citrate-soluble CaO and 1.5 % of citrate-soluble MgO compared to 0.8 % water-soluble SiO_2 and 5.3 % water-soluble K_2O. Long-term field trials demonstrate that on the average, application of this slow-released potassium silicate increased the yield of many crop species by an average of 10 % (Liu et al. 2011; see Chap. 10 for a complete list of species).

To increase potassium content in the fertilizer, Su et al. (2011) invented an improved hydrothermal chemical reaction method to produce potassium-rich silicate fertilizer (kalsilite) (a patented product) using feldspar powder. The hydrothermal chemical reaction is completed within 2–8 h at 220–280 °C. The content of citrate-extractable SiO_2 and K_2O is approximately 20 % and 24 %, respectively.

According to the field trials across China (Z. Yu and Y. Zhang 2014. personal communications), these potassium-containing calcium silicates can significantly increase the yield of a wide range of crops.

In 1986, a liquid potassium silicate, which is guaranteed of containing at least 12 % of water-soluble SiO_2 and 6.0 % of water-soluble potassium, appeared as a readily available silicate fertilizer in Japan. It was produced by diluting potassium silicate and potassium carbonate in water (Ma and Takahashi 2002).

12.4 Fused Potassium Magnesium Silicate

Fused potassium magnesium silicate appeared as a slow-releasing potassium magnesium silicate fertilizer around 2000 in China. It is produced by melting potash-rich minerals like feldspar together with magnesium sulphate in a furnace at around 900–1,300 °C, followed by cooling and grinding and/or pelletization. The content of citrate-extractable SiO_2, CaO, K_2O and MgO is approximately 25–30 %, 27–32 %, 4 % and 5 %, respectively. This potassium magnesium silicate is registered as soil amendment or soil conditioner. The results from field trials across China demonstrate that this soil amendment is effective in increasing the yield and quality of crops including rice, wheat (*Triticum aestivum*), maize (*Zea mays*), sugarcane, peanut (*Arachis hypogaea*), soybean (*Glycine max*) and vegetables and fruits and providing resistance to biotic (plant diseases and pest damage) and abiotic stresses.

12.5 Porous Hydrate Calcium Silicate

Porous hydrate calcium silicate (tobermorite) or calcium silicate hydrate used for light wall material in construction is manufactured by a chemical reaction with lime, quartz and cement at 180 °C at 10 atm pressure (Ma and Takahashi 2002). Because of the strict standards for wall material, a high percentage of nonstandardized products resulted as a waste by-product of the building industry. This waste material, which has Si solubility greater than slags from the iron and alloy mills, was used as a fertilizer in Japan (Gascho 2001; Ma and Takahashi 2002). In 1993, a 'light porous cement powder fertilizer' was recognized as a new silicate fertilizer with a new standard established. The content of 0.5 N HCl-soluble SiO_2 must be greater than 15 % and that of base component must be greater than 15 % (Ma and Takahashi 2002).

12.6 Soluble Silicate Fertilizer

Only sodium silicate and potassium silicate can be water soluble. In 1840, sodium silicate was first suggested by Liebig (1840) as a Si fertilizer and as the first source of Si for greenhouse experiments on sugar beets. A long-term field experiment, which has been conducted at the Rothamsted Experimental Station (England) since

1856, demonstrates that sodium silicate is effective in increasing grass productivity. Potassium silicate is even more effective in stimulating plant growth and productivity as it contains potash. However, both potassium silicate and sodium silicate are too expensive for field incorporation. These soluble silicates are available commercially, but in most cases, for economical and efficient considerations, they are used as a foliar fertilizer at a relatively low concentration. Water glass (sodium silicate) is reported to be effective in increasing crop yield and resistance to diseases and pest damage for both soil applications (Elawad et al. 1982; Ma et al. 1987b; Qian et al. 1988) and as foliar sprays (McAvoy and Bible 1996). However, it is not cost-effective for the rates recommended for soil application (Gascho 2001). Potassium silicates are applied to effectively control *Pythium* diseases in tomato and cucumber (*Cucumis sativus*) plants grown hydroponically (Adatia and Besford 1986) and powdery mildew in cucumber, muskmelon (*Cucumis melo*) and zucchini squash plants (*zucchini squash*) (Menzies et al. 1992) and in grape (*Vitis vinifera*) plants (Bowen et al. 1992). Potassium silicates are also too expensive for the rates recommended for soil applications (Gascho 2001). In the late 1980s, water glass was successfully used to manufacture a mixture of powdered sodium metasilicate and sodium disilicate in China (Ma et al. 1987a, b; 1992; Ma 1991). This fertilizer is a patented product and produced by centrifuging the water glass at a high speed, followed by spraying at a given rate and drying at a given temperature (Ma 1991). This fertilizer contains 50–55 % of water-soluble SiO_2 and 3–5 % water-soluble potassium; but it is not cost-effective for the rates recommended for soil applications. It is completely water soluble and can thus be used as a liquid fertilizer or drip irrigation fertilizer, but still not at a cost-effective concentration. Application of this soluble Si fertilizer at a low rate (e.g. 90–105 kg ha^{-1}) is reported to be effective in significantly increasing the yield of rice, sugarcane and wheat and providing resistance to lodging and diseases (Ma 1991; Ma et al. 1992; Shui et al. 1995; Ye and Xu 1995; Liang et al. 1994).

12.7 Silica Gel

Silica gel, which is usually used as a desiccating agent in the industry, is manufactured by neutralizing water glass, followed by gelling and dehydrating (Ma and Takahashi 2002). Si is required by rice at the seedling stage, but the silicate fertilizers commercially available (mainly slag-based Si fertilizer) in Japan are not suitable for use in nursery bed because they are too basic or alkaline (pH 10–12) and will raise the pH (Ma and Takahashi 2002). Silica gel was found to be a suitable Si source for use in rice nursery beds in Japan in 1999. As silica gel is not dissolved in hydrochloric acid, the content of 0.5 M sodium hydroxide-extractable SiO_2 in silica gel, according to the Japanese official standard, should not be lower than 80 %.

12.8 Other Sources

Apart from the sources of Si discussed above, some naturally occurring Si-containing minerals such as wollastonite and olivine ($MgSiO_3$) are used as sources of Si fertilizer used in agriculture after being pulverized and/or pelletized (Park 2001; Gascho 2001). In addition, diatomaceous earth is commercially used as a source of Si fertilizer by some companies. However, research is needed to clarify whether Si contained in diatomaceous earth is plant available during crop-growing seasons after it is mechanically pulverized and the agronomic effects reported are attributable to Si, but not to other nutrients contained in the diatomaceous earth.

References

Adatia MH, Besford RT. The effects of silicon on cucumber plants grown in recirculating nutrient solution. Ann Bot. 1986;58:343–51.

Bowen P, Menzies J, Ehret D, Samuels L, Glass AD. Soluble silicon sprays inhibit powdery mildew development on grape leaves. J Am Soc Hortic Sci. 1992;117:906–12.

Elawad SH, Gascho GJ, Street JJ. Response of sugarcane to silicate source and rate. I. Growth and yield. Agric J. 1982;74:481–4.

Gascho GJ. Silicon sources for agriculture. In: Datnoff LE, Snyder GH, Korndorfer GH, editors. Silicon in agriculture. Amsterdam: Elsevier Science B.V.; 2001. p. 197–207.

Han C. A hydro-thermal chemical reaction method of manufacturing potassium fertilizer using potash-rich rocks and calcium carbonate. Patent No. ZL01100474.6; 2005.

Haynes RJ, Belyaeva ON, Kingston G. Evaluation of industrial wastes as sources of fertilizer silicon using chemical extractions and plant uptake. J Plant Nutr Soil Sci. 2013;176:238–48.

Liang YC, Ma TS, Li FJ, Feng YJ. Silicon availability and response of rice and wheat to silicon in calcareous soils. Commun Soil Sci Plant Anal. 1994;25:2285–97.

Liebig J. Organic chemistry in its application to agriculture and physiology. From the manuscript of the author by Lyon Playfair. London: Taylor & Walton; 1840.

Liu JM, Han C, Sheng XB, Liu SK, Qi X. Potassium-containing silicate fertilizer: its manufacturing technology and agronomic effects. Oral presentation at 5th International Conference on Si Agriculature; 2011 September 13–18, Beijing; 2011.

Ma TS. Manufacture and application of silicate fertilizer. Chem Fertil Indust. 1991;6:24–6 (In Chinese).

Ma JF, Takahashi E. Soil, fertilizer, and plant silicon research in Japan. Amsterdam: Elsevier; 2002.

Ma TS, Wang DP, Cheng ZF, Tang JD, Chen YF. Feasibility of water glass as a soluble silicate fertilizer. J Nanjing Agric Univ. 1987a;2:129–30 (In Chinese).

Ma GR, He NZ, Shi WY. Investigation of sulphur, potassium and silicon concentrations and pH value of waters in main streams of Zhejiang province. Zhejiang Agric Sci. 1987b;1:17–9 (In Chinese).

Ma TS, Wang DP, Liang YC, Chen XH, Zhang FS, Wang XL, Qian ZR, Liu LR. Effect of high-efficient silicate fertilizer on rice. Soils. 1992;3:168–9 (In Chinese).

McAvoy RI, Bible BB. Silica sprays reduce the incidence and severity of bract necrosis in poinsettia. HortSci. 1996;31:1146–9.

Menzies J, Bowen P, Ehret D, Glass AD. Foliar applications of potassium silicate reduce severity of powdery mildew on cucumber, muskmelon, and zucchini squash. J Am Soc Hortic Sci. 1992;117:902–5.

Park CS. Past and future advances in silicon research in the Republic of Korea. In: Datnoff LE, Snyder GH, Korndörfer GH, editors. Silicon in agriculture. Amsterdam: Elsevier; 2001. p. 359–71.

Qian ZR, Zhang FS, Ma TS, Liu LR. Field trial efficacy of water glass as silicon fertilizer. J Nanjing Agric Univ. 1988;11:132–3 (In Chinese).

Shui MX, Du XF, Chen DF, Ye C, Li Q. Analysis of effect of high-efficient silicate fertilizer on resistance to rice blast. Acta Agric Zhejiangensis. 1995;4:42–5 (In Chinese).

Snyder GH, Jones DB, Gascho GJ. Silicon fertilization of rice on Everglades Histosols. Soil Sci Soc Am J. 1986;50:1259–63.

Su SQ, Ma HW, Yang J, Liu H. A hydrothermal chemical reaction method of manufacturing kalsilite using feldspar powder. Patent No. ZL 201010543249; 2011.

USEPA. The toxicity characteristic leaching procedure. US code of federal regulations. 40th ed., Part 261, Appendix II. USEPA, Washington, DC, USA; 1992.

Ye C, Xu J. Yied-increasing effect of high-efficient silicate fertilizer on sugarcane. Fujian Agric Sci Technol. 1995;4:13–4 (In Chinese).

Zhou YF, Haynes RJ. A comparison of inorganic solid wastes as adsorbents of heavy metal cations in aqueous solution and their capacity for desorption and regeneration. Water Air Soil Pollut. 2011;218:457–70.

Zippicotte J. Fertilizer. U.S. Patent No. 238240. Official Gazette of the United States Patent Office; 1881. pp. 19, 9, and 496.

Chapter 13
Summary and Perspectives

13.1 Summary

Chapters in this book summarize the current knowledge inherent to the importance of Si nutrition for plant growth and development, resistance to biotic and abiotic stress and yield production and quality. The readers should refer to the individual chapters for information on past, current and future research linking Si with soils, plants and fertilizers. In general, the chapters are closely connected thereby offering a comprehensive discussion about the specific topics while avoiding repetition.

The first chapter addresses the most important historical points and general introduction to Si research worldwide. Chapter 2 critically reviews the current analytical methods used for determination of Si in a wide variety of materials including soils, plants and fertilizers. Chapter 3 is more soil oriented and includes environmental chemistry of Si and its bioavailability for crops. Chapter 4 provides an overview of Si uptake and transport by higher plants highlighting the most recent achievements in cloning and characterization of different types of Si transporter in different plant species and plant tissues. Chapters 5, 6, 7, and 8 cover the broad issue of the complex role of Si in enhancing plant resistance and/or tolerance to abiotic stresses (e.g. mineral excess and nutrient deficiency, salinity, drought, low temperature, UV radiation and flooding) and suggest the possible mechanisms of Si-mediated antioxidative defence in plants. Chapters 9 and 10 review the role of Si against biotic stress in plants, focusing on physical, biochemical and molecular components of Si-mediated plant defence and priming. Chapter 11 deals with the effect of Si on growth, development, biomass production and yield quality of a wide range of crops. Chapter 12 examines the sources of Si in agriculture focusing on the manufacturing and application of the commercially most important environmentally friendly silicate fertilizers.

Y. Liang et al., *Silicon in Agriculture*, DOI 10.1007/978-94-017-9978-2_13

13.2 Perspectives

Recent advancements concerning uptake, transport and accumulation of Si in higher plants have broadened our understanding of the beneficial effects of this element. Accordingly, it is now well established that Si accumulation in plants is attributed to an effective uptake system mediated not only by passive diffusion but also by specific transporters. Although the regulation of the expression level of Lsi transporters has been clearly elucidated in high (e.g. rice, barley and maize) and moderate (e.g. pumpkin and soybean) Si-accumulating species, the whole picture of how these genes are regulated in many other agronomically important crop species and cultivars, and in Si-excluding species in particular, remains to be further elucidated. The work on cloning of genes involved in Si uptake and transport from other crop species should be continued with special focus on the identification and characterization of novel Si transporters.

Silicon supply and its subsequent accumulation in plants could be exploited as a strategy to improve crop health and productivity in stressful environments. However, little information is available on the response of Si transporters to plant stresses. More investigations are therefore needed to establish whether there is a relationship between Si transporters and the benefits of Si to plants subjected to multiple stresses. Finally, a comprehensive understanding of Si transport is important to draw novel breeding and sustainable crop management strategies to enhance productivity of crops growing under ever-increasingly stressful environmental conditions as a consequence of global climate change.

Recent studies have revealed the active role of Si in priming plants for a better defence response against biotic stress possibly through SA, JA, ethylene and/or ROS signalling pathways. However, whether and how Si acts as a secondary messenger in priming plants in response to biotic and perhaps abiotic stresses is still uncertain. Moreover, information is still scant on the metabolically active roles of Si in plants under abiotic stress conditions, particularly on the molecular aspects of Si-mediated nutrient and water use efficiency.

Combining different state-of-the-art analytical techniques (e.g. ICP-MS, XPS, AFM), He et al. (2013, 2015) have recently demonstrated that most of Si is naturally present in the cell walls as Si-hemicellulose complex rather than intra- or extracellular silica deposits. Therefore, such organic matrix-bound Si may play a significant role in maintaining cellular integrity and structure of the walls, thereby shedding more light on the biochemical and structural role of Si in preserving cell shape and mechanical properties as well as in defence against abiotic and biotic stress. On the other hand, the essentiality of Si for higher plants has not been confirmed yet. One of the main reasons is that a Si-free environment cannot be created currently due to technical problems as Si contamination from purified water, chemicals and dust cannot be avoided. It seems to suggest that progress in research on Si nutrition depends on the adoption of the cutting-edge technologies and advanced instruments.

Overall, further research directions should focus on accumulating more convincing evidence to unveil the molecular mechanisms of Si involved in plant resistance or tolerance to biotic and abiotic stresses as well as in the regulation of sensing, signal transduction pathways and gene expression involved in biosynthesis of key compounds related to plant growth.

References

He CW, Wang LJ, Liu J, Liu X, Li XL, Ma J, Lin YJ, Xu FS. Evidence for 'silicon' within the cell walls of suspension-cultured rice cells. New Phytol. 2013;200:700–9.

He CW, Ma J, Wang LJ. A hemicellulose-bound form of silicon with potential to improve the mechanical properties and regeneration of the cell wall of rice. New Phytol. 2015;206:1051–62

The manufacturer's authorised representative in the EU is Springer Nature Customer Service Centre GmbH, Europaplatz 3, 69115 Heidelberg, Germany. If you have any concerns regarding our products, please contact ProductSafety@springernature.com

Printed and bound by CPI Group (UK) Ltd, Croydon, CR0 4YY
30/09/2026
02225445-0005